ADVANCED CERAMIC PROCESSING
AND TECHNOLOGY

MATERIALS SCIENCE AND PROCESS TECHNOLOGY SERIES

Editors

Rointan F. Bunshah, University of California, Los Angeles *(Materials Science and Process Technology)*

Gary E. McGuire, Microelectronics Center of North Carolina *(Electronic Materials and Process Technology)*

DEPOSITION TECHNOLOGIES FOR FILMS AND COATINGS: by *Rointan F. Bunshah et al*

CHEMICAL VAPOR DEPOSITION IN MICROELECTRONICS: by *Arthur Sherman*

SEMICONDUCTOR MATERIALS AND PROCESS TECHNOLOGY HANDBOOK: edited by *Gary E. McGuire*

SOL-GEL TECHNOLOGY FOR THIN FILMS, FIBERS, PREFORMS, ELECTRONICS AND SPECIALTY SHAPES: edited by *Lisa A. Klein*

HYBRID MICROCIRCUIT TECHNOLOGY HANDBOOK: by *James J. Licari* and *Leonard R. Enlow*

HANDBOOK OF THIN FILM DEPOSITION PROCESSES AND TECHNIQUES: edited by *Klaus K. Schuegraf*

IONIZED-CLUSTER BEAM DEPOSITION AND EPITAXY: by *Toshinori Takagi*

DIFFUSION PHENOMENA IN THIN FILMS AND MICROELECTRONIC MATERIALS: edited by *Devendra Gupta* and *Paul S. Ho*

SHOCK WAVES FOR INDUSTRIAL APPLICATIONS: edited by *Lawrence E. Murr*

HANDBOOK OF CONTAMINATION CONTROL IN MICROELECTRONICS: edited by *Donald L. Tolliver*

HANDBOOK OF ION BEAM PROCESSING TECHNOLOGY: edited by *Jerome J. Cuomo, Stephen M. Rossnagel,* and *Harold R. Kaufman*

FRICTION AND WEAR TRANSITIONS OF MATERIALS: by *Peter J. Blau*

CHARACTERIZATION OF SEMICONDUCTOR MATERIALS—Volume 1: edited by *Gary E. McGuire*

SPECIAL MELTING AND PROCESSING TECHNOLOGIES: edited by *G.K. Bhat*

HANDBOOK OF PLASMA PROCESSING TECHNOLOGY: edited by *Stephen M. Rossnagel, Jerome J. Cuomo,* and *William D. Westwood*

FIBER REINFORCED CERAMIC COMPOSITES: edited by *K.S. Mazdiyasni*

HANDBOOK OF SEMICONDUCTOR SILICON TECHNOLOGY: edited by *William C. O'Mara, Robert B. Herring,* and *Lee P. Hunt*

HANDBOOK OF POLYMER COATINGS FOR ELECTRONICS: by *James J. Licari* and *Laura A. Hughes*

ADVANCED CERAMIC PROCESSING AND TECHNOLOGY—Volume 1: edited by *Jon G.P. Binner*

Related Titles

ADHESIVES TECHNOLOGY HANDBOOK: by *Arthur H. Landrock*

HANDBOOK OF THERMOSET PLASTICS: edited by *Sidney H. Goodman*

SURFACE PREPARATION TECHNIQUES FOR ADHESIVE BONDING: by *Raymond F. Wegman*

ADVANCED CERAMIC PROCESSING AND TECHNOLOGY

Volume 1

Edited by

Jon G.P. Binner

Department of Materials Engineering and Materials Design
University of Nottingham
Nottingham, England

 **np** NOYES PUBLICATIONS
Park Ridge, New Jersey, U.S.A.

Copyright © 1990 by Noyes Publications
No part of this book may be reproduced or utilized in
any form or by any means, electronic or mechanical,
including photocopying, recording or by any information storage and retrieval system, without permission
in writing from the Publisher.
Library of Congress Catalog Card Number: 90-7766
ISBN: 0-8155-1256-2
Printed in the United States

Published in the United States of America by
Noyes Publications
Mill Road, Park Ridge, New Jersey 07656

10 9 8 7 6 5 4 3 2 1

Library of Congress Cataloging-in-Publication Data

Advanced ceramic processing and technology / edited by Jon G.P.
Binner.
p.   cm.
Includes bibliographical references and index.
ISBN 0-8155-1256-2 (v. 1) :
1. Ceramics. I. Binner, J.
TP807.A334   1990
666--dc20   90-7766
                                             CIP

To the memory of

William J. Knapp
(1916–1986)

Preface

Advanced ceramics and ceramic matrix composites are finding increasing use in modern technological applications as ever more stringent demands are placed upon material's properties. This has led to a significant expansion over the past two decades, in terms of research and development into optimising the properties of these generally brittle and unforgiving materials. However, whilst advanced ceramics and ceramic composites have many potentially useful properties, they can be extremely difficult to fabricate into usable artefacts. This has resulted in an ever increasing emphasis being placed on advanced ceramic processing and technology.

As long ago as 1972 Stuijts emphasised the need for precise control of microstructure as a means of achieving control of the properties of the final component. This approach, which is strongly reflected throughout the current book, must begin with the precursor powders and continue through to green body formation and the densification of the body via some sintering mechanism. Not only must accidental variations in the microstructure be avoided, but the design of the microstructure must be optimised with the final application of the component in mind—and then achieved.

This book, the first of two volumes, contains a series of independent chapters, each focussing on a different aspect of ceramics processing. It is not intended that these chapters should form a complete portfolio of all the possible techniques currently available for fabricating ceramics; such an approach would be more at home in a ceramics encyclopedia. Rather the aim is to offer the views of leading experts as to the current state-of-the-art of a number of ceramics processing options and, most importantly, the future directions which they see their fields taking. The two volumes, then, are aimed at the materials engineer who already has a grasp of the fundamentals underlying ceramic science and engineering and who is now looking to expand his or her knowledge of processing techniques and their underlying philosophies.

For a number of reasons this text has been a long time in the making and I would like to extend my heartfelt thanks to all the authors (and the publisher) who, without exception, have shown great patience. In particular, I should like to thank those authors who met the original manuscript deadlines and then found themselves, some time later, having to significantly update their chapters. Finally, I should like to thank Pam and Elaine for their excellent help in typing some of the incoming manuscripts.

This volume is dedicated to the memory of Professor Bill Knapp, former member of the Department of Materials Science and Engineering at the University of California at Los Angeles. Originally to be Bruce Kellet's co-author for the first chapter, Bill was tragically killed in a hit-and-run accident whilst out jogging one morning in late 1985. Bill was a very fine ceramist, but more importantly, he was a very special man.

Nottingham, England   Jon G.P. Binner
June, 1990

Contributors List

Jon G.P. Binner
University of Nottingham
Nottingham, England

Anselmo O. Boschi
Federal University of Sao Carlos
Sao Carlos SP, Brazil

Ramesh C. Budhani
Brookhaven National Laboratory
Upton, New York

Rointan F. Bunshah
University of California,
   Los Angeles
Los Angeles, California

David S. Cannell
Morgan Matroc Unilator Division
Clwyd, England

Stephen C. Danforth
Rutgers University
Piscataway, New Jersey

Julian R.G. Evans
Brunel University
Middlesex, England

Eric Gilbart
The University of Leeds
Leeds, England

Steven N. Heavens
Chloride Silent Power Limited
Cheshire, England

Bruce J. Kellett
Ecole Polytechnique Federale
   de Lausanne
Lausanne, Switzerland

Fred F. Lange
University of California at
   Santa Barbara
Santa Barbara, California

Andrew C. Metaxas
University of Cambridge
Cambridge, England

Kevin J. Nilsen
Dow Chemical Company
Midland, Michigan

Roy W. Rice
W.R. Grace and Company
Columbia, Maryland

Richard E. Riman
Rutgers University
Piscataway, New Jersey

Paul Trigg
Filtronic Components Ltd.
West Yorkshire, England

Walter T. Symons
AC Rochester
Flint, Michigan

NOTICE

Contents

1

Advanced Processing Concepts for Increased Ceramic Reliability

B.J. Kellett [†] and F.F. Lange [*]

† Ecole Polytechnique Federale de Lausanne, Department des Materiaux, Laboratoire de Ceramiques, 34 ch. de Bellerive, CH-1007 Lausanne, Switzerland.

* Materials Department, College of Engineering, University of California at Santa Barbara, Santa Barbara, CA 93106, USA.

1. INTRODUCTION

Man's skill in processing functional ceramics dates back many millennia, preceding the introduction of more formable and less brittle materials, viz. metals, that have since received more economic, technological and scientific attention due to their deserved engineering importance. Ceramic materials, with their multiplicity of elemental combinations and structural arrangements, produce a multitude of unique properties, which are still being uncovered. Today, advanced ceramics are finding potential applications ranging from advanced heat engines to communication and energy transmission and they are emerging as the leading class of materials needed to implement many advanced technologies.

Engineering implementation of advanced ceramics is still hindered by their formability and brittle nature; however ceramic processing technology has advanced little beyond the needs associated with functional, traditional ceramics. Such traditional approaches inherently lack a clear methodology for controlling microstructural heterogeneities and uniformity. This lack of processing control leads to property variability and consequent uncertain engineering reliability.

1

The objective here is to review new approaches to powder processing that minimize heterogeneities common to this 'many bodied' problem. The review will start by outlining other approaches to ceramic processing. New thinking concerning densification will set the stage for discussions concerning new approaches to powder preparation and consolidation that emphasize the colloidal approach.

2. PROCESSING METHODS

Although powder methods dominate ceramic forming, ceramics can also be formed by glass-ceramic and gelation methods.

2.1 Glass-Ceramic Methods

Glass-ceramic methods can be used for compositional systems with relatively small free energies of crystallization (e.g. silicates) so that solidification occurs before crystallization. Direct crystallization from the melt must be avoided because very large grains result which produces a relatively friable material. Shapes are formed by conventional high temperature glass processing to take advantage of Newtonian rheology. Crystallization is induced by a two-step nucleation/growth process at moderate temperatures; however complete crystallization is rarely (if ever) achieved. Hence ceramics produced by the glass-ceramic method contain a residual glass phase which degrades the mechanical properties at high temperatures. Calculations[1,2] suggest that residual glassy pockets within a polycrystalline material can be thermodynamically stable. Many advanced structural ceramics (e.g. silicon nitride and carbide) decompose before melting, whereas others crystallize too readily for use of this method. The glass-ceramic process is thus limited to materials that melt and do not readily crystallize.

2.2 Gelation Methods

Gelation methods are analogous to the glass-crystallization method in that processing starts with a metastable system. With this method, soluble metal-organic precursors are 'gelled' (e.g. hydrolysis of alkoxides). After the liquid is removed, organic residuals are removed by heat treatment preceded by crystallization and densification. Unlike the glass-ceramic method, compositions are not restricted to those that are glass formers at high temperature, viz. crystallization is induced by heating and not cooling.

One of the major attributes of the gelation method is that multi-element, metastable systems (intimately mixed at the molecular level) can be produced at low temperatures. Phase partitioning from these metastable systems can be used to control microstructures by heat treatment at higher temperatures.

Removal of the liquid is one major limitation of the gelation method. Capillary pressure causes the low density network to shrink during drying[3]. Shrinkage initiates at the surface and generates stresses that usually cause the drying system to break apart into small granules (analogous to the mud crack pattern observed on a drying lake bed). The shrinkage stresses can be reduced by extremely slow drying to produce sound, monolithic bodies, but these drying periods (of the order of weeks) are not practical. Surface tension and thus capillary pressures on the network can be completely eliminated by removing the fluid phase at temperatures and pressures above the fluid's critical point (i.e. super critical drying)[4]. Super critical drying (used for more than 40 years to produce aero-gels) results in very low density networks (relative densities <0.2) and thus a large degree of shrinkage occurs during heat treatments that produce crystallization and densification. Low densities and hence large shrinkages are the second major drawback of the gelation method. A third, but lesser problem is the elimination of the organic radicals bound to the polymer networks which must be carefully controlled to avoid gas entrapment, etc. These limitations become greater with component size.

Gelation methods are generally limited to the processing of thin films, fibres and powders.

2.3 Powder Methods

Powder methods are used to fabricate most advanced ceramics. They involve powder manufacture, preparation of the powder for consolidation, consolidation of the powder into a shape and densification (elimination of the void phase). Post-densification heat treatments can develop specific microstructures to optimize certain properties.

Although powder methods are much less restrictive than those discussed above, its 'many-bodied' nature makes it prone to heterogeneities. One of the major causes of these heterogeneities is the powder itself. Nearly all current powders are agglomerated, i.e. groups of either weakly or strongly bonded particles. Agglomerates pack together during consolidation to produce compacts with differentials in packing density leading to poor densification and the formation of crack-like voids which can be a major strength degrading flaw population[5].

It is common practice to reduce agglomeration size by attrition milling. Studies[6] have shown that milling has a low probability of eliminating all agglomerates. It also introduces contaminates and large inclusions not acceptable for the fabrication of reliable, advanced ceramics. Various organics can be added as helpful binders and/or lubricants during consolidation. For rheological consolidation methods (e.g. slip casting, tape casting of thin sheets, extrusion and injection moulding), the non-volatile residual polymer content of the system can be between 40 and 50 volume percent. Elimination of this polymer (e.g. through pyrolysis) is not only time consuming (days) but can also produce disruptive phenomena.

The most common method of consolidating powders is via dry pressing in which forces are applied to powder contained within a die. Since dry powders are naturally agglomerated and the consolidation of agglomerates must be avoided to produce reliable ceramics, this technique is undesirable. In addition, since dry, fine powders do not flow to uniformly fill pressing dies, powder slurries with polymer additions are currently spray dried to produce large, flowable particles (>50 μm agglomerates). These massive particles (agglomerates) produce larger separating forces due to differential acceleration which can overcome attractive (e.g. Van der Waals) forces during flow. Thus, although spray drying is helpful in producing a flowable powder, it can introduce large agglomerates and thus produce large crack-like voids during densification.

Powders are also consolidated from slurries. Current methods include: filtration (slip casting), electrophoresis, evaporation (casting of thin sheets, i.e. tape casting), extrusion and injection moulding. Although these slurry state consolidation methods are adaptable to new colloidal methods discussed below, each has limitations concerning kinetics, shapes and/or polymer content.

New consolidation technologies must emerge from basic scientific reasoning. As discussed below, colloidal methods of preparing agglomerate free powder must be combined with new slurry state consolidation methods based on the basic understanding of how particles rearrange under the action of both applied and interparticle forces, i.e. basic rheology studies of highly filled slurry states are required.

Powder compacts are made dense by heat treatments at temperatures where mass transport can occur to eliminate the void phase. Densification is driven by the free energy decrease associated with the powder's surface area. Differential surface curvature causes mass to 'fill' the contact region between touching particles. This process, known as sintering, has been extensively studied for the case of two touching particles. Liquid phases, produced by reactions between different constituents in the powder (e.g. impurities, added sintering aids) can aid mass transport. Quoting Rhines[7]: "..sintering is one of the major processes in nature that is provided by such processes as the flow of liquids and solids, phase changes, and so on. When we know enough about sintering we will greatly enhance our understanding of all of nature."

Sintering theories concerning two contacting particles have been applied to predict the densification behaviour of powder compacts. Despite our extensive theoretical knowledge concerning two particle systems, Exner[8] summarizes his review with the statement, "The quantitative understanding of sintering processes is still very incomplete in spite of the great number of experimental and theoretical investigations." It is thus understandable why fabricators of today's advanced ceramics pay little attention to theory, but instead, use an Edisonian approach to find the right conditions (temperature, time, sintering aids etc.) to manufacture dense ceramics.

Current theories do not consider the effect of particle packing structure (i.e. bulk density) on sintering behaviour. Particle packing is a major process variable. One example concerning the inconsistency of theory and reality concerns the theoretical result that densification kinetics should increase with decreasing particle size. Common experience indicates that powders with a very small crystallite size (e.g. <0.1 μm) can be very difficult to pack and densify. It is now commonly accepted that strong, partially sintered agglomerates, common to the chemically derived powders that make up powders of very small crystallite size, are the main reason for poor sinterability.

Rhodes[9] was the first to report empirical relations between agglomerates and sinterability. He showed that the density achieved by a fixed heating schedule was inversely proportional to the agglomerate size. When combined with other studies, these results suggest that particle packing has a great influence on densification. Mounting evidence suggests that the two particle models, which lack pores, are insufficient. It must be concluded that since the practical aim of densification is to rid the powder compact of its void phase, the void phase produced by particle arrangement must be studied along with the mass transport produced by the particles.

Microstructures that control properties are primarily developed during densification; post-densification heat treatments can be used to develop specific microstructures to optimize a specific property. In all cases the average microstructure and heterogeneities within the microstructure are a direct consequence of all processing stages starting with the chemistry and characteristics of the powder.

Most advanced ceramics are not single phase materials. Processing aids are added to improve densification (but may also produce unwanted second phases), control microstructure and produce desired properties. Thus, powders used to fabricate advanced ceramics may contain more than one powder phase, presenting the problem of phase homogeneity during powder preparation. Control of phase homogeneity in powders is critical in controlling reaction kinetics during densification and phase homogeneity in the dense material.

Grain size distribution and morphology are related to a variety of processing variables. It is known that second phases, introduced and distributed during powder preparation, can be used to control grain size and abnormal grain growth; distribution of the second phase is critical to grain size distribution[10].

It is well known[11] that the high fracture toughness of polyphase silicon nitride materials is due to its fibrous grain morphology. This morphology is developed during densification; the aspect ratio and diameter of the fibrous grains can be controlled by the phase content and particle size respectively, of the starting powder[12]. If one knew why Si_3N_4

develops a fibrous microstructure, one might develop similar tough microstructures in other materials. These and other correlations have little or no basic understanding and thus are not intentionally used in either designing new materials or in processing advanced ceramics.

Microstructural heterogeneities are introduced during processing. These heterogeneities include: crack-like voids produced by the differential sintering of agglomerates, abnormally large grains, inorganic inclusions that produce residual stresses during cooling, organic inclusions that produce irregular shaped voids, non-uniformly distributed second phases, etc. It is now recognized that each type of heterogeneity is a potential strength degrading flaw population. That is, a ceramic component will contain many different flaw populations, each related to a different type of heterogeneity introduced during some stage of processing. Inadvertently, the processor introduces a variety of flaw populations that limit the potential strength of the product. Slight changes in processing variables can, unknowingly, produce the same material with a different set of flaw populations and thus, what appears to be the same material with different strength statistics. New processing methods with a high probability of either controlling and/or eliminating microstructural heterogeneities must be developed to ensure structural reliability.

3. DENSIFICATION CONCEPTS

3.1 Two Particle Concepts

Traditional sintering theories are based on the kinetics of mass transport motivated by differential surface curvature as described by the Gibbs-Kelvin equation:

$$\mu = \gamma_s \Omega (1/r_1 + 1/r_2) \qquad\qquad 1$$

which relates chemical potential (μ) to local surface curvature, expressed by the principal radii of curvature r_1 and r_2 at the contact region between two particles, the surface energy per unit area (γ_s), and the atomic volume (Ω). A powder compact lowers its free energy by promoting mass transport to particle-particle contact regions (as has been described in all sintering theories) and to larger particles causing particle coarsening, i.e. grain growth. Since the free energy decrease attributed to grain growth is usually significantly less than that for neck growth, most researchers have neglected the effect of grain growth on densification. As discussed below, the interaction of grain growth and neck growth leads to new and interesting conclusions.

Experimental evidence of neck growth was first demonstrated by Kuczynski[13], who in 1949 sintered large polycrystalline particles onto flat polycrystalline substrates. Theories, based on a two sphere model (each a single crystal), were developed to determine the rate of neck growth and the rate at which particle centres approach one another. These theories conclude that the rate of neck growth is inversely proportional to particle size raised to a power that depends on the mass transport path. Many mass transport paths were subsequently considered, bulk diffusion, surface diffusion, grain boundary diffusion, viscous flow, evaporation-condensation, liquid solution-reprecipitation, and dislocation motion (see reference 8 for a review).

Sintering theories based on the neck growth between two particles were used to explain the initial sintering of powder compacts. As necks grow, the morphology of the compact changes from one of touching particles and interconnected porosity to one of (closed) pores. Many theories have also been developed for the closed pore (final) stage of densification. These theories predict that pores, assumed to be spherical and located at 2 and 3, or 4 grain junctions, continuously shrink at a rate controlled by the kinetics of either volume and/or grain boundary diffusion.

3.2 Multiparticle Concepts

Studies performed on the packing of monosized spheres have shown that particle packing rarely results in periodic arrangements as implied when two particle neck growth models are applied to the initial sintering stage of powder compacts. The random, dense packing of monosized spheres has been reviewed by Frost[14] as an arrangement of different, irregular polyhedra consisting of a 'pore' surrounded by touching spheres (particles) at the vertices. The structure of the powder compact is formed by the polyhedra joined at faces, edges and corners. The number of touching particles define the pore's coordination number; as shown below this is an important physical property which helps to define the thermodynamic stability/instability of pores within polycrystals.

It was first pointed out by Coble[15], and later in more detail by Kingery and Francois[16], that isolated pores found in the latter stage of sintering can only be spherical if they reside within a single grain of isotropic surface energy.   As shown in Figure 1, the surface of a pore curvature is controlled by the dihedral angle and the number of the surrounding grains, i.e. its coordination number.   Kingery and Francois concluded that a pore would either grow or shrink depending on whether its curvature is either concave or convex (as viewed from within the pore), respectively. In an unpublished draft, Cannon[17] discussed the consequences of mass transport between pores and grain boundaries to conclude that, depending on the dihedral angle, highly coordinated pores shrink to equilibrium size, whilst pores of lower coordination shrink and disappear.

Hoge and Pask[18], in earlier work, reached a less specific but similar conclusion concerning pore stability by determining the lowest free energy configuration developed by three dimensional periodic arrangements of spheres (i.e. cubic, face-centred cubic, body-centred cubic, etc.). They showed that

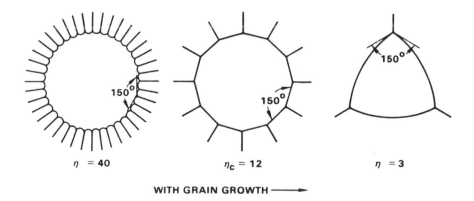

$\eta = 40$ $\eta_c = 12$ $\eta = 3$

WITH GRAIN GROWTH ⟶

Figure 1: Illustration of pore curvature as the coordination number decreases.

an equilibrium periodic structure of interpenetrating spheres can develop without complete densification. Although Hoge and Pask did not develop a general criterion for pore closure, they were the first to suggest that, under certain conditions, equilibrium pores may exist.

3.3 Minimum Energy Configurations of Particle Arrays

To further address the effect of particle packing on sintering behaviour and to develop a more general relationship between pore coordination number and pore stability, Kellett and Lange[19] determined the minimum energy configuration of symmetric particle arrays containing a single pore. Particle energy was determined by summing the energies associated with the particle's surface area (A_s) and grain boundary area (A_b):

$$E = A_s\gamma_s + A_b\gamma_b \qquad\qquad 2$$

where γ_s and γ_b are the energies per unit area associated with the surface and the grain boundary respectively and are assumed to be isotropic. Young's equation was used to relate γ_s and γ_b with the dihedral angle, ψ_e, viz. cos $\psi_e/2 = \gamma_b/2\gamma_s$. The minimum in the free energy function ($dA_s/dA_d = -2\cos \psi_e/2$) was used to determine the equilibrium configuration of the array.

Arrays composed of identical symmetrically arranged particles do not experience a driving force for interparticle mass transport, i.e. no one particle grows at the expense of another. It was therefore assumed that particles conserve their mass to produce multi-particle equilibrium configurations by only intraparticle mass transport. The case for an infinite linear array of identical touching cylinders, with an initial radius r_i, is shown in Figure 2. It is assumed that centres approach one another as cylinders penetrate one another. The interpenetrated mass is uniformly redistributed over each cylinder to increase their radii (r) during interpenetration. Using the geometrical variables shown in Figure 2b and equation 2, the energy per unit length of cylinder, normalized by the initial energy ($2\pi r_i\gamma_s$), was determined as a function of three variables, viz. the contact angle (ϕ) (angle formed where the surface tangents meet the grain boundary), the dihedral angle, and the normalized radius ($R = r/r_i$). It can be shown[19] that the energy per unit length of cylindrical particle can be expressed as:

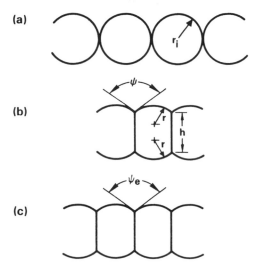

Figure 2: a) Initial, b) generalized and c) equilibrium configuration of a linear array of cylinders.

$$E = 2\pi r_i \gamma_s \left[\frac{\pi - \phi}{\pi} \left[1 - \frac{1}{2} \frac{\cos \frac{\phi_e}{2}}{\cos \frac{\psi}{2}} \right] R + \left[\frac{1}{R} + \frac{R}{\pi} \sin \phi \right] \frac{\cos \frac{\phi_e}{2}}{\cos \frac{\phi}{2}} \right] \qquad 3$$

The equilibrium structure[1] is determined by minimizing particle energy with respect to the geometrical variables ϕ and R (assuming constant dihedral angle (ϕ_e)), i.e.

$$\delta E/\delta \phi = \delta E/\delta R = 0 \qquad\qquad 4$$

Solving these equations (for constant particle volume) it can be shown that the configuration of minimum energy occurs when:

---

[1] It can also be shown that the second derivative is >0, as required for a minimum energy condition.

$$\phi = \phi_e \qquad\qquad 5a$$

$$R = [\pi/(\pi - \phi_e + \sin\phi_e)]^{\frac{1}{2}} \qquad\qquad 5b$$

The equilibrium structure is shown in Figure 2c. Figure 3 illustrates the particle energy (per unit length of cylinder, normalized by its initial energy, $2\pi r_i \gamma_s$) as a function of the contact angle ϕ. As shown, the energy per particle exhibits a minimum when the contact angle is equal to the dihedral angle. Also, larger dihedral angles result in a larger free energy decrease. The same conclusions are reached with arrays formed with spheres[19].

For the calculations discussed above, particle centres were allowed to approach one another (e.g. mass transport paths involving grain boundary and/or volume diffusion). Similar calculations can be performed for the case where the particle centre distance remains unchanged (e.g. mass transport paths involving evaporation-condensation and/or surface diffusion). It can be shown[19] that the minimum energy achieved for both centre approach and fixed centre distance is nearly identical for dihedral angles ≤100°, and the centre approach case becomes increasingly favourable with increasing dihedral

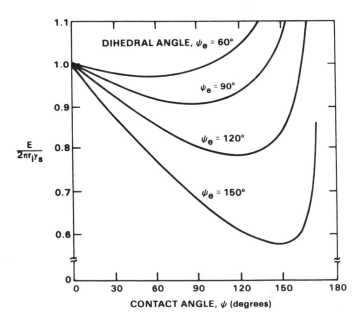

Figure 3: Normalized particle energy per unit length of cylinder as a function of the contact angle (ψ) for different dihedral angles (ψ_e).

angles.   This result suggests that much of the excess energy needed for
densification via particle centre approach may be dissipated if the initial
mass transport path favours evaporation-condensation and/or surface
diffusion.   This is relevant to ceramic processing since it is generally
believed that evaporation-condensation and surface diffusion may dominate at
low temperatures, i.e. during them initial heating to sintering temperatures.

Closed arrays that contain a single pore better represent conditions
within a powder compact.   The pore within the array is defined by the number
of coordinating particles (pore coordination number, n) and its size (R_p)
defined as the radius of the circumscribed circle (or sphere) as shown in
Figure 4.

Using the same method defined by equation 2, the energy per particle
can be calculated for rings of cylinders, rings of spheres and polyhedra
formed with spheres[19].   The energy per particle for the ring array formed
with identical cylinders is plotted in Figure 5 as a function of the pore
size for pores coordinated by 5, 10 and 20 particles with a dihedral angle
of 150°.   As shown, larger coordinated pores shrink to an equilibrium size,

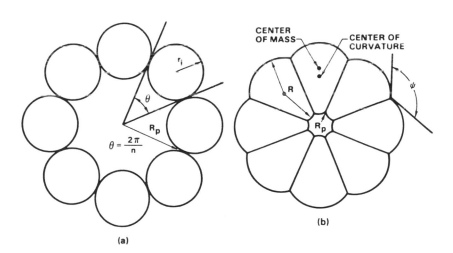

Figure 4: Schematic of a ring of cylinders (or spheres) of coordination
number, n = 8. a) Initial configuration, and b) intermediate (or final)
configuration as $\phi \rightarrow \phi_e$.

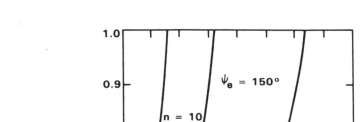

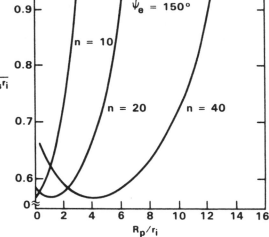

Figure 5: Normalized particle energy per unit length of cylinder, as a function of pore radius.

whilst others disappear. It can be shown[19] that a critical coordination number exists (n_c = $2\pi/(\pi-\psi_e)$) such that when n > n_c, the pores are thermodynamically unstable and will disappear. The same conclusions can be reached for rings of spheres and polyhedra formed with spheres.

3.4 Stability Conditions for Isolated Pores

The change in free energy of an isolated pore within a large, but finite body can be determined with respect to its volume change, dV_p, in a similar manner as described above[20]:

$$dE/dV_p = \gamma_s[dA_s/dV_p + 2(dA_b/dV_p)\cos(\phi_e/2)]$$

6

As shown in Figure 6, the exterior dimension of the cylindrical body is described by R_{ext} and contains pores surrounded by identical grains.

As detailed elsewhere[20], it can be shown that:

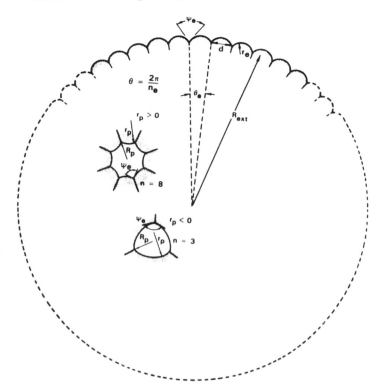

Figure 6: Schematic of two isolated pores with different surface curvatures within a finite polycrystalline body.

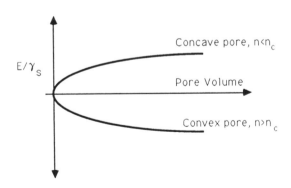

Figure 7: Energy of isolated pores vs pore volume for $n > n_c$ and $n < n_c$.

$$dE_p/dV_p = -\gamma_s/r_p \tag{7}$$

where r_p is the radius of curvature of the pore's surface.

Appropriately, equation 7 is the Gibbs-Kelvin equation. The free energy of a pore is schematically illustrated in Figure 7 as a function of its volume. This figure suggests that pores with convex surfaces decrease their energy by growing, and pores with concave surfaces decrease their energy by disappearing. These conclusions, initially reported by Kingery and Francois[16], are however incorrect since one must also include the free energy change of the body's external surface before drawing conclusions concerning pore stability/instability conditions.

If mass is transported from the external surface to the pore then the external surface will contract decreasing both surface area and grain boundary area. The differential energy change in the external region (dE_e) with volume change (dV) is:

$$dE_e/dV = \gamma_s[dA_s/dV + 2(dA_b/dV)\cos(\phi_e/2)] \tag{8}$$

Using the geometry shown in Figure 6, it can be shown that[20]:

$$dE_e/dV = -\gamma_s/r_e \tag{9}$$

Since the exterior surface curvature is always positive, the exterior surface energy always increases with increasing volume.

If we assume that the change in pore volume is equal to the change in specimen volume ($dV_p = dV$), then the total change in system energy with respect to pore volume is:

$$dE_t/dV = \gamma_s(1/r_e - 1/r_p) \tag{10}$$

(the sum of equations 7 and 9), and is a function of both the curvature of the pore and the grains on the external surface.

As shown in Figure 7 concave pores ($r_p < 0$, $n < n_c$) increase their energy with pore volume, implying that these pores will continuously shrink or disappear, mass transport permitting. Convex pores ($r_p > 0$, $n > n_c$) will either shrink or grow depending on the relative curvature values, i.e. when

$r_e > r_p$ pore growth decreases system energy and for $r_e < r_p$ pore shrinkage decreases system energy. Equilibrium ($dE_t/dV = 0$) occurs when:

$$r_e = r_p \qquad\qquad 11$$

suggesting that pores with convex surfaces ($n > n_c$) will be stable when their curvature equilibrates with that of the exterior surface.

Equating the curvature of grains on the surface to the grain size, D, (see Figure 6):

$$r_e = D/[2\cos(\phi_e/2)] \qquad\qquad 12$$

it can be shown that at equilibrium, the pore size is dependent on the grain size and the pore coordination (n):

$$R_p = D\left[\frac{\cos\left[\dfrac{\phi_e}{2} + \dfrac{\pi}{n}\right]}{2\cos\dfrac{\phi_e}{2}\sin\dfrac{\pi}{n}}\right] \qquad\qquad 13$$

Equation 13 shows that when $n \to n_c = 2\pi/(\pi - \psi_e)$ the pore disappears ($R_p \to 0$) and when $n > n_c$ the pore is stable with a finite size ($R_p > 0$).

3.5 Grain Growth and Densification

With these new ideas in mind, it may be concluded[21] that grain growth, which is expected to reduce the pore coordination number of stable pores, may be both desirable and necessary for complete densification. As evidence, grain growth has been observed during all stages of densification. For example, Greskovich and Lay[22] observed grain growth in very porous alumina compacts containing 70% void space whilst Gupta[23] analyzed previous data on Cu, Al_2O_3, BeO and ZnO to show that grain size increased linearly, though very slowly, with density until the latter reached 90% of theoretical. At these higher densities the grain size increased very rapidly. With the hypothesis that grain growth (or coarsening) during densification must be linked to the densification process itself, both a theoretical and experimental study are required[24]; the first task being to determine how grains coarsen within a partially dense network.

If the assumption is made that the initial particles within a compact are spherical, but differ in size, then it may be reasoned that, because of their different radii, a driving force exists for interparticle mass transport (i.e. coarsening or grain growth) as well as for intraparticle transport to the region where the particles contact (i.e. sintering). It may be further assumed that because i) the differential surface curvature between the contact region and either adjacent, touching particle is much greater than that between the particles themselves, and ii) because the diffusion distance required to increase the contact area could be much smaller than the average diffusion distance between the particles, neck growth (sintering) will initially dominate mass transport phenomena relative to interparticle mass transport (coarsening).   Thus, although sintering and coarsening are concurrent phenomena, it may be assumed that sintering occurs first, followed by grain growth.

Figure 8 illustrates the sintering and coarsening of a three particle array where the smaller particle is sandwiched between two identical, larger particles.   Sintering without interparticle diffusion (each particle retains

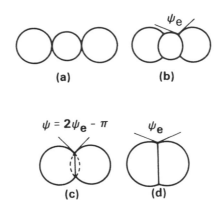

Figure 8: Configurational changes produced during sintering and interparticle mass transport for three co-linear particles.   Note that disappearance of smaller grain (c) reinitiates sintering.

their initial mass) results in the configuration shown in Figure 8b. It can be shown[24] that this is the configuration where sintering between the three particles is complete, i.e. where the driving force for neck growth has diminished to zero. Note that the grain boundaries exhibit curvature because of the different radii of adjacent particles. In dense materials, grain boundaries will move toward their centre of curvature by very short range diffusion (hopping of atomic species across the boundary). However, the grain boundaries shown in Figure 8b can not simply move as they would in a dense material without increasing their area and thus the free energy of the particle array. That is, if either of the two boundaries were to move into the smaller grain they would increase their area and encounter an energy barrier for further motion.

Once interparticle mass transport is 'turned on', the smaller grain shown in Figure 8b will decrease its size as the size of the two adjacent grains increases. During this period it is assumed that the dihedral angle is maintained between the grain boundary and the surface of the particle. The atom movement will eventually lead to the configuration shown in Figure 8c, where the two larger grains now touch one another and the grain boundaries (shown by broken lines) that still define the smaller particle move together and decrease their energy to form a single grain boundary (solid line). Up to this configuration, boundary motion without changes in the surface configuration will always be impeded by an energy barrier. For this reason, one might conclude that grain boundary motion in a partially dense network will be constrained by the surface configuration and that grain coarsening will require interparticle mass transport. It can then be reasoned that longer range, interparticle mass transport might govern grain growth in a porous material whilst much shorter range transport governs boundary motion after densification. This reasoning is consistent with the observations of Gupta[23].

The relation between grain growth and densification first became obvious by recognizing the implications of Figure 8c. If it is assumed that the two grains in the figure are identical, interparticle mass transport is terminated once the sandwiched, smaller grain disappears. Of significance is the fact that once the two larger particles touch one another the angle between the grain boundary and the surface is less than the dihedral angle. The new local configuration that has arisen reinitiates intraparticle mass transport to the contact region and ultimately leads to the configuration shown in Figure 8d. Similar conclusions may be reached for more complex particle arrays and networks.   Simply stated, grain growth reinitiates sintering.

Now that a relation between grain growth and sintering has been established it is pertinent to establish a relation between grain growth and densification.   Assuming that sintering produces shrinkage, analytical expressions may be established[24] for simple one-, two- or three-dimensional particle arrays where the changing distance between the grain centres is related to grain growth. For the purposes of illustrating these calculations and results, consider a symmetric ring containing two different sets of identical spheres with nearly the same initial radii. It is assumed that the spheres within the ring undergo the sequential cycles of sintering, grain growth and sintering in a similar manner to those described above. Once this array has undergone one full cycle the number of spheres in the ring will be reduced by a factor of 2.   The array is then subjected to further cycles with the assumption that every other grain is somewhat smaller than its neighbour and will disappear via interparticle mass transport.   The number of cycles required to close the pore within the ring will depend on the number of initial particles within the ring and the dihedral angle as discussed in the previous section. The relative grain size can be determined after each cycle since the number of grains is reduced by a factor of two. The shrinkage of the ring, and thus the density of the disc that defines the ring, can also be calculated after each cycle.   It can be shown that this relative density (ρ_{m+1}) after each cycle of grain growth and sintering (defined as 'm') can be expressed as a function of the new grain size after each cycle (D_m) and the dihedral angle (ψ_e) as:

$$\rho_{m+1} = \frac{1}{3}\left[\frac{2r_i}{D_m}\right]^3 \cos\left[\frac{\psi_e}{2}\right]\left[3 - \cos^2\left[\frac{\psi_e}{2}\right]\right]\left[\cos\left[\frac{\psi_e}{2}\right]\left[1 + \left[\sin\frac{2^m\pi}{n_o}\right]^{-1}\right]\right]^{-2} \qquad 14$$

where n_o is the initial number of spheres of radius r_i in the ring, and D_m = $2r_i(2^{m/3})$. Figure 9 illustrates a plot of equation 14 for 3 different dihedral angles, showing that the relative density/grain size relation is a near linear function until relative densities > 90%, consistent with Gupta's experimental review.   Similar relations are obtained for three-dimensional arrays[23].

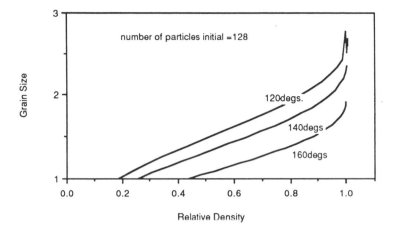

Figure 9: Density of disc containing ring of spheres as a function of grain size.

It is significant to note that the contribution of grain disappearance to shrinkage is very small relative to that for sintering and is equal to zero when the dihedral angle is > 130°. The reason is that the larger, neighbouring grains grow as fast as the smaller, sandwiched grain shrinks, leading to little or no displacement of the larger grain centres. This surprising result was not predicted by intuition.

In summary, it can be shown that grain growth in a partially sintered particle network reinitiates the sintering process which leads to further shrinkage of the network and densification. Although the grain growth phenomenon does not directly contribute to shrinkage, per se, it is the interparticle mass transport and its kinetics which is responsible for densification once the initial particles have formed a sintered network. According to this theory, the kinetics of densification should be related to grain growth kinetics rather than those associated with sintering.

3.6 Experiments Relating Grain Growth to Pore Disappearance

Theory outlined in the previous sections strongly suggests that unless identical particles are periodically packed such that the coordination number

of all pores is less than the critical number, grain growth will be necessary
to fully densify a powder compact.   To test this hypothesis, Al_2O_3/ZrO_2
composite powders were consolidated containing a small volume fraction
(<0.01) of identical, plastic spheres (either 1μm, 2μm or 4μm diameters)
which produced the only remnant porosity observed after densification[20].
Specimens were heat treated for different periods at a sintering temperature
of 1600°C to induce grain growth.   Grain size, pore size and pore number
density measurements were made after each heat treatment.   As shown in
Figures 10 and 11, respectively, the number density of the pores decreased
with increasing grain size, whereas the size of the pores that remained after
heat treatment was unchanged.   These data are not consistent with a kinetic
view of densification in which the number density would remain nearly
constant, whilst the pore size would decrease with the heat treatment period.
Rather, the data are consistent with the thermodynamic view outlined in the
previous sections which suggest that pores will disappear once their
coordination number is lowered below the critical value by grain growth and
that the size of the remaining pores should be relatively unchanged by the
heat treatment.

4. COLLOIDAL POWDER PROCESSING

4.1 Heterogeneities Associated with Powder Processing

The previous sections conclude that the sinterability of a given powder
can best be optimized by packing the particles to produce pores with
coordination numbers as low as possible.   Unfortunately, powder processing
is a 'many bodied' problem prone to the introduction of heterogeneities,
where the major cause of these heterogeneities can be the powder itself.

Nearly all current powders of interest are agglomerated, i.e. dry
powders with average particle sizes of about 1 μm are composed of large, weak
agglomerates consisting of particles held together by capillary forces due
to condensed moisture, Van der Waals forces, or other short-range adhesive
forces.   Most powders also contain strong agglomerates comprising chemically
bonded particles.   These are formed either by partial sintering during
pyrolysis of precursors (e.g. salts) or by 'cementation' (e.g. bonding due
to hydroxides) during powder drying.   Agglomerates pack together during

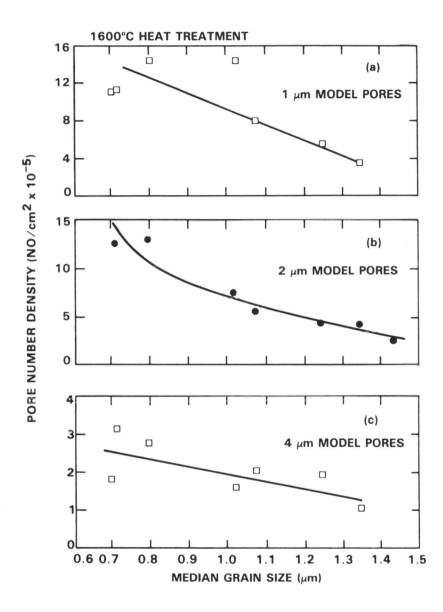

Figure 10: Number of pores observed per unit area in three materials containing pores produced with plastic spheres after post densification heat treatment to produce grain growth.

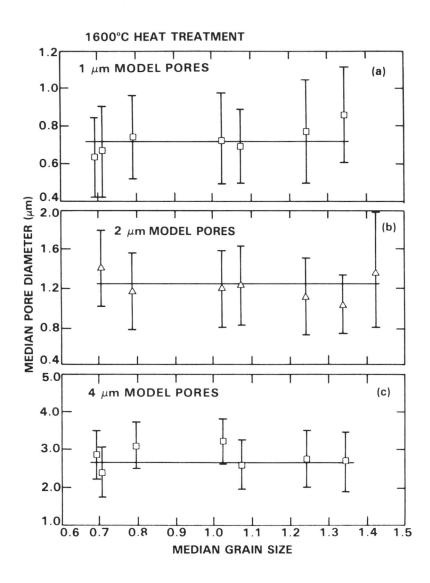

Figure 11: Size of pores observed in three materials containing pores produced with plastic spheres after post densification heat treatment to produce grain growth.

consolidation to produce compacts with large differential packing densities leading both to poor densification for reasons discussed above and the formation of crack-like voids that become a major strength degrading flaw population[5].

Besides the crack-like voids formed during densification, conventional powder processing is subject to the introduction of many other heterogeneities that include: inorganic and organic inclusions, poorly distributed second phases, particle packing that encourages abnormal grain growth, etc. It is now recognized that each type of heterogeneity is a potential strength degrading flaw population. Inadvertently, typical processing procedures introduce a variety of flaw populations that limit the potential strength of the product. Slight changes in processing variables can, unknowingly, produce a different flaw population. New processing methods with a high probability of either controlling and/or eliminating microstructural heterogeneities must be developed in order to ensure engineering reliability.

Heterogeneities introduced during processing are best uncovered by observing fracture origins. The flaw must be identified and related to a processing step and new processing methodology implemented to eliminate it. This achieved, new fracture origins will uncover a new, but less severe heterogeneity which in turn must be identified, related to processing and removed by changes to the production route. This is therefore an iterative process.

Figure 12 presents the chronological increase in the mean strength for three transformation toughened materials as a function of such iterative processing changes. As reported elsewhere[5], large two phase agglomerates that produce crack-like voids during densification were observed at fracture origins after dry pressing (Process Step 1). These large agglomerates were produced after drying milled powder slurries. They were eliminated by removal of the drying step and consolidating directly from the slurry by slip casting[25] (Process Step 2). Smaller agglomerates were then observed at subsequent fracture origins, which were later identified as hard, partially sintered agglomerates present in one of the two powders (the zirconia) used to produce the two phase material. To remove this problem, all agglomerates larger than approximately 1 μm were fractionated from both powders by sedimentation prior to powder mixing and slip casting (Process Step 3)[26]. Irregularly shaped voids were then discovered as the fracture origins and

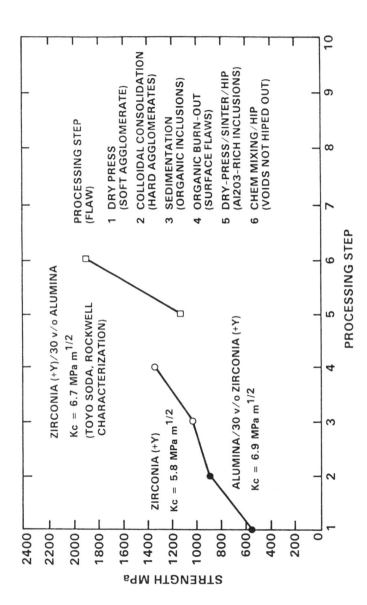

Figure 12: Mean strength vs processing changes for three transformation toughened materials.

were found to be caused by organic inclusions which burnt out during sintering. These were eliminated by burning out the organics present in the powder compact at a low temperature, followed by isostatically pressing the compact at room temperature (Process Step 4) prior to densification at high temperatures[26].

Hot isostatic pressing (HIPing) can be used as a post-densification processing step to close voids left after pressureless sintering[27]. However, whilst HIPing can be effective in removing some detrimental voids, such as crack-like voids produced by differential densification of agglomerates[5], it is ineffective in eliminating other heterogeneities. The zirconia/alumina material processed by Toyo Soda Ltd[2] is post-sintered HIP-treated to increase strength by eliminating detrimental voids (Process Step 5). Large alumina-rich, two-phase inclusions were observed as the fracture origin in this material. This heterogeneity stems from poor mixing of the two powder phases and as such can not be removed by HIPing. It was removed when Toyo Soda researchers improved mixing procedures (Process Step 6) which resulted in a significant strengthening.

4.2 Colloidal Methods for Preparing and Consolidating Powders

Many of the processing methods used to eliminate heterogeneities discussed in the previous section involved various aspects of what has become known as colloidal processing. This approach involves the manipulation of interparticle forces. Colloidal methods being used, for example, to; break apart weakly bonded agglomerates, eliminate strongly bonded agglomerates by sedimentation, fractionate desirable particle size distributions, eliminate inorganic and organic inclusions greater than a given size, homogeneously mix two or more powders, store powder slurries without mass segregation and consolidate powders to very high bulk densities. The potential usefulness of colloidal methods in developing reliable processing methods for advanced ceramics can not be over-emphasized (see also Chapter 3).

Colloidal phenomena involve interparticle forces in liquids, liquid mixtures, or vapours. Certain aspects of the colloidal processing of ceramics require repulsive interparticle forces whilst others require attractive forces. For example, repulsive forces produce pourable slurries

---

2 Toyo Soda Mfg Co Ltd, Tokyo, Japan.

containing up to 60 volume percent solids, whereas attractive forces cause the same slurry to behave nearly like a solid. A number of basic interactions can be used to alter interparticle forces. These include attractive Van der Waals forces, repulsive electrostatic forces, solvation (i.e. short range hydration or hydrophobic) forces and steric forces, in addition to attractive capillary (Laplace) forces[28]. The solvation and steric forces can be attractive or repulsive in nature.

With the electrostatic approach, ions or charged molecules are attracted to or dissociated from the particle surfaces to produce a system of similarly charged particles. When the repulsive 'double-layer' electrostatic forces between the particles are greater than the attractive Van der Waals forces, the particles repel to produce a dispersed system. The net interparticle force (in aqueous solutions) can be altered by changing the type of concentration of the ionic species as well as the pH. When the particle charge approaches zero, the particles floc and eventually produce a very open network of touching particles. The zeta potential provides a convenient experimental measure of such forces.

With the steric approach, bi-functional macromolecules attach themselves to the particles. The macromolecular additives are usually completely soluble in the dispersing fluid, but are designed with certain functional groups to bind them to the particles. The particles then repel one another once their separation distance is less than the radius of gyration of the macromolecule[29]. Flocculation usually occurs when the conditions are changed to make the polymer insoluble in the solvent medium. This can be controlled by changing the temperature, by adding another fluid to the system, by changing the polymer molecular weight and, of course, by changing the polymer.

Polyelectrolytes are multi-charged macromolecules that may absorb or anchor to surfaces of ceramic particles in a polar (often aqueous) medium. Observations suggest that certain polyelectrolytes can be most effective in dispersing a wide range of ceramic powders. Although we lack the basic understanding of why some are quite so effective, it is known that they can produce an electrostatic 'double-layer' surface charge, thereby opposing the attractive Van der Waals forces, in addition to offering the possibility of steric interaction. The interparticle forces produced by polyelectrolytes may be controlled by altering pH, ionic strength, temperature, molecular weight, etc.

While the colloidal method may appear straightforward, its success depends on an understanding of interparticle forces and how to manipulate them. Although such forces have a strong theoretical base, verified through direct surface force measurements[28], the choice of the best surface active agent to use to control the forces is still a matter of trial and error for most ceramic systems.

The application of the colloidal method is certainly not new to ceramic powder preparation, for example, ancient applications include the ageing of clays for hand moulding. In addition, colloidal methodology has been applied for more than one hundred years to prepare clay slurries for casting into porous moulds, usually plaster-of-paris, to form thin walled, complex shaped bodies. Colloidal methods are also commonly used to fractionate ceramic abrasive media to obtain a desired narrow particle size distribution, however their use prior to consolidation of a ceramic body to eliminate or reduce in size the common heterogeneities associated with ceramic powders, such as agglomerates and inclusions, is relatively rare.

Figure 13 shows schematically one colloidal approach to treat and store ceramic powders prior to consolidation[26]. As received, dry powders are

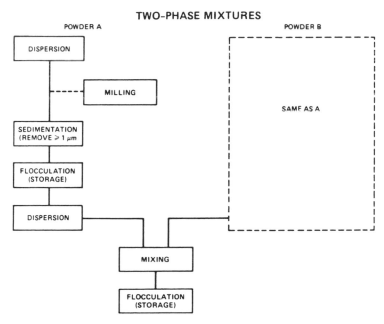

Figure 13: Colloidal method for treating powders to fractionate, mix and store.

dispersed in an appropriate fluid with a surfactant that produces interparticle repulsive forces which break apart the weak agglomerates. Partially sintered and other agglomerates, which are too strong to be broken apart by the surfactant, and inorganic inclusions greater than a desired size, are eliminated by sedimentation. This step can be automated by centrifuging. After removal of undesired particulates, the remaining dispersed slurry is adjusted to contain the desired powder fraction and is flocced by changing the interparticle forces from repulsive to attractive. Floccing concentrates the particles to form a continuous 'touching' network which consolidates under its own weight, partially separating the particles from the fluid phase. Flocced slurries can also be washed to remove excess salts and/or surfactants used to manipulate the interparticle forces. Centrifuging can further concentrate this particle network. Floccing also prevents further mass segregation during storage even when acted upon by centrifugal forces[30].

Figure 13 also shows that two or more powder phases, separately treated as summarized above, can be mixed together to form multiphase slurries. If the phases are colloidally compatible, i.e. do not floc one another, each flocced slurry can be redispersed (by again adding the proper surface agent that produces repulsive interparticle forces) and mixed. More commonly, the phases are not compatible, but because flocced mixtures can be mechanically redispersed by means of a device that produces a high shear-rate (an ultrasonic horn, high speed rotors, etc), the different flocced slurries can be mixed within such a high shear rate field. The mixture subsequently flocs to form a new mixed particle network which prevents phase separation during storage and further processing.

The question of how uniform the mixture is and what tools can be used to define this property has recently been addressed[31]. The method used for phase uniformity is quite simple[31], can be directly related to the rheology of the slurry mixture[32] and can be used as an inline processing sensor to assess mixing uniformity.

When a multiphase body is observed in a scanning electron microscope, the non-dispersive X-rays can be collected to obtain an energy (EDX) spectrum representing the different elements. This spectrum can be analyzed to quantitatively define the atomic fraction of each element and thus, if different elements are associated with different phases, the content of each phase within the area scanned by the electrons. At low magnifications, the

EDX spectrum defines the phase content within the large body.   With reasonable counting periods, the standard deviation for different areas examined at low magnifications is low (usually <3%) but is associated with counting errors.   At very high magnifications, the area examined may not be representative of the large body and the deviation of the spectrum relative to the large body can be very large.   At some intermediate magnification, the standard deviation will begin to depart from that produced by counting errors.   At this magnification, the area scanned is statistically identical (within an acceptable standard deviation somewhat larger than that due to counting errors) to the large body.   The size of this area thus defines the smallest area that contains the same phase distribution as the whole body. This area (A_{um}) can be defined quantitatively and used to represent the phase uniformity after the phases are mixed together during processing.   The better the mixing, the smaller the value for A_{um}.   This parameter is thus an extrinsic property of the multiphase material that depends solely upon processing.

Values of A_{um} can be related to different mixing methods, different mixing periods, e.g. different resident periods for mixed slurries within a high shear rate field, and to the properties of the mixed slurry itself.   It may also be related to the viscosity of flocced, two phase slurries[34] and it is possible that an inline viscosity measurement could be used to determine phase uniformity (as defined by a number, that is A_{um}) during processing.

4.3 Consolidating Powders Consistent with the Colloidal Method

If colloidally treated powders are dried, weak to moderately strong agglomerates will reform.   In addition, drying can re-expose the powder to unwanted environments (e.g. dust) which are expensive to control.   That is, dry pressing methods of consolidating colloidally treated powders is subject to great uncertainty that old problems associated with heterogeneities will rise again to plague processing reliability.   Thus, colloidally treated powders should never be dried and never exposed to an uncontrolled environment until the consolidated body has been formed.   This can simply be accomplished by piping the slurry directly to the consolidation machine.

The current consolidation methods listed in Table 1 use slurries and are based on some knowledge of interparticle forces[3]. With modifications, all could be adapted to the fractionation procedures and stringent environmental control requirements outlined above. Unfortunately, each suffers from some limitation, also briefly indicated in the table. For this reason, new consolidation methods must be developed that are consistent with the colloidal powder processing methods.

Method	Major Limitation
Slip casting (fluid removed by filtration)	Parabolic kinetics (Darcy's Law) too slow; shapes limited to thin walls
Tape casting (fluid removed by evaporation)	Limited to thin sheets, large residual polymer content
Electrophoresis (charged particles attracted to an electrode)	Shape, wall thickness very limited
Injection moulding (highly filled dispersion frozen during injection into a die)	40 to 50 volume percent polymer must be eliminated prior to densification: limits thickness of object

Table 1: Current state of slurry consolidation techniques.

New consolidation methods are being developed using the understanding of how interparticle forces control rheology and how rheological behaviour can be changed with applied pressures. Directions leading to these new consolidation methods might be conceptually envisaged by examining the rheological behaviour of dispersed and flocced slurries containing a high volume fraction of the particulate phase. Pourable, dispersed slurries can contain up to 60 volume percent of the particulate phase. The viscosity of these highly filled, dispersed slurries increases with shear rate (they flow under gravitation forces, but become 'rigid' when quickly sheared). Conversely, the viscosity of highly-filled, flocced slurries, produced by centrifuging and pressure filtration, decreases with increasing shear rate.

---

[3] It is interesting to note that the additives used to increase particle loading and decrease viscosity of injection moulding slurry systems are rarely defined as agents that alter surface forces, nor is the system viewed in terms of colloid chemistry (34).

The pseudo-plastic behaviour of flocced slurries can be advantageously used for pressure moulding.

With this rudimentary knowledge, it is apparent that interparticle forces might be manipulated (or applied stresses used) to dramatically change rheology, with the objective of forming complex shapes having mechanical integrity prior to drying. Several examples can be suggested:

1)   Dispersed slurries can be poured into a complex mould and then 'solidified' by changing the interparticle forces from repulsion to attraction, viz. by in-situ flocculation,

2)   Dispersed slurries can be injected into a complex die and then made more 'rigid' by increasing the strength of the attractive interparticle forces, e.g. by gelation[34],

3)   Highly-filled, flocced slurries with a 'yield stress' suitable for shape retention can be liquified when coupled to the high amplitude vibrations produced by an ultrasonic device (the slurry then 'freezes' once the device is turned off), etc.

Pressure filtration also produces dramatic changes in rheology as the fluid phase is removed. Unlike injection moulding, ceramic slurries used for pressure filtration fill the die cavity prior to the application of pressure. This attribute of pressure filtration avoids many of the flow sensitive problems (and thus flaw populations and non-uniform microstructures) inherent with injection moulding. Preliminary experiments have shown that the layer of packed particles that consolidate as fluid passes through the filter can fill a complex shaped die cavity to produce a complex shaped engineering component[35]. In addition, pressure filtration devices can be designed so that the fluid phase is removed by in situ percolation after consolidation. These combined attributes strongly suggest that pressure filtration can become the new method for consolidating reliable ceramic components.

Pressure filtration experiments have been conducted with Al_2O_3 powder in which conventional methods (controlled pH) are used to alter interparticle forces[36]. Conditions that produce the maximum interparticle repulsive forces result in an extremely high particle packing density (relative densities up to 70%) which is relatively insensitive to the applied pressure. On the other hand, the packing density for conditions that produce the maximum attractive interparticle forces is very pressure sensitive. These results suggest that repulsive interparticle forces would be desirable. However, two significant problems exist. First, the consolidated layer produced by the dispersed slurry is dilatant and does not retain its shape after consolidation. Second, because of the very high packing density, the permeability is more than an order of magnitude lower relative to the lower density compacts (60% relative density) produced at the same pressure when the interparticle forces are attractive. These results show that interparticle forces control packing density, pressure filtration kinetics and mechanics, and the rheology of the consolidated compact.

These same pressure filtration studies have shown that one problem of pressure filtration is that time-dependent stress redistribution occurs on pressure removal[35]. That is, relief on the strain energy stored in the connective particle network during pressure filtration is time dependent due to fluid flow. Stresses that arise within the consolidated body upon pressure removal can cause cracks to form in the compact after the body is removed from the moulding die. It was shown that this cracking problem can be avoided with small additions (about 1 volume %) of certain polymers.

5. CONCLUDING REMARKS

Within the last few years, new concepts have emerged relating particle arrangement to densification and the control of particle arrangement through colloidal methods. The new concepts concerning densification suggest that the microstructure development during densification, viz. grain growth, can control the thermodynamics and kinetics of the densification process itself. They explain how packed agglomerates produce pores within the powder compact that can be thermodynamically stable unless grain growth can lower the coordination number to allow the pore to disappear. They explain how forces develop between particles that can cause rearrangement when the particles are not identical and periodically arranged. In essence, these new concepts emerged by examining the pores within a powder compact as well as the particles. The new concepts concerning colloidal methods were developed out of a need to eliminate and/or minimize heterogeneities common to the powders. These concepts are not really new, but have been borrowed from related sciences and technologies and adapted to advanced ceramic processing.

Innovative slurry consolidation methods must be developed to implement these new colloidal methods which have the potential to increase processing reliability and thus property reliability. This will be the major challenge for tomorrow's process engineer.

REFERENCES

1. Raj, R., Morphology and stability of the glass phase in glass ceramic systems. J. Amer. Ceram. Soc. 64 [5] 245-8 (1981).

2. Raj, R. and Lange, F.F., Crystallization of small quantities of glass (or a liquid) segregated in grain boundaries. Acta Met. 29 [12] 1993-2000 (1981).

3. Scherer, G.W., Drying gels: I. General theory. J. Non-Cryst. Sol. 87 199 (1986).

4. Zarzycki, J., Monolithic xero and aerogels for gel-glass process. In Ultrastructure Processing of Ceramics, Glasses and Composites; Hench, L.L. and Ulrich, D.R. (Eds), Wiley (1984).

5. Lange, F.F., Processing related fracture origins: I. Observations in sintered and isostatically hot-pressed composites. J. Amer. Ceram. Soc. 66 [6] 398-9 (1983).

6. Lange, F.F., Davis, B.I. and Aksay, I.A., Processing related fracture origins: Part III. Differential sintering of ZrO_2 agglomerates in Al_2O_3/ZrO_2 composites. J. Amer. Ceram. Soc. 66 [6] 407-8 (1983).

7. Rhines, F.N., In Science of Sintering and its Future; Ristic, M.M. (Ed), International Institute for the Science of Sintering, Belgrade (1975).

8. Exnor, H.E., Principles of single phase sintering. Reviews on Powder Metallurgy and Physical Ceramics 1 [1-4] 1-251 (1979).

9. Rhodes, W.H., Agglomerate and particle size effects on sintering yttria-stabilised zirconia. J. Amer. Ceram. Soc. 64 [1] 19 (1981).

10. Lange, F.F. and Hirlinger, M.M., Hindrance of grain growth in Al_2O_3 by ZrO_2 inclusions. J. Amer. Ceram. Soc. 67 [3] 164 (1984).

11. Lange, F.F., Fracture toughness of silicon nitride as a function of the initial alpha phase content. J. Amer. Ceram. Soc. 62 [9-10] 428 (1979).

12. Iskoe, J.L., Lange, F.F. and Diaz, E.S., Effect of selected impurities on the strength of Si_3N_4. J. Amer. Ceram. Soc. 11 908 (1976).

13. Kuczynski, G.C., Self-diffusion in the sintering of metallic particles. AIME 185 169 (1949).

14. Frost, H.J., Overview 17: Cavities in dense random packing. Acta Met. 30 [5] 899-904 (1982).

15. Coble, R.L., Diffusion sintering in the solid state. pp147-63 in Kinetics of High Temp Proc., Ed by W.D. Kingery. Joint Publ. of the Technology Press of MIT and J. Wiley, NY. (1959).

16. Kingery, W.D. and Francois, B., Sintering of crystalline oxide, I. Interaction between grain boundaries and pores. In Sintering and Related Phenomena; Kuczynksi, G.C., Hooten, N.A. and Gilbon, G.F. (Eds), 471-98, Gordon Breach (1967).

17. Cannon, R.M., On the effects of dihedral angle and pressure on the driving force for pore growth or shrinkage. Unpublished manuscript (1981).

18. Hoge, C.E. and Pask, J.A., Thermodynamics and geometric considerations of solid state sintering. Ceramurgia International $\underline{3}$ [3] 95-9 (1977).

19. Kellett, B.J. and Lange, F.F., Thermodynamics of densification, Part I: Sintering of simple particle arrays, equilibrium configurations, pore stability and shrinkage. J. Amer. Ceram. Soc. $\underline{72}$ [5] 725-34 (1989).

20. Kellett, B.J. and Lange, F.F., Thermodynamics of densification, Part III: Experimental relation between grain growth and pore closure. J. Amer. Ceram. Soc. (in review).

21. Lange, F.F., Sinterability of agglomerated powders. J. Amer. Ceram. Soc. $\underline{67}$ 83 (1984).

22. Greskovich, C. and Lay, K.W., Grain growth in very porous Al_2O_3 compacts. J. Amer. Ceram. Soc. $\underline{55}$ [3] 142-6 (1972).

23. Gupta, T.K., Possible correlations between density and grain size during sintering. J. Amer. Ceram. Soc. $\underline{55}$ [5] 176 (1972).

24. Lange, F.F. and Kellett, B.J., Thermodynamics of densification, Part II: Grain growth in porous compacts and relation to densification. J. Amer. Ceram. Soc. $\underline{72}$ [5] 735-41 (1989).

25. Aksay, A.K., Lange, F.F. and Davis, B.I., Uniformity of Al_2O_3-ZrO_2 composites by colloidal filtration. J. Amer. Ceram. Soc. $\underline{66}$ C-190 (1983).

26. Lange, F.F., Davis, B.I. and Wright, E., Processing-related fracture origins: IV. Elimination of voids produced by organic inclusions. J. Amer. Ceram. Soc. $\underline{69}$ [1] 66-9 (1986).

27. Engle, V. and Hubner, H., Strength improvement of cemented carbides by hot isostatic pressing. J. Mat. Sci. $\underline{13}$ [9] 2003-13 (1978).

28. Israelachvili, J.N., Intermolecular and surface forces. Academic Press, London, (1985).

29. Israelachvili, J.N., Tandon, R.K. and White, L.R., J. Colloid Interface Sci, $\underline{78}$, 430, (1980).

30. Lange, F.F., US Patent 4,624,808, Forming a ceramic by flocculation and centrifugal casting.

31. Lange, F.F. and Hirlinger, M.M., Phase distribution studies using energy dispersive X-ray spectral analysis. J. Mat. Sci. Let. $\underline{4}$ 1437-41 (1985).

32. Lange, F.F. and Miller K.T., A colloidal method to ensure phase homogeneity in beta"-Al_2O_3/ZrO_2 composite systems. J. Amer. Ceram. Soc. (in press).

33. Edirisinghe, M.J. and Evans, J.R.G., Review: Fabrication of engineering ceramics by injection moulding. I. Materials Selection. Int. J. High Tech. Ceram. 2 1-31 (1986).

34. Rivers, R.D., US Patent 4,113,480, Method of injection moulding powder metal parts, Sept. 12, 1978.

35. Lange, F.F. and Miller, K.T., Pressure filtration: Kinetics and mechanics. Bull. Amer. Ceram. Soc. (in press).

2

Processing of Silicon Nitride Powders

S.C. Danforth, W. Symons[+], K.J. Nilsen[*] and R.E. Riman

Center for Ceramics Research, Rutgers, The State University Piscataway, NJ 08855-0909, USA

+ AC Rochester, Flint, MI 48556, USA.
* Dow Chemical Co., Midland, MI 48674, USA.

1. INTRODUCTION

With the advent of many new ceramic materials, as well as methods of forming and densifying them, there has been keen interest in the potential for using ceramics in many high technology applications. These include areas such as: insulators, electronic substrates, high Tc superconductors, tool bits, wear surfaces, heat exchangers, turbine and automotive engine components, etc. While the market potential for these applications is very high, there are still several difficulties that must be overcome before the full potential of structural ceramics is realized. One of the most critical problems has been the lack of suitable raw materials (powders) from which ceramic components could be fabricated. Most traditional methods of powder synthesis yield powders with inherent "defects". These defects can often be traced to resultant microstructural flaws which ultimately limit the properties and reliability of the ceramic material[1].

Conventional powders (in this author's terminology) have the following general characteristics:
1) a large average size, >1-10 microns, with some particles in the 20-100 micron size range,
2) a very wide distribution of particle sizes, with R_{max}/R_{ave} >5,
3) a high degree of agglomeration, both weak and strong inter-particle bonds,
4) irregular morphologies,
5) impurity levels in excess of desirable limits, and

6) a lack of control over phase chemistry and crystal structure.

It is now very clear that powders with these characteristics often require special procedures to achieve full densification[2,3]. These include, but are not limited to the following: ball milling, sieving, purification, mixing, spray drying, addition of sintering aids, hot pressing, hot isostatic pressing, and use of very high sintering temperatures or long sintering times.   As increasing demands are made on properties and reliability, however, newer methods must be developed for: powder synthesis and dispersion, consolidation/shape forming, and densification.

As indicated a number of years ago, there is a great advantage to processing ceramics from powders with an idealized set of physical and chemical characteristics[1]. Such ideal characteristics include:
1) a small size, generally less than 1 micron,
2) a narrow size distribution, where $R_{max}/R_{ave} <3$,
3) an equiaxed morphology, tending towards spherical,
4) no agglomeration, or very weak agglomerate bonds which can be broken during processing, and
5) a very high degree of chemical and cyrstal phase purity.

There have been numerous investigations that have clearly demonstrated the value of working with such powders. Some of the advantages claimed include: reduced sintering times and temperatures, reduced grain size and distribution, higher sintered densities, reductions in the use of pressures or sintering aids, increased dimensional tolerances etc., all of which lead to improved property reliability[4-11]. It should be noted however, that use of such ideal powders alone, is insufficient to insure the realization of the potential benefits listed above. Appropriate methods for dispersing and subsequently consolidating these powders, that will result in green bodies that have the maximum degree of microstructural uniformity, must also be developed.

It should be noted that this set of ideal powder characteristics has been found to be generally useful for many ceramic systems. One must, however, be careful in determining the exact set of characteristics that are appropriate for the particular application at hand, i.e., the exact characteristics for a Si_3N_4 powder for structural applications will be different from those required for a 1-2-3 high T_c superconducting powder.

At least two schools of thought have developed in terms of the proper approach to take toward processing science.   One school states that the majority of technical problems can be solved through the use of such idealized powders, without the use of additives and external pressures.   The other school advocates the use of sintering additives and external pressures (where needed), owing to the high degree of difficulty (and cost) of working with such ideal powders.   Harmer appropriately states that it is not simply a question of one approach or the other[12].   No matter how perfect the powder may be, the statistical probability of forming a packing defect or other microstructural non-uniformity in the green body is so high, that one may still need small amounts of uniformly dispersed additives and/or external pressures to reach   >99% theoretical density, especially for covalent ceramics or reinforced composites.   The competition between these two schools of thought, and the worlds of practical and theoretical ceramics are shown schematically in Figure 1.   It is clear that a balanced prospective may indeed be the wisest.

There has been considerable interest in developing ceramic materials for use in advanced heat engines.   Possessing low density and excellent thermomechanical properties, ceramics provide a means for producing heat

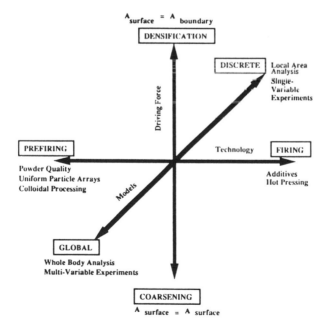

Figure 1: Sintering space.   A concise summary of the science and technology of sintering.

engines with efficiency ceilings far above what is presently possible with today's super alloys. One material, silicon nitride, has received considerable attention due to its high decomposition temperature (approximately 1880°C) as well as superb intrinsic creep properties[13,14].

Exploitation of silicon nitride's intrinsic properties, however, is not presently possible. Current practice relies on the addition of one or more oxide sintering additives (ie Y_2O_3, MgO, CeO, Al_2O_3, and BeO) to promote densification[15-17]. By reacting with the native SiO_2 layer (typically 0.5-3wt%) present on the starting alpha-Si_3N_4 powders, densification is accomplished through a liquid phase sintering process. The SiO_2 reacts with the sintering aids to form a liquid, which then dissolves the α-Si_3N_4 and subsequently precipitates β-Si_3N_4. Processing of Si_3N_4 in this manner results in the formation of a residual glassy phase in the grain boundaries. This phase softens at the desired elevated use temperatures, thus compromising the thermomechanical properties of conventionally processed material. As an example, the creep rate of Si_3N_4 can be reduced by up to 5 orders of magnitude if the grain boundaries are very clean (as in reaction bonded Si_3N_4)[18].

Obviously, in order to take advantage of silicon nitride's intrinsic properties, it is highly desirable to densify Si_3N_4 without sintering additives. The sintering difficulties associated with Si_3N_4 which require the use of these additives are attributed to various factors. Silicon nitride has very low bulk diffusion coefficients which result from strong directional covalent bonding. Sublimation introduces problems of decomposition and variable stoichiometry as processing temperatures are increased in an effort to overcome the low diffusion coefficients. The net result is a high ratio of non-densifying sintering mechanisms (i.e. surface diffusion and evaporation-condensation) to those which promote densification (i.e. volume and grain boundary diffusion). Thus, without oxide sintering additives, compacts of Si_3N_4 powder typically coarsen with little shrinkage.

Due to the difficulties associated with sintering "pure" Si_3N_4 mentioned above, the material has been labeled "unsinterable". Silicon nitride, however, is not alone in this category. Silicon and SiC, also covalent in nature, exhibit similar sintering difficulties. It has been shown, however, that Si and SiC can be densified without developing residual glassy grain boundary phases, if the particle size is small enough, or if appropriate sintering aids are used[19-21]. In addition, pure Si, 55 nm in size, has been found to sinter provided initial green densities are sufficiently high i.e. >42% theoretical density[22,23].

While "pure" Si_3N_4 has been labeled as being unsinterable, there is no apparent thermodynamic barrier to its densification[19]. Utilizing Si as a model powder for the sintering of covalent materials, it was determined that extensive sintering should develop in powder compacts when the ratio of bulk (or boundary) diffusivity to particle radius is approximately 4×10^{-6} cm^{-1}[19,24]. Assuming that the ratio of diffusivity to particle radius for other covalent materials must equal or exceed silicon's value, Si_3N_4 powders should sinter without liquid forming sintering additives, if they are less than 25 nm in size.

Densification of covalent materials may be further assisted through the application of an external pressure during the sintering cycle. It has been demonstrated that through pressure application, the energy available for densification can be increased by more than 20 times[25-28]. Hot pressing (HP) and hot isostatic pressing (HIPing) are typical methods for applying pressure and temperature simultaneously. In the case of Si_3N_4, the use of an over pressure of N_2 0.7-70MPa (~100-10,000psi) has been found to provide both a compressive force for densification, as well as an increased thermodynamic barrier to decomposition.

Miyamoto et al., have recently demonstrated the densification of "undoped" Si_3N_4 powders by HIPing[29]. Resultant bulk specimen properties (density, microhardness and flexural strength) were very good, both at room and elevated temperatures. Starting powders, however, were processed in ambient atmospheric conditions prior to HIPing, no doubt resulting in significant levels of oxygen.

Due to the higher free energy of formation possessed by SiO_2, Si_3N_4 will oxidize or hydrate in atmospheres of oxygen and water vapor[30,31]. Recent work has demonstrated that the hydroxilation kinetics of an amorphous laser derived Si_3N_4 powder, when exposed to various relative humidity atmospheres at room temperature, are extremely fast[32]. However, on exposure to N_2-20% O_2 atmospheres containing extremely low concentrations of water vapor, the kinetics of oxidation were negligible over 72 hours. Water vapor attack on Si_3N_4 powders results in Si-OH layer formation. This layer may act as a sintering aid during heat treatment. The fact that Miyamoto et al. reported the occasional identification of a secondary oxynitride phase in their HIPed samples seems to support this suggestion.

In order to densify "pure" Si_3N_4, it seems imperative to:
1) synthesize a powder <25 nm in size,
2) pack powders uniformly into high green density compacts,
3) process the material to prevent the formation of an oxide layer, and
4) assist densification and limit decomposition through the application of an external pressure.

The objective of the experimental work described herein is to study the effect of oxygen contamination on the dispersion and subsequent densification behavior of ultra-high purity, laser-derived Si_3N_4 powders. In particular, stoichiometric, amorphous powders approximately 17 nm in size, both exposed and unexposed to the ambient environment, were dispersed in a non-aqueous solution in an attempt to form high green density, uniformly packed green bodies with little or no O_2 contamination.

A dispersion of Si_3N_4 powder in Unamine-O^{TM}-hexane solutions was selected as the best system for creating high purity, high density, uniformly packed green compacts for the following reasons: (1) the dispersions have a low relative viscosity, (2) the solvent and dispersant both have a low oxygen content, limiting possible oxygen contaminiation, (3) the Unamine^{TM} dispersant exhibits the high surface activity essential for good dispersions, (4) Unamine^{TM} is available commercially, and is commonly prepared by the reaction of oleic acid with 2-amino-ethanolamine, and (5) Unamine^{TM} should exhibit good burnout behavior[32].

HIPing of the consolidated laser synthesized Si_3N_4 powders was carried out in sealed ampules that maintained the exposed or unexposed status of the powders. No sintering aids were used in this investigation.

2. EXPERIMENTAL PROCEDURE

2.1 Powder Synthesis and Characterization

Si_3N_4 powders were produced via laser-driven gas phase reactions. Details of the process have been reported elsewhere, and therefore are only summarized here[20,33-35]. Si_3N_4 powders were produced by passing electronic grade SiH_4 and NH_3 orthogonally through a CO_2 laser beam. Due to the strong absorption of the 10.6 µm CO_2 emission by SiH_4, the reactants heated rapidly and subsequently decomposed and reacted to form Si_3N_4 powders. By careful manipulation of process variables (ie flow rates, reactor pressure, gas ratios and laser intensity) powders of controlled characteristics (size, stoichiometry and crystallinity) were produced. All post-synthesis processing was done in atmospheres containing a combined oxygen-water environment <10 ppm, or the ambient atmosphere. Powders processed with and without atmospheric exposure are termed exposed or unexposed respectively.

Synthesized powders were subjected to the following battery or characterization methods: X-ray and electron diffraction, He-pycnometry, single point BET gas absorption, TEM, XPS analysis based on a 101.3 eV silicon 2p binding energy for stoichiometric Si_3N_4 powders[36], neutron activation, and wet chemical analysis. The activity of Si_3N_4 powders, both exposed and unexposed, was determined by measuring the surface area of powders annealed in nitrogen as a function of temperature.

2.2 Powder Dispersion

For the dispersion studies, all powders referred to as unexposed were processed in a glove box, while powders referred to as exposed were introduced to air at room temperature ~50% RH for at least 72 hr. Powders referred to as O_2 exposed were subjected to a gas stream of 21% O_2 in nitrogen mixture at a flow rate of ~0.75 1/min. ACS grade hexane was used as received. Unamine O^{TM} was also used with no further purification. Solutions of imidazoline surfactant in hexane were prepared in air and dried over a freshly activated molecular sieve.

2.3 Adsorbate Surface Studies

The quantity of chemisorbed and physisorbed water on Si_3N_4 powder was determined by a modified Karl Fisher titration technique used for determining the hydroxyl content of amino alcohols[37,38]. The powder used for the study was an exposed powder that had been heated and evacuated at 200°C and 0.8 torr vacuum for 24 hours (to remove physisorbed H_2O), and subsequently stored in a glove box. Weight gain experiments were used to quantify the degree of hydroxylation and hydration of the Si_3N_4 powder using an apparatus similar to that of Wentzel[39]. In a separate experiment, the oxidation of the Si_3N_4 powder was measured by weight gain of an unexposed powder in an ultra high purity 21% O_2 in nitrogen gas stream at a flow rate of approximately 0.75 l/min.

Diffuse reflectance infra-red spectroscopy (DRIFT) was used to characterize the powder surface chemistry and powder surface-dispersant interaction. The unexposed samples were transferred to the FTIR ultra high purity nitrogen purged sample chamber with minimal air exposure. The exposed samples were run under dry air conditions. O_2-exposed powder was analyzed with the same procedure used for unexposed powder.

2.4 Dispersion Properties

The rheological properties of the dispersion systems were assessed using relative viscosity measurements, where the relative viscosity is defined as the viscosity of the suspension divided by the viscosity of the solvent-dispersant system. The measurements of shear stress vs shear rate were performed on dispersions with 1 vol% solids, which were mechanically agitated for 72 hr at ~20°C. Dispersions referred to as water added followed the procedure outlined above, except that 0.15 vol% water was added to unexposed slurries after 72 hours of agitation in a glove box.

2.5 Powder Consolidation

In order to evaluate the consolidation behavior of these dispersed powders, unexposed and exposed Si_3N_4-Unamine O^{TM}-hexane dispersions were first prepared (as above) and subsequently dried in a glove box. The dried mass was removed from the glove box and granulated using a mortar and pestle in air, followed by die pressing in air at pressures from 69-346 MPa. The density of the green compacts were calculated based on measurements of sample dimensions and weight.

2.6 HIPing

Pellets for use in the HIPing studies were formed in a two stage pressing process. Synthesized powder was first hand pressed at pressures less than 350 kPa, followed by cold isostatic pressing at 283 MPa. Pellets for HIPing were then encapsulated in high purity boron nitride powder inside Pyrex^{TM} tubes. The samples were outgassed at 5×10^{-6} torr and 600°C for 24 hours. All sample preparation was carried out in an effort to maintain the sample's exposure status, ie unexposed or exposed. Sealed cans were subsequently placed into a 600°C oven and cooled slowly.

Hot isostatic pressing was conducted using an argon gas HIP. All samples were first heated to 830°C, the softening point for Pyrex^{TM}, at a rate of 30°/min under 602 kPa pressure. After a 15 minute soak at 830°C, pressure and temperature were ramped to desired values.

3. RESULTS AND DISCUSSIONS

3.1 Powder Synthesis and Characterization

The synthesis conditions utilized to produce Si_3N_4 powders by the laser process are listed in Table 1. The properties of powders produced under such conditions are presented in Table 2. Figure 2 is a TEM micrograph of the

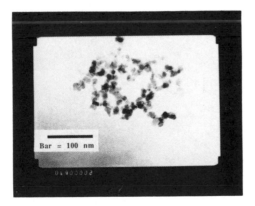

Bar = 100 nm

Figure 2: TEM micrograph of as synthesized Si_3N_4 powders. Bar = 100 nm.

Silane	72 cm^3min^{-1}
Ammonia	320 cm^3min^{-1}
Argon: annulus	600 cm^3min^{-1}
chamber	1.5 $lmin^{-1}$
Reactor pressure	720 torr
Flame temperature	1110°C
Laser intensity	5.6×10^3 Wcm^{-2}

Table 1: Si_3N_4 powder synthesis conditions.

Surface area		122 m^2g^{-1}
Particle size:	BET-ESD	17.0 nm
	TEM	16.0 nm
He-pycnometric density		2.9 gcm^{-3}
Crystallinity		Amorphous
Chemistry		Stoichiometric, (No Si detected by X-ray and XPS).

Table 2: Si_3N_4 powder characteristics.

Element	Wt%
C	0.072
Cl	<0.010
Ta	<0.008
W	<0.008
S	0.007
B	0.003
Cu	0.001
Fe	0.001
Al	<0.001
Ca	<0.001
Cr	<0.001
Mg	<0.001
Ni	<0.001
Na	<0.001
Y	<0.001

Table 3: Chemical analysis of Si_3N_4 powders.

as synthesized Si_3N_4. Table 3 lists the trace impurities of the powder as determined by wet chemical analysis. Particularly notable are the low iron, calcium, aluminium and carbon contents of these powders. Neutron activation analysis indicates that exposed powders possess an oxygen content of ~4 wt% while unexposed powders have only 0.1-0.2 wt%. Results indicate that extremely "pure" Si_3N_4 powders of controlled characteristics may be produced by laser-driven gas phase reactions.

3.2 Surface Chemistry of Unexposed and Exposed Powders

A review of silicon chemistry facilitates understanding and optimization of the dispersion behavior of the Si_3N_4-imidazoline-hexane system. The surface chemistry of Si_3N_4 powder prepared in various fashions and processed in aqueous suspensions or ambient air has been found to consist of an equilibrium between amino and silanol (Si-OH) groups[40,41]. Recently, Busca has shown that the surface of an air exposed Si_3N_4 consists primarily of Si-NH groups and silanol groups[42]. The Si_3N_4 powder surface species and their reactivity (without exposure to air or water), have not been widely studied.

Silica surface studies give an indication of the reactivity of silanol groups on Si_3N_4 surfaces. Room temperature silica surface studies indicate that condensation reactions between aliphatic alcohols and silanol groups on Si_3N_4 are highly unlikely. Instead, the adsorbate alcohol will probably hydrogen bond to the silanol sites[43]. Based upon the literature for nonaqueous and aqueous solution chemistry of organosilylamines (viz. R_3-Si-NH_2) and silazanes (viz R_3-Si-NH-Si-R_3), any amino groups would be expected to undergo hydrolysis and alcoholysis quite readily at room temperature. Many times, however, an acid initiator is essential for silazane alcoholysis. Silazanes have not been observed to undergo hydrolysis when the molecule was insoluble in neutral or basic aqueous solutions[44]. Other functional groups such as: C=C bonds, and carbonyl groups (C=O), and amine groups (NH_x) have been shown to adsorb on to silica surface silanol groups[45-47].

Exploring the chemistry of the functional groups present in the surfactant aids in understanding the interaction of Si_3N_4 with the Unamine O^{TM} surfactant.   Unamine O^{TM} is a 1-hydroxy-ethyl 2-heptadecenyl 2-imidazoline which consists of a five member heterocyclic ring with an alcohol at the one position and an aliphatic chain at the two position as depicted in the reaction below[48]:

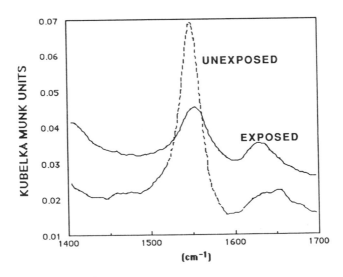

$$(1)$$

In acidic aqueous solutions at room temperature, the imidazoline has not been observed to undergo hydrolytic ring fissure[49]. The imidazoline ring is extremely reactive in the presence of water in non-aqueous and aqueous solutions that are neutral or basic. The imidazoline undergoes a base-catalyzed hydrolisis ring opening resulting in an amide reaction product (Reaction 1)[50].

The surface chemistry of the Si_3N_4 surfaces and their interactions with water and the Unamine O^{TM} dispersant were examined to understand the dispersion and pressing properties of exposed and unexposed powder. The unexposed powder surface chemistry was characterised to understand the intrinsic, oxidative, and hydrolytic nature of the Si_3N_4 powder. Interpretation of the DRIFT spectra of the Si_3N_4 powder indicates that the surface of the unexposed powder is amino in nature. In Figure 3, Kubelka

Figure 3: FTIR DRIFT spectra of unexposed and exposed Si_3N_4 powders from 1700 cm^{-1} to 1400 cm^{-1}.

Munk absorption peaks at 1650 cm^{-1}, 1630 cm^{-1}, and 1550 cm^{-1} are assigned to an NH$_2$ bending, a free NH$_3$ bending and an NH bending mode respectively[42,51,52]. The absorption of the NH mode is dominant even though this mode has a weaker absorption coefficient[52] than the other two modes. This indicates that there is a predominance of silazane (Si$_2$-NH) species and a negligible concentration of Si-NH$_2$ on the unexposed Si$_3$N$_4$ powder surface.

Since triply bonded nitrogen surface species (Si$_3$N$_4$) are not directly detectable by infrared spectroscopy, the possibility exists that this specie is also present on the surface of the unexposed Si$_3$N$_4$. This is highly unlikely, however, due to the strained nature of this group and the reaction temperature (~1110°C) during powder synthesis. The powder density (2.92 gcm^{-3}, ~92% theor) is close to that for crystalline Si$_3$N$_4$ and there is no evidence of internal porosity in TEM. The hydrogen content measured for these powders is 0.15 wt%. A calculation of the expected H content was made, based on the powder surface area of 120m^{2}g^{-1} and the Si$_2$NH surface chemistry. The calculation was based on a site density of 4 surface hydrogens per square nanometer. This indicates that a powder whose surface is made up of Si$_2$NH sites would have 0.19 wt% H, in close agreement with the measured value. The combination of the high density and the evidence that the 0.15 wt% H resides at the surface, is strong evidence that amino groups reside only on the powder surface, and not in the bulk of the powder. The surface of an unexposed powder can be summarized as being predominantly silazane groups (Si$_2$-NH) with a minor constituent of silylamine groups (Si-NH$_2$) and some small amount of adsorbed ammonia.

Weight gain experiments were carried out in an attempt to characterize the reactivity of the Si$_3$N$_4$ powder's surface to the atmosphere. The results (Figure 4) indicate that the unexposed Si$_3$N$_4$ powder reacts very quickly with water vapor. The initial rapid weight gain regime (~1.5wt%) can be attributed to the hydroxylation of silazane species following the reaction:

$$Si_2-NH + 2H_2O = 2Si-OH + NH_3 \qquad\qquad 2$$

The hydroxylation of the silazane surface species to a level of 1.5 wt%, results in a calculated Si$_2$-NH site density of 4 sites/nm^2. Beyond 1.5 wt%, a slower weight gain regime results. This is attributed to physically (or hydrogen bonded) adsorbed water. The amount of physically adsorbed water is seen to be proportional to the relative humidity.

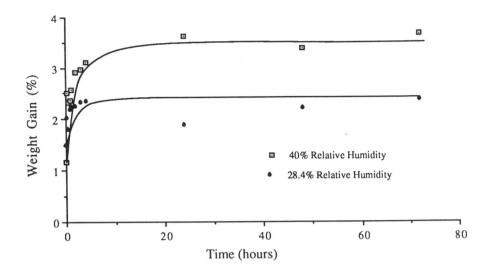

Figure 4: Weight gain vs time of Si_3N_4 powders exposed to different relative humidities.

The Karl Fisher measurements of the heated evacuated powder yield a total water content of 3.73 wt%, while an exposed powder has a total water concentration of 5.72 wt%. Titration measurements indicate that the physisorbed water content of the heated evacuated powder is 1.73 wt%. Thus, the remaining 2.00 wt% corresponds to the concentration of Si-OH sites. This is considered to be in good agreement with the weight gain data (described above).

The infrared spectrum of the O_2 exposed powder in the region of 1700 cm^{-1} (Figure 5) is quite different from the unexposed powder spectrum. The NH_2 and NH_3 absorption peaks present at 1650 cm^{-1} and 1630 cm^{-1} for the unexposed powder, have disappeared for the O_2 exposed powder, but the NH absorption at 1550 cm^{-1} has not. This indicates that the small concentration of $Si-NH_2$ species present on the unexposed powder reacts to form Si-O groups after O_2 exposure. However, no characteristic Si-O stretching was seen at 1190 cm^{-1} for the O_2 exposed, which is plausible, since very few $Si-NH_2$ species exist on an unexposed powder surface.

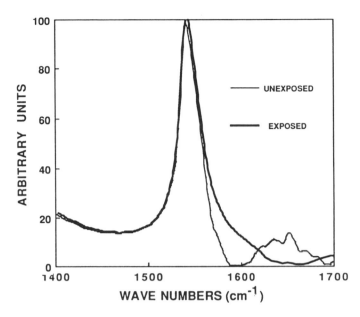

Figure 5: FTIR DRIFT spectra of unexposed and dry air exposed Si_3N_4 powders from 1700 cm^{-1} to 1400 cm^{-1}.

The DRIFT spectrum of the exposed Si_3N_4 powder (Figure 6) is very different from the unexposed powder spectrum as shown in Figure 3. The stretching regime of the unexposed powder shows the absorption due to N-H stretch at 3490 and 3395 cm^{-1}, whereas the exposed powder displays a freely vibrating Si-OH peak at 3760 cm^{-1} and a broad peak at 3350 cm^{-1} due to a combination of H_2O stretch, NH stretch, and associated Si-OH absorption. The presence of molecular water for the exposed powder is confirmed by the appearance of a peak at 1635 cm^{-1} (Figure 6). The peak at 1550 cm^{-1} indicates that silazane groups are present on the exposed Si_3N_4 powder surface. However, this peak was much less intense that the peak found in the unexposed powder spectrum indicating that reaction 2 has taken place resulting in a lower concentration of silazane groups. This spectral data confirms the hydrolysis chemistry inferred from KFT and weight gain measurements.

3.3 Imidazoline-Si_3N_4 (Exposed and Unexposed) Surface Chemistry

DRIFT spectra were obtained on both exposed and unexposed powders with

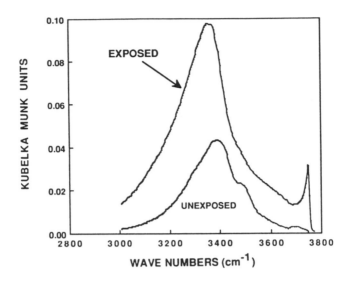

Figure 6: FTIR DRIFT spectra of unexposed and exposed Si_3N_4 powders from 3800 cm^{-1} to 3000 cm^{-1}.

adsorbed Unamine O^{TM} (Figures 7 and 8). These spectra exhibit features identical to the characteristic of the carbon backbone of the dispersant molecule[51-53]. However, the spectral features exhibited by the nitrogen bearing groups indicate that the dispersant interacts differently with exposed powder than it does with unexposed powder (Figures 7 and 8 respectively). The 1607 cm^{-1} spectral absorption in the unexposed case corresponds to the C=N bond[54,55] in the imidazoline ring structure. The presence of this absorption indicates that the ring structure of the dispersant has not changed upon adsorption onto the unexposed powder. In contrast, the spectra of exposed powder does not show a peak at 1607 cm^{-1}, but instead, shows two absorptions at 1645 cm^{-1} and 1545 cm^{-1} indicative of a secondary amide[51]. The conclusion that these changes in the spectral features are due to the hydrolysis of the imidazoline ring was confirmed by DRIFT spectra of a reaction product derived from Unamine O^{TM} - ammonia-ethanol solutions.[56]

Based on the literature discussed ealier, the adsorption phenomenon observed in both the exposed and unexposed systems is believed to occur through hydrogen bonding. For the unexposed system, the hydrogen bonding is

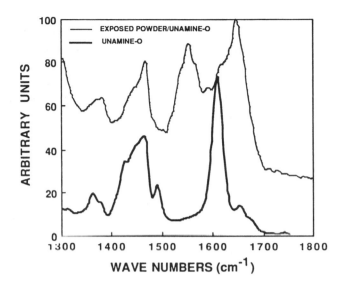

Figure 7: FTIR DRIFT spectra of exposed Si_3N_4 powder with adsorbed Unamine O^{TM} and pure Unamine^{TM} from 1800 cm^{-1} to 1300 cm^{-1}.

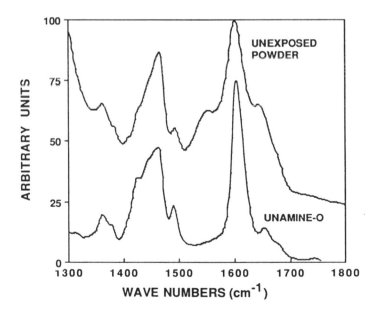

Figure 8: FTIR DRIFT spectra of unexposed Si_3N_4 powder with adsorbed Unamine O^{TM} and pure Unamine O^{TM} from 1800 cm^{-1} to 1300 cm^{-1}.

believed to take place between the alcohol group of the Unamine O^{TM} and the silazane groups of the powder surface. The exposed powder-adsorbate system is complicated by the hydrolysis reaction which produces various powder surface sites.   In addition, ring fissure of the imidazoline results in a secondary amide molecule.   It is possible that this amide molecule could interact with the exposed powder surface in a much different manner than the imidazoline does with an unexposed surface.

3.4 Dispersion Properties

A series of visual observations on the powder dispersions combined with rheological data summarize the impact that water contamination has on the Si_3N_4 dispersion properties.   Unexposed powder dispersions showed no visible sedimentation in a 0.5 hr time frame, while water contaminated (or exposed) systems exhibited rapid sedimentation (Table 4).   The unexposed system had a lower relative viscosity than the exposed system (Figure 9).

Possible reasons for the dispersion behavior to be adversely affected by water are: (1) incomplete coverage of the dispersant molecule due to

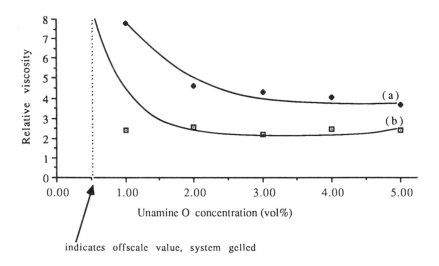

indicates offscale value, system gelled

(a) EXPOSED Si_3N_4 POWDER

(b) UNEXPOSED Si_3N_4 POWDER

Figure 9: Silicon nitride viscosity vs Unamine O^{TM} concentration.

Slurry type	Viscosity (cP) Unamine 0^{TM} content		30 min observation (3 vol% solution)
	1 vol%	3 vol%	
Exposed	7.79	4.27	Settled into loose, large flocs.
Unexposed	2.40	2.20	Did not settle.
Water added	4.90	2.45	Settled into loose, large flocs.

Table 4: Si_3N_4 - imidazoline - hexane dispersion properties.

competitive adsorption by water, (2) multiple attachment of the dispersant molecule, and (3) a change in the adsorbed layer thickness or a possible reduction in Hamaker shielding of the particles due to the break up of the dispersant molecule and/or modification of the surface groups on the powder. Adsorption isotherm data does not show significant differences in surface coverage between these two systems[56].

3.5 Processing

In an effort to determine the effectiveness of the hexane-Unamine 0^{TM} system, exposed and unexposed powders were dry pressed as described earlier. For the unexposed powder, extensive ammonia evolution due to hydrolysis (Reaction 2) was detected by the pungent odor of ammonia during the granulation process carried out in air. On the other hand, the exposed powder exhibited no noticeable odor during granulation. Compacts prepared from the exposed-adsorbate powder were easily removed from the die with no endcapping, whereas all but the low pressure unexposed powder exhibited extensive endcapping. This behavior was believed to be due to the ammonia gas evolution.

The results of these pressing studies are seen in Figure 10, where the theoretical density is plotted against pressure for unexposed-adsorbate, and exposed-adsorbate powder. The results indicate that the exposed system has slightly higher densities than the unexposed system. Based on the rheological and visual data, this behavior was unexpected. Ammonia release

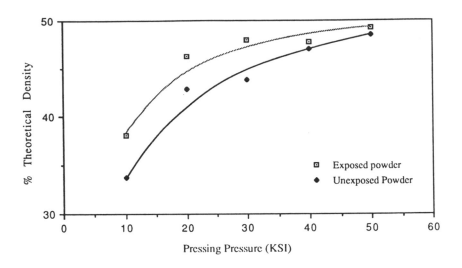

Figure 10: Packing density of Si_3N_4 powders that were dispersed, dried and die pressed at various pressures.

during pressing and variations in the granulation process may have caused these discrepancies.   To avoid the problem of ammonia evolution during processing, unexposed powder must clearly be processed in inert atmospheres. The potential for forming uniformly packed green bodies from this sytem can, however, clearly be seen from the green density data presented.   Without dispersion processing, unexposed and exposed powders cold isostatically pressed at a pressure of 347MPa only reach densities in the range of 35-40%[57].   On the other hand, for both the unexposed or exposed powders, use of a dispersion process followed by drying and die pressing yields green densities as high as ~50%.   These processing procedures are far from optimized, and it is anticipated that even higher green densities may be realized by using more advanced powder consolidation methods.

3.6 Powder Activity

While the FTIR and weight change studies discussed above indicate how atmospheric exposure significantly affects the oxygen content of synthesized powders, annealing studies were used to indicate how exposure affects the

Si_3N_4 powder's activity. Figure 11 shows the activity of an unexposed and
two exposed Si_3N_4 powders. The difference between the two exposed powders
is that one was exposed to the atmosphere for only 48 hours prior to
annealing (termed short exposure) while the other was exposed for a period
of greater than two months (termed long exposure). It is seen that exposure
of the powders to ambient surroundings dramatically affects their activity.
Unexposed samples are extremely active, displaying a strong temperature
dependence as well as a large activation energy for the mechanism responsible
for the measured surface area reduction. However, when samples are exposed
to the ambient environment, the temperature dependence for surface area
reduction decreases. The decrease in temperature dependence is also a
function of exposure time indicating that the chemistry and or structure of
the exposed amorphous Si_3N_4 powder may be changing over long periods of time.
X-ray diffraction   and   TEM   analysis   indicated   conclusively   that
crystallization was responsible for the measured surface area reductions.

Particularly interesting results were found when powders were annealed
~20°C-50°C below the onset temperature for surface area reduction.   In
particular, powder compacts were observed to shrink during these soaks.
Figure 12 shows the linear shrinkage exhibited by short term exposure powders

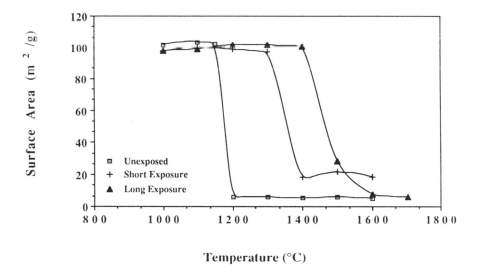

Figure 11: Surface area vs temperature for Si_3N_4 powders.

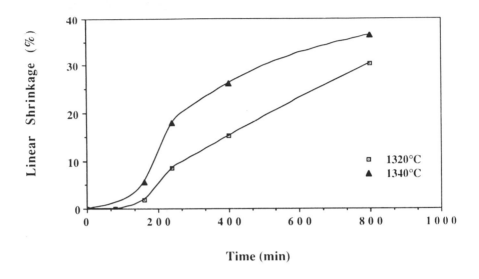

Figure 12: Linear shrinkage vs time for exposed Si_3N_4 powders.

as a function of time at temperature for several annealing temperatures. Similar results were found for long term exposure powders as well as unexposed samples. The following list contains possible explanations for the observed behavior;

1) weight loss; unlikely since the highest recorded weight loss of all samples tested was 5%,

2) coarsening; this mechanism is capable of surface area reduction but cannot account for shrinkage,

3) rearrangement; due to the lack of mechanical stress to assist densification it is doubtful this would contribute to shrinkage,

4) crystallization; while samples crystallized during annealing, conversion from 2.9 g/cm^3 to 3.18 g/cm^3 cannot account for the observed shrinkage, for 10%-15% shrinkage was observed in samples prior to crystallization,

5) oxygen presence; possibly attributable to exposed samples but not in the case of unexposed samples,

6) viscous sintering; it is possible that the small particle size and amorphous nature of the starting powder may permit sintering by a densifying mechanism. In particular, viscous sintering may be responsible for observed shrinkages prior to crystallization.

Presently it is unclear what particular mechanism is responsible for the shrinkages observed during the annealing of laser powders. The results are very encouraging if they do in fact represent the possibility of sintering Si_3N_4 (and perhaps other covalent ceramics) in an amorphous state.

3.7 Hot Isostatic Pressing (HIPing)

The final densities of HIPed samples are shown in Figure 13 as a function of temperature at pressure for exposed and unexposed samples. A pressing pressure of 198 MPa and hold time of 60 minutes was used. Density was determined by mercury displacement. As may be seen the contamination of laser derived powders with oxygen from the ambient environment significantly alters the densification behavior of the powder. Unexposed samples attained a maximum density of 67% theor. at 2050°C, while exposed samples reached a significantly higher density of 91% theor. at 2050°C. X-ray diffraction analysis of HIPed samples indicates that both exposed and unexposed samples attain higher percentages of the beta phase as the HIPing temperature increases. The kinetics of the phase transformation (alpha to beta) for exposed and unexposed samples, however, appear to be different.

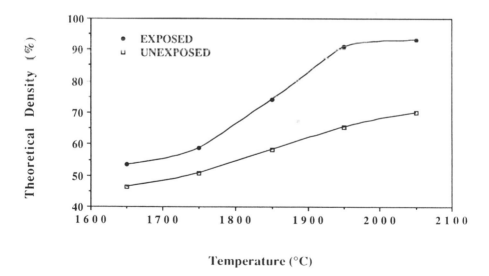

Figure 13: Density vs temperature for exposed and unexposed Si_3N_4 powders HIPed at 198 MPa for 60 min.

Quantification of phase contents and conversion kinetics is currently being carried out.

Densification kinetics of exposed and unexposed samples at a temperature of 1950°C and pressing pressure of 198 MPa were investigated.   Figure 14 shows HIPed final densities at 1950°C for pressing times of 1-180 minutes. Exposed samples are observed to be extremely active compared to unexposed samples.   For both cases, the rate of densification falls off significantly after a time of approximately 20 minutes.   X-ray diffraction analysis again indicates that the beta-Si_3N_4 content of samples increases with time at temperature, and that the kinetics of the phase conversion from α to β-Si_3N_4 differ for exposed and unexposed powders.   Unexposed samples were observed to convert to 100% β-Si_3N_4 after a time of 15 minutes while exposed samples became a 100% β-Si_3N_4 only after an hour.

Figures 15 and 16 show the microstructural evolution of exposed and unexposed specimens respectively.   After one minute at temperature, Figures 15a and 16a, individual particles are easily discernable for both exposed and unexposed samples.   Particle sizes have increased to an average size of

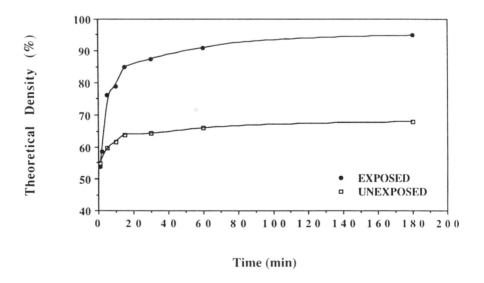

Figure 14: Densification kinetics for exposed and unexposed Si_3N_4 powders HIPed at T = 1950°C and p = 198 MPa.

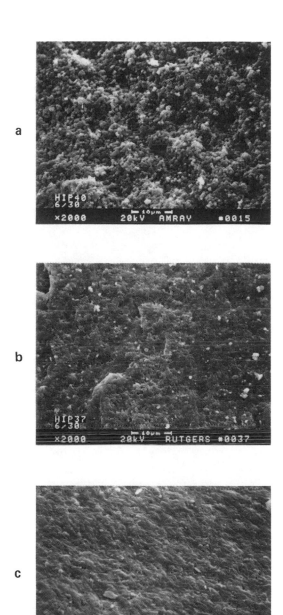

Figure 15: Microstructures of exposed Si_3N_4 powders HIPed for various times at T = 1950°C and p = 198 MPa.   a) 1 min, b) 10 min and c) 60 min.

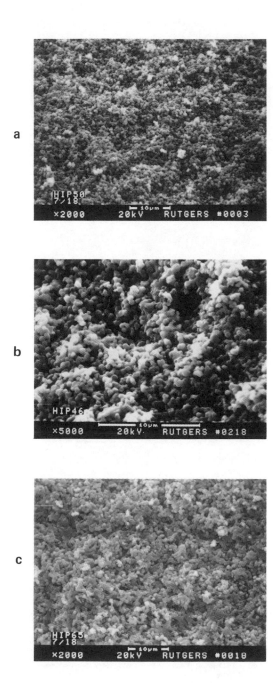

Figure 16: Microstructures of unexposed Si_3N_4 powders HIPed for various times at T = 1950°C and p = 198 MPa.   a) = 1 min, b) = 10 min and c) = 60 min.

~0.5μm with some localized sintering observed in the exposed sample. After ten minutes at temperature, there is a large deviation between the microstructures of exposed and unexposed samples (Figures 15b and 16b). Individual particles are still quite discernable in the unexposed sample, while particle individuality is no longer apparent in the exposed sample. After one hour, Figures 15c and 16c, little change has occurred in the microstructure of the unexposed sample, while the exposed sample has entered final stage sintering.   It is interesting to note that some localized sintering has occurred in the unexposed samples. Since the unexposed samples possess significantly less O_2 than the exposed samples do, this is indeed an encouraging result.

Considering the starting bulk density of samples (36% for unexposed and 42% for exposed) it is not surprising that the unexposed samples did not attain the high density of the exposed. This is especially true when one considers the extremely low bulk diffusion coefficients which operate in Si_3N_4. Obviously, the contamination of powders on exposure to the ambient atmosphere plays an important role in increasing the rate at which the dominate densification mechanisms operate. However, the need to create green compacts of increased green density, possessing unagglomerated microstructures is quite apparent. However, the current results indicate that laser derived Si_3N_4 powders can be densified. While exposure of synthesized powders to the ambient atmosphere enhances the densification kinetics, unexposed powders are observed to display a limited amount of sintering. Through improved processing (ie the consolidation of stabilized dispersions of unexposed as well as exposed powders), green bodies of improved microstructures may be utilized in HIPing to deliver superior HIPed densities as well as microstructures[56].

4. SUMMARY AND CONCLUSIONS

Silicon nitride powders possessing unique characteristics are synthesized via laser-driver gas phase reactions. Synthesized powders have the following (nearly ideal) characteristics: 1) 17 nm in size, 2) spherical, 3) loosely agglomerated, 4) narrow size distribution, 5) amorphous, and 6) stoichiometric. These laser synthesized Si_3N_4 powders are extremely reactive with water (but not oxygen) in the ambient environment. FTIR spectra of unexposed powders show that the surface consists primarily of silazane groups (viz Si_2-NH). Weight gain experiments indicate that the unexposed surface reacts quickly with water. FTIR, KFT, and weight gain data indicate an equilibrium is established for an exposed powder, consisting of silanol, silazane, and adsorbed water species. FTIR spectra of O_2 exposed powder indicate that only the small constituent of $Si-NH_2$ species oxidize, whereas the predominant Si_2-NH surface species do not. The unexposed Si_3N_4 powder hexane-Unamine O^{TM} dispersions displayed significantly better rheological and flocculation behavior than the exposed powder system. The enhanced dispersion properties of the unexposed system (in comparison to the exposed system) are related to the chemical modification of the imidazoline molecule at the exposed powder surface. Processing results show that the dispersion system has great potential for yielding high green density compacts. A great deal of care must be taken in processing the unexposed system so as to avoid problems associated with hydroxylation of the powder surface.

During nitrogen annealing of synthesized powders, exposed as well as unexposed samples were observed to display linear shrinkages of up to 40%. This result is scientifically very interesting, if it can indeed be proven to result from a viscous sintering mechanism. The possibility of being able to synthesize and easily densify covalent (non-oxide) ceramics in the amorphous state, followed by subsequent crystallization is extremely exciting.

Samples for HIPing studies were prepared by die and cold isostatic pressing followed by encapsulating in evacuated, BN lined, pyrex ampules. Hot isostatic pressing of laser synthesized Si_3N_4 powders exposed to the ambient environment produced rapid initial densification and high final densities, ~91% theor. The microstructures showed that the discrete nature of the crystalline particles is lost even after 10 minutes at temperature. In constrast, the unexposed powders were found to only attain a final density of ~64% theor. for the same HIPing conditions. Microstructural evidence shows that the discrete nature of the particles was retained even after 60 minutes at temperature for the unexposed samples. In addition, there was very little coarsening that occurred between 1 and 60 minutes for the unexposed case. It should be noted, however, that some localized sintering did occur for the unexposed samples after 60 minutes. This suggests that higher final densities may be attainable if unexposed powders can be packed more efficiently and uniformly into higher green density samples. It is also possible that higher densities may be achieved if higher sintering temperatures and pressures are used.

As stated earlier, the green compacts for HIPing experiments were not prepared via colloidal dispersion techniques. As a result, there are certainly large packing heterogeneities in these green bodies which will limit end point densities. Based on our previous processing experience we anticipate that if samples were prepared via colloidal routes, it is anticipated that the exposed and unexposed samples could be HIPed to densities of 99% theor. and 85% theor. respectively.

This research effort has clearly demonstrated that O_2, whether from water vapor or oxygen in the atmosphere, plays a critical role in determining the Si_3N_4 powder surface chemistry, powder surfactant interactions, the consolidation behavior, and the densification kinetics and end point HIP density. It will be very interesting to learn exactly what the minimum oxygen content is, that is needed to reach full density via HIPing. It is anticipated that these materials should exhibit extremely good high temperature properties.

ACKNOWLEDGEMENTS

The authors would like to gratefully acknowledge the financial support of the New Jersey Commision on Science and Technology and the Center of Ceramics Research at Rutgers, The State University of New Jersey.

REFERENCES

1. Bowen, H.K., Basic Research Needs on High Temperature Ceramics for Energy Applications. Mater. Sci. Eng. 44 [1] 1-56 (1980).

2. Brook, R.J., Tuan, W.H. and Xue, L.A., Critical Issues and Future Directions in Sintering Science. In Cer. Trans. Vol 1, Ceramic Powder Science; Messing, G.L., Fuller, E.R. Jr. and Hausner, H. (Eds), American Ceramic Society, (1988).

3. Harmer, M.P., Science of Sintering as Related to Ceramic Powder Processing. In Cer. Trans. Vol 1, Ceramic Powder Science; Messing, G.L., Fuller, E.R. Jr. and Hausner, H. (Eds), p 824-39, American Ceramic Society, (1988).

4. Rhodes, W.H., Agglomerates and Particle Size Effects on Sintering Yttria Stabilized Zirconia. J. Amer. Ceram. Soc. 64 19 (1981).

5. Barringer, E.A. and Bowen, H.K., Formation, Packing and Sintering of Monodisperse TiO_2 Powders. J. Amer. Ceram. Soc. 65 C199 (1982).

6. Lange, F.F., Sinterability of Agglomerated Powders. J. Amer. Ceram. Soc. 67 [2] 83-89 (1984).

7. Askay, I.A., Principles of Ceramic Shape Forming with Powder Systems. In Cer. Trans. Vol 1, Ceramic Powder Science; Messing, G.L., Fuller, E.R. Jr. and Hausner, H. (Eds), p 663-74, American Ceramic Society, (1988).

8. Haggerty, J.S., Garvey, G.J., Flint, J.H., Sheldon, B.W., Okuyama, M., Ritter, J.E. and Nair, S.V., Processing and Properties of Reaction Bonded Silicon Nitride and Sintered Silicon Carbide Made from Laser Synthesized Powders. In Cer. Trans. Vol 1, Ceramic Powder Science; Messing, G.L., Fuller, E.R. Jr. and Hausner, H. (Eds), p 1059-68, American Ceramic Society, (1988).

9. Danforth, S.C. and Haggerty, J.S., Mechanical Properties of Sintered and Nitrided Laser Synthesized Silicon Powder. J. Amer. Ceram. Soc. 66 [4] 273-275 (1983).

10. Sacks, M.D. and Tseng, T.Y. Preparation of Silica from Model Powder Compacts: I, Formation and Characterization of Powders, Suspensions and Green Compacts, and II: Sintering. J. Amer. Ceram. Soc. 67 [8] 526-637 (1984).

11. Gauthier, F.G.R. and Danforth, S.C., Packing of Bimodal Mixtures of Colloidal Silica. In Cer. Trans. Vol 1, Ceramic Powder Science; Messing, G.L., Fuller, E.R. Jr. and Hausner, H. (Eds), p 709-15, American Ceramic Society, (1988).

12. Harmer, M.P., Science of Sintering as Related to Ceramic Powder Processing. In Cer. Trans. Vol 1, Ceramic Powder Science; Messing, G.L., Fuller, E.R. Jr. and Hausner, H. (Eds), p 824-39, American Ceramic Society, (1988).

13. Pasto, A E., Causes and Effects of Fe-Bearing Inclusions in Sintered Si_3N_4. J. Amer. Ceram. Soc. 67 [9] 178-180 (1984).

14. Yeckley, R M., Stress Rupture Life of HIP Silicon Nitride. Talk 83-C-88F, presented at 12th Annual Conference on Composites and Advanced Ceramics, Cocoa Beach, FL, USA, (1988).

15. Yamada, T., Shimada, M., and Koizumi, M., Fabrication of Si_3N_4 by Hot Isostatic Pressing. J. Amer. Ceram. Soc. 60 [11] 1225-1228 (1981).

16. Terwillinger, G R., and Lange, F F., Pressureless Sintering of Si_3N_4. J. Mater. Sci. 10 [7] 1169-1174 (1975).

17. Clarke, D R., Zaluzec, N J., and Carpenter, R W., The Intergranular Phase in Hot-Pressed Silicon Nitride: I, Elemental Composition. J. Amer. Ceram. Soc. 64 [10] 601-607 (1981).

18. Seltzer, M., High Temperature Creep of Silicon-Based Compounds. Amer. Ceram. Soc. Bull. 56 4 (1977).

19. Greskovich, C. and Rosolowski, J H., Sintering of Covalent Solids. J. Amer. Ceram. Soc. 9 [7-8] 336-343 (1976).

20. Danforth, S.C. and Haggerty, J.S., Synthesis of Ceramic Powders by Laser Driven Reactions. Energy Laboratory Report, MIT-EL-81-003, (1980).

21. Haggerty, J.S., Sinterable Powders from Laser Driven Ractions. Energy Laboratory Report, MIT-EL-82-002, (1981).

22. Moller, H.J. and Welsch, G., Sintering of Ultrafine Silicon Powders. J. Amer. Ceram. Soc. 68 [6] 320-325 (1985).

23. Danforth, S.C. and Haggerty, J.S., Mechanical Properties of Sintered and Nitrided Laser-Synthesized Silicon Powder. J. Amer. Ceram. Soc 66 [4] C58-59 (1983).

24. Sawhill, H T., Crystallization of Ultrafine Amorphous Si_3N_4 During Sintering. Masters Thesis, MIT, Cambridge, MA, USA, (1981).

25. Hermannsson, L., Nystrom, L., and Adlerborn, J., Hot Isostatic Pressing of Silicon Carbide With No Sintering Agents. In Ceramic Materials and Components for Heat Engines; Bunk, W. and Hausner, H. (Eds), p 353-60, Deutsche Keramische Gesellschaft (1986).

26. Prochazka, S., and Rocco, W A., High Pressure Hot Pressing of Silicon Nitride Powders. High Temperature, High Pressures 10 [1] 87-95 (1978).

27. Yamada T., Shimada M., Koizumi, M., Densification of Si_3N_4 by High Pressure Hot Pressing. Amer. Ceram. Soc. Bull. 60 [12] 1281-1288 (1981).

28. Richardson, D W., In Modern Ceramic Engineering Properties, Processing and Use in Design; Boothroyd, G. and Dieter, G. (Eds), Marcel-Dekker Inc., New York, (1982).

29. Miyamoto, Y., Tanaka, K., Shimada, M., and Koizumi, M., Survey of HIP Sintering Condition and Characterization of Dense Silicon Nitride Without Additives. In Ceramic Materials and Components for Heat Engines; Bunk, W. and Hausner, H. (Eds), p 271-78, Deutsche Keramische Gesellschaft, (1986).

30. Peuckert, P., and Greil, P., Oxygen Distribution in Silicon Nitride Powders. J. Mater. Sci. 22 3717-3720 (1987).

31. Singhal, S.C., Thermodynamic Analysis of the High-Temperature Stability of Silicon Nitride and Silicon Carbide. Cream. Int. 2 123-130 (1976).

32. Nilsen, K.J., Riman, R.E., and Danforth, S.C., The Effects of Moisture on the Surface Chemistry and Dispersion Properties of Silicon Nitride in Imidazoline-Hexane Solutions. In Ceramic Powder Science II; Messing, G.L., Fuller, E.R. and Hausner, H. (Eds), p 469-76, American Ceramic Society, (1988).

33. Cannon, W.R., Danforth, S.C., Flint, J.H., Haggerty, J.S., and Marra, R.A., Sinterable Ceramic Powders from Laser-Driven Reactions: I, Process Description and Modeling. J. Amer. Ceram. Soc. 65 [7] 324-330 (1982).

34. Cannon, W.R., Danforth, S.C., Flint, J.H., Haggerty, J.S., and Marra, R.A., Sinterable Ceramic Powders from Laser-Driven Reactions: II, Powder Characterization and Process Variables. J. Amer. Ceram. Soc. 65 [7] 330-335 (1982).

35. Symons, W., and Danforth, S.C., Synthesis and Characterization of Laser-Synthesized Silicon Nitride Powders. In Advances in Ceramics, Vol. 21, Ceramic Powder Science; Messing, G.L., Mazdiyasni, K.S., McCauley, J.W. and Haber, R.A. (Eds), p 249-56, The American Ceramic Society (1987).

36. Vasquez, R.P., et al., X-ray Photoelectron Spectroscopy Study of the Chemical Structure of Thermally Nitrided SiO_2. Appl. Phys. Lett. 44 [10] 969 (1984).

37. Smith, D.M. and Mitchell, J. Jr., Aquametry, Part III; Wiley (1977).

38. Smith, D.M., Mitchell, J. Jr. and Hawkins, W., Analytical Procedures Employing Karl Fisher Reagent X. The Determination of Aliphatic Hydroxy Amines (Amino Alcohols). J. Amer. Chem. Soc. 66 715-716 (1944).

39. Wentzel, An atmosphere Producer for Laboratory Use, in 'Humidity and Moisture: Measurement and Control in Science and Industry, vol 3, Wexler, A. & Wildhack, W., Reinhold Publ Corp, NY, 1963.

40. Whitman, P.K. and Fede, D.L., Colloidal Characterization and Modification of Silicon Nitride and SiC Dispersions. Amer. Ceram. Soc. Bull. 65 [2] 89 (1986).

41. Shaw, T.M. and Pethica, B.A., Preparation and Sintering of Homogeneous Silicon Nitride Green Compacts. J. Amer. Ceram. Soc. 69 [2] 89 (1986).

42. Busca et al., FT-IR Study of the Surface Properties of Silicon Nitride, Materials Chem. and Phys. 14 123 (1986).

43. Hair, M.L., Infrared Spectroscopy in Surface Chemistry; Marcell Dekker (1967).

44. Earborn, C., Organosilicon Compounds; Academic Press (1960).

45. Marshall, K. and Rochester, C.H., Infrared Study of the Adsorption of Oleic and Linolenic Acids onto the Surface of Silica Immersed in Carbon Tetrachloride. J. Chem. Soc., Faraday Trans. I [71] 1754 (1975).

46. Mills, A.K. and Hockey, J.A., Selective Adsorption of Methyl Esters on n-Fatty Acids at the Silica/Benzene and Silica/Carbon Tetrachloride Interface. J. Chem. Soc., Faraday Trans. I. [71] 2398 (1975).

47. Low, M.J.D. and Lee, P.L., Infrared Study of Adsorption in situ at the Liquid-solid Interface. J. Coll. and Interface Sci. [45] 148 (1973).

48. Product Information, Tertiary Amines; Lonza Corp, Fair Lawn, NJ, USA.

49. Hofman, K., Imadazole and its Derivitives, Part 1; Interscience (1953).

50. Harnsberger, B.G. and Riebsomer, J.L., The Influence of Alkyl Substituents on the Rates of Hydrolysis of 2-Imidazolines. J. Heterocyclic Chem. 1 188-192 (1964).

51. Bellamy, L.J., The Infrared Spectra of Complex Molecules; Wiley (1975).

52. Nakanashi, K. and Soloman, P.H., Infrared Absorption Spectroscopy (2nd ed); Holden-Day (1977).

53. Nilsen, K.J., Wautier, H. and Danforth, S.C., Dispersion of Laser-Synthesized Si_3N_4 Powder. In Advances in Ceramics, Vol. 21, Ceramic Powder Science; Messing, G., Mazdiyazni, K., McCauley, J. and Haber, R., p 537-48, The American Ceramic Society (1987).

54. Rosen, M.J. and Goldsmith, H.A., Systematic Analysis of Surface Active Agents (2nd ed); Wiley (1972).

55. Katrisky, A.R., Physical Methods in Heterocyclic Chemistry, Vol. IV; Academic Press (1971).

56. Nilsen, K.J., The Effect of Moisture on the Surface Chemistry and Non-Aqueous Dispersion Properties of Laser Synthesized Silicon Nitride. Ph.D. Thesis, Rutgers University, (1988).

57. Symons, W. The Influence of Oxygen Exposure on the HIP Characteristics of Laser Synthesized Si_3N_4. Ph.D. Thesis, Rutgers University, (1989).

3

Wet Forming Processes as a Potential Solution to Agglomeration Problems

A.O. Boschi[†] and E. Gilbart[*]

† Materials Engineering Department, Federal University of Sao Carlos, Cx. Postal 676, Sao Carlos SP, Brazil.

* Division of Ceramics, School of Materials, The University of Leeds, Leeds, LS2 9JT, UK.

1. INTRODUCTION

The aim in any component manufacture is to make a product which has the desired shape, dimensions and properties. The first two characteristics are determined mainly by the forming process, whilst the properties of the final product, which in most cases is a polycrystalline (and often multiphase) material, are dependent on other parameters, as indicated in Figure 1.

The intrinsic properties are defined by the choice of the chemical system to be used. Subsequently, any control over the final properties can be achieved only by controlling the product microstructure, by regulating the different stages of the ceramic fabrication process. Therefore, once the desired microstructure is known, the ceramist has to consider how to adjust the fabrication conditions to produce it. The success of this approach depends on a considerable understanding of the effect of the various steps of the fabrication process on the microstructure and the interrelationship between them.

Nearest to the final microstructure is the firing operation, and most attention was initially concentrated on it. However, as a result of great advances in understanding how the green microstructure evolves into the final microstructure, as well as the recognition of the limitations of sintering

73

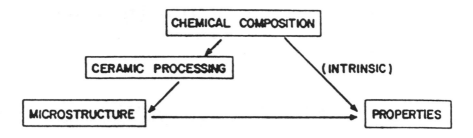

Figure 1: The properties of ceramic products.   After Stuijts[1].

alone to control the characteristics of the final microstructure, a series
of recommendations for all the other steps of the process has been issued
(see, for example, reference 2).   Initially, most of them were concerned with
the quality of the powder.   Subsequent advances, now from the two extremes
of the fabrication process (powder, right at the beginning, and sintering,
the last operation), have demonstrated the central importance of the green
compact[3,4].   According to this view, the most profitable way forward would
involve a greater level of attention being given to the green compact (Figure
2).

2. THE GREEN MICROSTRUCTURE

The characteristics of the green compact are a consequence of the powder
characteristics and the manner in which the forming process assembles the
particles together to give them the required shape and approximate
dimensions.   Two of the most important characteristics of the green body are
density and homogeneity (both chemical and physical)[5].   Green density is

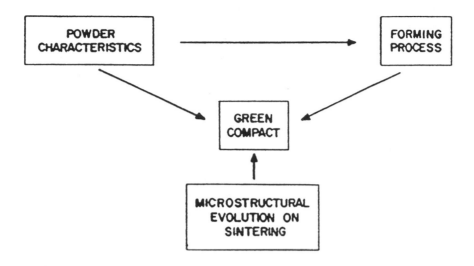

Figure 2: Relationship between different parts of the ceramic process.

a consequence of the solid's density and the volume of pores per unit volume of sample. It does not take into consideration the sizes nor the physical distribution of the porosity. However, if on sintering all pores are to disappear, the shrinkage must be equal to the volume of pores in the green body. Therefore, if the volume of pores (the density) varies from place to place in the green body, the extent of shrinkage will also vary. This process will lead to the development of stresses in the body which will make sintering to the theoretical density a much more difficult, if not impossible, task. Flaws critical to the strength and reliability of the component may remain.

Variations in size and/or physical distribution of pores will also lead to different sintering rates in different regions of the same body, due to the variations of diffusion distances. Stresses will also be developed in these cases.

The presence of agglomerates and/or aggregates in a powder is widely recognized as one of the major causes of physical inhomogeneities in green compacts[6-8]. These clusters of primary particles are usually formed during the manufacture of the powder and they may or may not survive the forming

operation.   But in cases where they do survive they usually form a bimodal
pore distribution with pores inside the agglomerates having a smaller size
than those between them[7].   During firing these agglomerates sinter to
nearly theoretical density at relatively low temperatures or at the early
stages of sintering, while the pores between agglomerates require for their
elimination higher temperatures and/or longer sintering times (see also
Chapter 1).

The above can be explained from fundamental sintering theory where the
rates of densification are inversely related to the scale of the structure
to sinter (the scaling law).   Therefore, the small pores within the
agglomerates disappear before the larger ones between them.

Rhodes[6] has demonstrated the dramatic effects of eliminating
agglomerates on the powder sinterability (Figure 3).   Zirconia powder in
which the agglomerates were eliminated by sedimentation reached full density
at temperatures much lower than the original powder, and presented a fine
grained and uniform microstructure.

It is important to mention here that the better sinterability of the

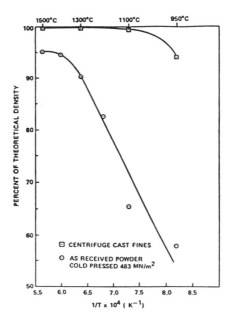

Figure 3: Effect of eliminating agglomerates on sintering of yttria-stabilised zirconia.   After Rhodes[6].

finer powder, freed of agglomerates by sedimentation, was also a consequence of the different forming process used. To avoid the formation of new agglomerates on drying, this powder was given shape by centrifugal casting. This process, apart from facilitating agglomeration control, can generate a more uniform distribution of particles and pores in the green compact.

As a result of the high specific surface energy of very fine powders, which is the driving energy for sintering, and the scaling law effect[9], which indicates that atoms have shorter distances to diffuse in compacts consisting of smaller particles (and thus require shorter sintering times), there has been a trend to work with ever finer powders. However, as particles become smaller and smaller, the gravitational force acting over them decreases very rapidly (as the cube of the diameter), whilst the natural surface adhesion forces decrease approximately by the first or second power of the diameter. This natural tendency of fine particles to stick together forming agglomerates creates serious handling problems which may offset the advantages predicted by the scaling law.

It is usually found that when particles become smaller, the size of the inhomogeneities in the green compact do not decrease proportionally with the particle's diameter. This may lead to the situation in which inhomogeneities much larger than the primary particles themselves control the evolution of the microstructure during sintering.

3. ORIGIN AND NATURE OF AGGLOMERATES

The bonds between primary particles may be classified by the medium through which the linkage is made:

1)   in gas (electrostatic and van der Waals)
2)   in liquid (liquid bridges and capillary liquid)
3)   in solid (solid bridges).

Rumpf and Schubert[10] suggested that, as a first approximation, the adhesion forces increase from 1 to 3 above. Only the bonding due to "solid" bridges resulting from the use of viscous binders occurs intentionally, all the other mechanisms usually being a consequence of small size particles and the manufacturing process.

A popular route for producing ceramic powders is chemical precipitation from homogeneous solutions. In this process the powder, or its precursor, is obtained during precipitation in a liquid medium. The drying operation that follows, as well as the calcination, if required, is usually most critical for control of the characteristics of the clusters of primary particles formed[11]. The liquid medium itself, in the drying stage, as well as the drying conditions, has been shown to be of great importance in determining the characteristics of the agglomerates, e.g. for ultrafine grained zirconia powders Haberko[12] has found that the agglomerates formed on drying from aqueous suspensions were much harder than when ethyl alcohol was used. Roosen and Hausner[7] have demonstrated the effects of different drying methods, with freeze drying giving a powder free of agglomerates and easier to sinter.

Liquid bridges can hold the particles together by capillary action. Very little liquid is needed for this type of bonding because it is concentrated at the contact points between primary particles. These capillary forces may be able to give the compact good mechanical strength when the particles are very small, as a consequence of the larger number of contact points per unit volume of the compact.

Solid bridges may be formed during drying by the crystallization of salts dissolved in the liquid of the original solution. During drying the concentration of these salts increases and the remaining liquid, in which they are dissolved, is driven to the contact points between primary particles by capillary action. Finally, the concentration becomes too high for the salt to remain in solution and precipitation starts forming crystals which touch both particles linking them together.

Solid bridges can also be formed during calcination, as a result of local sintering. Agglomerates formed before calcination will become harder after this stage. In this respect the temperature and duration of calcination is critical.

Therefore, there are only three possible alternatives to avoid the presence of aggregates in ceramic powders:

a)   to avoid their development during powder fabrication,

b)   to eliminate them by sedimentation, and

c)   to try to crush them by milling.

Most powder preparation routes used nowadays for advanced ceramics produce a suspension which has to be dried. Even if one, at this stage, separates the larger particles and agglomerates by sedimentation or any other technique, they may be formed again during drying. One possibility is to avoid drying the suspension before forming by using processes such as slip casting. In this manner it is possible both to separate the undesirable agglomerates and to produce a uniform green compact.

4. DEVELOPMENT OF THE GREEN MICROSTRUCTURE

As shown in Figure 2 the characteristics of the green compact are a consequence of the powder characteristics and the forming process. The forming processes most used in the ceramic industry can be divided into two general groups: dry and wet. Among the dry, die pressing is one of the most used, and in the wet group slip casting is one of the most popular.

4.1 The development of the green microstructure in dry pressing[13-15]

Because of the poor flowability of very fine powders and the deleterious effects of a nonuniform filling of the die, it is usual to go through a process of granulation of the powder before actually introducing it into the die. The granulated powder flows better mainly because the granules are much bigger than the particles giving a smaller number of contact points (points of friction) per unit volume and the granules have a round shape. The most popular of these granulation processes is spray drying.

The changes that take place in the internal structure of the compact during compaction are usually divided into three stages:

STAGE 1:   Initially, the granules each containing thousands of primary particles linked together and behaving as if they were one single larger particle, are poured into the die and distribute themselves in such a manner as to achieve an equilibrium. This equilibrium is established through the balance of the granule weight and the forces acting at the contact points with other granules and/or the die wall. During this stage the arrangement of primary particles in the granules does not change. The only variations are due to changes in the arrangement of the granules.

STAGE 2: At the beginning of this stage the granules are locked together as shown in Figure 4, and start to resist the application of pressure. The concentration of force at the contact points, resulting in high local pressures, leads to the fracture of some granules which makes the system unstable and produces rearrangement of the granules. The new arrangement will have a higher contact area between granules, and thus will resist stronger forces before the contact points are broken again and more rearrangement takes place.

STAGE 3: Now no further rearrangement of granules can take place and they are disintegrated by the application of forces strong enough to break the bonds between the primary particles. Ideally at this stage we should have a redistribution of the primary particles in such a manner as to form a uniform structure. However, since the primary particles are usually very small and as a consequence the compact has a very high number of contact

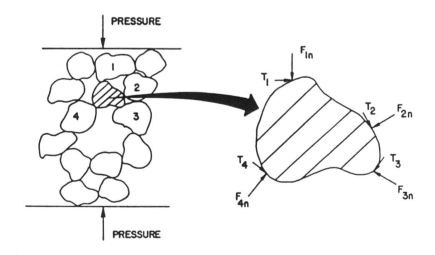

Figure 4: Forces acting on a particle during compaction, after Reed and Runk[13]. F and T represent normal and shear forces, respectively.

points per unit volume, friction inhibits the movement of the particles and makes the achievement of a uniform distribution of particles and pores quite difficult.   There is also the natural attraction forces acting between particles which become more important the smaller the particle.

If the powder used has agglomerates with different strengths a structure of locked particles may develop which will protect from fracture the harder agglomerates situated within this matrix of softer agglomerates and fine particles.   Under these conditions the agglomerates will constitute a non-uniformity in the green compact which, through differential densification rates during sintering, may build up enough stress to produce flaws which will limit the use of the ceramic part[16].

4.2 The development of the green microstructure in slip casting

In slip casting, initially, it is necessary to make up a suspension of the powder in a liquid and then, by a sort of filtration, a compact layer of the powder is deposited on the surface of a porous mould (Figure 5).   The green microstructure in this case is basically developed as the liquid is

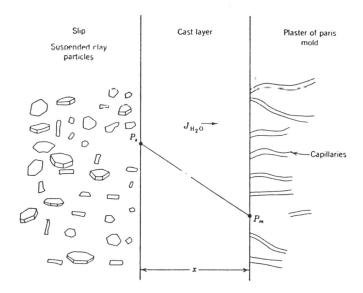

Figure 5: Schematic representation of the casting process. P represents the pressure at the slip-cast (s) and cast-mould (m) interfaces, and J the flux of liquid flowing through the cast layer of thickness x.

drawn into the mould by capillary action and the characteristics of the slip are known to play a major role[17].

Since the fine dry powder agglomerates spontaneously, and for slip casting it is necessary to make up a reasonably stable suspension, a dispersion process is necessary. Parfitt[18] has divided the overall process of dispersion into three stages: wetting of the powder, breaking of the clusters and stabilization.

The wetting process depends on the nature of the liquid phase, the character of the solid surface, the dimensions of the interstices in the clusters, and the nature of the mechanical process used to produce the wetting of the particles. Once the particles are wet some form of mechanical energy is usually required to bring about their effective separation. The state of agglomeration of the primary particles in the liquid will depend on the stability of the dispersion against coagulation.

When two particles approach each other, due to Brownian motion, sedimentation or processing forces, they flocculate or repel depending on the character and magnitude of the interparticle forces. The general equation describing these forces consists of attractive and repulsive terms:

$$V_t = V_a \text{ (van der Waals)} + V_r \text{ (electrostatic)} + V_r \text{ (steric)} + V_r \text{ (solvation)}$$

where V stands for interaction energy and the subscripts t, a and r for total, attractive and repulsive, respectively. In this equation the terms V_a (van der Waals) and V_r (solvation) are characteristics of the system and generally cannot be changed. Usually, the solvation forces are relatively small, although in some non-aqueous solvents of moderate or low dielectric constant, they may contribute significantly to the stability.

The only possible way of avoiding coagulation is by developing repulsive forces between particles, either V_r (electrostatic) or V_r (steric), which are strong enough to leave a repulsion resultant as V_t. This can be done by pH control, in the first case, and by using macro-molecules which may be adsorbed on to the particle's surface in the second[19,20].

By controlling the agglomeration state of the particles in suspension we are able to adjust the characteristics of the powder which will be driven towards the mould wall by the flow of fluid into the mould, and which will pack together to build up the green compact.

5. THE FORMING PROCESS AND GREEN MICROSTRUCTURE HOMOGENEITY

As mentioned before, the characteristics of the green compact are a consequence of the powder characteristics and the forming operation (Figure 2), and uniformity is one of the most important of these characteristics[4]. The presence of agglomerates in the powder, which are not eliminated before or during the forming operation, is one of the main causes of inhomogeneity in the green compact. To avoid this, one can control the powder fabrication process, grind the powder, or separate out the agglomerates. The effectiveness of the dry grinding in eliminating agglomerates is poor. As agglomerates may be formed during drying, they could be avoided by avoiding drying. Also, as sedimentation is one possibility for eliminating agglomerates, forming directly from suspension would allow the separation of the larger particles, and the agglomerates with them, before forming straight from suspension by slip casting.

To illustrate some of the ideas given above, the results of a group of experiments conducted to distinguish between the sort of structure produced when slip casting and die pressing have been used to shape the same powder, as well as their evolution on sintering, are quoted below.

6. EXPERIMENTAL PROCEDURE

The powder used was titania (rutile, R-SM2 from Tioxide, Cleveland, UK). Two forming methods were used: dry pressing and slip casting. For pressing, a pressure of 64 MPa was applied, and for slip casting the dispersion stability was adjusted by pH control using NH_4OH additions to achieve pH = 9.2 and 10.0. The suspensions with pH = 9.2 and 10.0 had densities of 1.64 and 1.13 gcm^{-3} (solid contents of 51 and 16.4 weight %) and high and low casting rates, respectively. The slips were ultrasonicated before and after pH adjustment. Compacts were heated at 10°C per minute for sintering at 1200°C for periods of 0 (to indicate changes occurring as a

result just of the heating process), 30, 90 and 300 minutes, followed by
direct removal from the furnace. The green microstructure and its evolution
on sintering was characterised by mercury porosimetry and observation of
fracture surfaces with the Scanning Electron Microscope (SEM).

7. RESULTS AND DISCUSSION

Figure 6 shows the fracture surface of green compacts of the same powder
formed by pressing and by slip casting. The differences in regularity are
striking. As the crack responsible for the fracture always follows the
weakest region near its tip, the more regular fracture surface of both
samples produced by slip casting can be seen as an indication of better
uniformity.

Chappell et al[21] have suggested that for an ordered array to be
formed, particles have either to reach the growing compact at an appropriate
position or they have to diffuse to such a position after arrival. For all
the particles to arrive at the appropriate position is thermodynamically
improbable, so they have to diffuse. For this to be possible the particles
arriving need to have freedom of movement, time to reach the position of
optimum fit, and finally have to be held in that position. Consider
producing not a perfect array of particles, but a random array instead; there
are two basic requirements: a) a slip containing a uniform (random)
distribution of primary particles (and only primary particles), and b) a
mould with a homogeneous capillary action.

The uniformity of the distribution of primary particles in the slip is
directly related to its stability and therefore requires strong uniform
repulsion between particles.

In the experimental procedure described above, the pH = 10.0 slip is
closest to the ideal situation. It has the strongest repulsion forces, which
gives the particles the necessary freedom of movement on arrival at the
growing cast, because they are repelled by those particles already forming
it, and its lower solid content (density) gives the particles more time to
search for the best fitting positions before they are trapped by the
subsequently arriving particles.

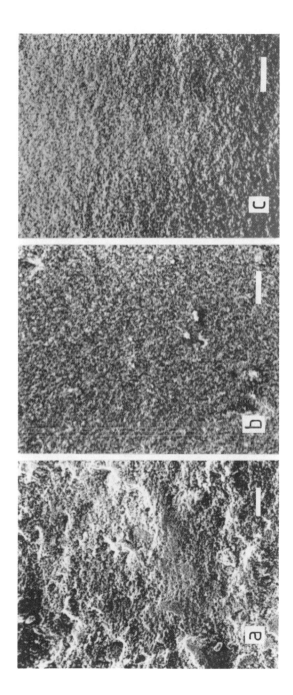

Figure 6: Fracture surfaces of green compacts formed by pressing (a) and by slip casting, at pH = 9.2 (b) and pH = 10.0 (c). Bar = 50 μm.

The uniformity of the green compact produced by pressing could certainly be improved by the use of binders, a better controlled granulation process and isostatic pressing. However, even then it would be very difficult to achieve a degree of uniformity similar to that achieved by the slip cast samples.

Figure 7 shows the pore size distributions for the pressed and slip cast green compacts. Here it is interesting to notice that both the total volume of porosity and the mean pore radius vary, such that the better dispersed and more dilute suspension, pH = 10.0, presents the smallest pores and the highest green density, as predicted by the reasoning given above. Other experiments[22] using a similar powder, but with a different particle size distribution, obtained by removal of some of the coarser fraction from the as-received powder by sedimentation, have shown that under similar dispersion conditions the pore size distribution of the green compacts were almost identical to those of Figure 7. Therefore, the dispersion conditions are seen to be the main feature in determining the characteristics of the green compact.

The variations of the mean pore aperture radius during isothermal sintering at 1200°C as a function of sample density are shown in Figure 8,

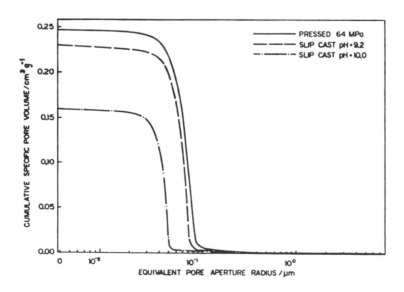

Figure 7: Effect of the forming process on the pore size distribution.

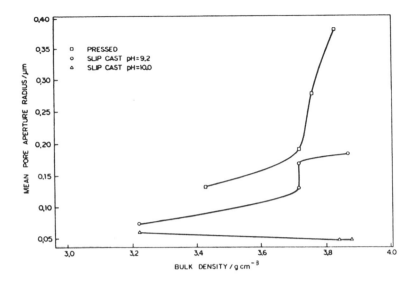

Figure 8: Variations of the mean pore aperture radius with bulk density for samples at 1200°C.

the corresponding sintering times are given in Table 1. The pressed sample exhibited the greatest pore growth rate and the slip with pH = 10.0 shows a decrease of the mean radius. In fact, since the size of the primary particles was the same and the mean pore aperture radius resulting from the various forming conditions was different, the ratio of actual pore/grain radius has been varied, and in accordance with the model developed by Kingery and Francois[23] it is only below a certain value of this ratio, which is a function of the (unchanged) dihedral angle, that conditions become favourable for pores to shrink. For oxides, the dihedral angle lies in the range 140–160°, and at 150°, the critical ratio is about 1.45[23].

The pore radius given by the porosimeter represents the "penetration radius" of the aperture through which mercury must penetrate to fill the pore, which is smaller than the actual pore radius which would be more critical in relation to the grain radius as regards coarsening rather than shrinking. Random packing of particles with pores the same size of the particles gives a predicted packing density near 0.6[23], and since the fractional green density of slip cast sample of pH = 10.0 is about this value, the assumption is made that for this sample, the pore and grain radii are equal, i.e. for the grain radius of 0.12 µm, the pore/grain radius ratio is 1.0. This also provides a correction factor for the other samples to obtain their respective pore/grain radius ratios from the porosimeter penetration radius, which are calculated to be 1.53, for the sample slip cast at pH = 9.2, and 1.65 for the die pressed compact[22]. So, only the pH = 10.0 sample fulfils the requirement that the pore/grain radius ratio is less than 1.45, and indeed only this sample exhibits a decrease in mean pore radius on sintering.

The observed behaviour of the slip cast sample with pH = 9.2, where densification temporarily ceases, could not yet be properly examined, but very similar results were obtained for samples made with a similar powder with a different particle size distribution[22].

Figure 9 shows the fracture surfaces of samples partially sintered. The differences in grain sizes are quite clear. The grain growth rate is strongly influenced by the presence of pores. To analyse the observed behaviour in a more quantitative way a simple geometric model was developed[22]. The model consists of a cubic array formed by equispaced intersecting cylinders of a radius equal to the mean pore radius, which represent the pores. This model uses a geometry which is similar to that employed to transform the results from mercury porosimetry (originally pressure and volume of mercury) into the so-called pore aperture size distribution. The spacing between the cylindrical pores is adjusted until the volume of the cylinders corresponds to the total pore volume as given by the porosimeter (or from the bulk density). The spacing in the green compacts may then be compared with the particle size, and the evolution on sintering of the calculated pore spacing and the calculated pore density for each forming process evaluated on the basis of this idealised model.

Table 1 gives the variations in mean pore spacing and pore density (number of pores per unit volume of sample) with time at 1200°C, for the samples formed under different conditions. The interaction between grain boundaries and pores is very important in establishing microstructural evolution (grain growth versus densification). Because pores usually cannot move as fast as grain boundaries they slow down coarsening[25]. Also, for a pore to be eliminated it has to remain near a boundary which acts as a source of atoms (ions) and sink of vacancies. Therefore, if a pore is left in the centre of a grain it will limit the maximum density to be achieved. Thus, in an ideal situation one should have a large number of pores which

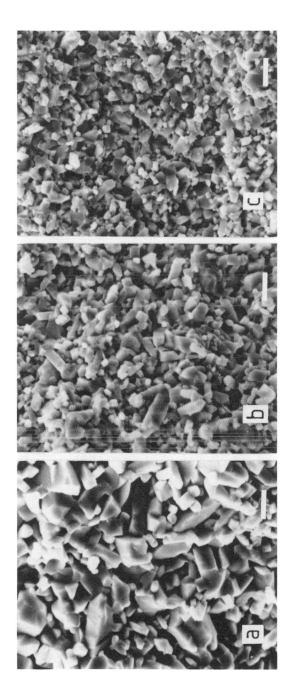

Figure 9: Fracture surfaces of samples partially sintered at 1200°C for 90 minutes, with similar densities. Samples were pressed (a), slip cast pH = 9.2 (b) and slip cast pH = 10.0 (c). Bar = 5 μm.

Forming	Time	Bulk density	Spacing	Pore density
	min	gcm^{-3}	μm	cm^{-3}
Pressed	G*	2.07	0.33	8.6E13
64 MPa	0	3.42	0.96 (66)	3.4E12 (-96)
	30	3.59	1.54 (38)	8.3E11 (-76)
	90	3.72	2.53 (39)	1.9E11 (-76)
	300	3.80		
Slip cast	G*	2.15	0.31	1.0E14
pH = 9.2	0	3.22	0.46 (66)	3.0E13 (-71)
	30	3.72	1.18 (61)	1.8E12 (-94)
	90	3.72	1.47 (20)	9.4E11 (-49)
	300	3.87		
Slip cast	G*	2.47	0.22	2.8E14
pH = 10.0	0	3.22	0.38 (42)	5.6E13 (-80)
	30	3.84	0.49 (23)	2.5E13 (-55)
	90	3.88	0.52 (6)	2.1E13 (-17)

Table 1: Values calculated from the cylindrical pore model[22] for mean pore spacing and pore density (number of pores per unit volume of sample).   The numbers in brackets are the percentage increase between successive values. Mean particle diameter = 0.24 μm.   * Green samples.

remain attached to the boundary during sintering. This is the case of sample pH = 10.0. The green pore spacing corresponds almost exactly to the mean particle diameter, implying a regularity of pore/grain association not achieved in other samples. It presents the smallest decrease of the number of pores per unit volume of sample during sintering (Table 1), the smallest grains (Figure 9) and the highest final density. This was certainly a consequence of the smaller pores in the green compact and the good uniformity of particle packing given by slip casting as well dispersed powder.

8. CONCLUSIONS

From the point of view of producing a dense and uniform microstructure with fine grains, the best results in the present work were achieved by slip casting. However, what lies behind the different forming techniques is the degree of uniformity of the green microstructure they can produce. Thus, in the conditions used here slip casting leads to a higher uniformity. However, this may not always be the case. If pressing is used carefully, with a free flowing powder free of agglomerates, the appropriate binders and lubricants in the correct amount, uniformly distributed, and the design of the body and the application of pressure is such that there is not a significant variation in the distribution of particles and pores in the green body, results similar to those achieved for slip casting could be expected. However, the finer the powder, the more difficult it becomes to achieve the conditions just described. Therefore, it becomes difficult to take full advantage of the high sinterability of fine powders. On the other hand, slip casting, and more generally the wet forming processes, are suitable for producing reasonably uniform green microstructures free of flaws even with very fine powders, and allows the elimination of agglomerates and larger particles by sedimentation and/or milling, before the actual shaping takes place. Using colloidal chemistry concepts a uniform and dense green microstructure, containing pores of the same order of magnitude as the particles, can be produced. This type of structure favours densification and inhibits grain growth, keeping a large number of pores attached to the grain boundaries.

ACKNOWLEDGEMENT

The practical work was supported by CAPES-Brazil and undertaken in the Division of Ceramics, School of Materials at the University of Leeds, UK, under the supervision of Dr. W.E. Worrall. Thanks are due to Professor R.J. Brook for his critical advice and constant encouragement.

REFERENCES

1. Stuijts, A.L., Ceramic microstructures. In Ceramic Microstructures '76; Fulrath, R.M. and Pask, J.A. (Eds), Westview Press, Boulder, p 1, 1977.

2. Yan, M.F., Sintering of ceramics and metals. In Advances in Powder Technology, Chin, G.Y. (Ed), American Society for Metals, p 99, 1982.

3. Onoda, G.Y. and Hench, L.L., Ceramic Processing before Firing, Wiley, New York, 1978.

4. Roosen, A. and Bowen, H.K., Influence of various consolidation techniques on the green microstructure and sintering behaviour of alumina powders. J. Amer. Ceram. Soc. 71 250 (1988).

5. Kingery, W.D., Firing - the proof test for ceramic processing, In Ref. 3, p 291.

6. Rhodes, W.H., Agglomerate and particle size effects in sintering yttria-stabilized zirconia. J. Amer. Ceram. Soc. 64 19 (1981).

7. Roosen, A. and Hausner, H., Low temperature sintering of zirconia. Ceramic Forum Int. 4/5 184 (1985).

8. Pampuch, R., ZrO_2 micropowders as model systems for the study of sintering. Science and Technology of Zirconia II, Claussen, N., Rühle, M. and Heuer, A.H. (Eds), Advances in Ceramics 12 773 (1984).

9. Herring, C., Effect of change of scale on sintering phenomena. J. Appl. Phys. 21 301 (1950).

10. Rumpf, H. and Schubert, H., Adhesion forces in agglomeration processes, In Ref. 3, p357.

11. Roosen, A. and Hausner, H., Techniques for agglomeration control during wet-chemical powder synthesis. Adv. Cer. Matls. 3 131 (1988).

12. Haberko, K., Characteristics and sintering behaviour of zirconia ultrafine powders. Ceram. Intl. 5 148 (1979).

13. Reed, J.S. and Runk, R.B., Dry Pressing. In Ceramic Fabrication Processes, Wang, F.F.Y. (Ed), Treatise on Matls. Sci. and Techn. 9 p71, Academic Press, (1976).

14. Macleod, H.M., Compaction of ceramics. In Enlargement and Compaction of Particulate Solids, Stanley-Wood, N.G. (Ed), p 241, Butterworths, (1983).

15. Frey, R.G. and Halloran, J.W., Compaction behaviour of spray-dried alumina. J. Amer. Ceram. Soc. 67 199 (1984).

16. Lange, F.F. and Metcalf, M., Processing-related fracture origins: 11, Agglomerate motion and crack-like internal surfaces caused by differential sintering. J. Amer. Ceram. Soc. 66 398 (1983).

17. Worrall, W.E., Flow properties of acid-flocculated alumina slips. Trans. Brit. Cer. Soc. 62 659 (1963).

18. Parfitt, G.D. (Ed), Dispersion of Powders in Liquids, Applied Science, (1981).

19. Overbeek, J.Th.G., Strong and weak points in the interpretation of colloid stability. Adv. Colloid and Interface Science 16 17 (1982).

20. Sato, T. and Ruch, R., Stability of Colloidal Dispersion by Polymer Adsorption, Marcel Dekker, New York, (1980).

21. Chappell, I.S., Birchall, J.D. and Ring, T.A., The origin of defects arising in colloidal processing of submicron, monosize powder. Novel Fabrication Processes and Applications, Davidge, R.D. (Ed), Brit. Cer. Proc. 38 49 (1986).

22. Boschi, A.O., Effects of Different Forming Processes on the Sinterability of Rutile. Ph.D. Thesis, University of Leeds, (1986).

23. Kingery, W.D and Francois, B., The sinterability of crystalline oxides. I. Interactions between grain boundaries and pores. Sintering and Related Phenomena. Kuczynski, G., Hooton, N.A. and Gibbon, C.F. (Eds) p 471, (1967).

24. Boschi, A.O., Gilbart, E. Worrall, W.E. and Brook, R.J., Pore stability during the sintering of TiO_2. High Tech Ceramics, Vincenzini, P. (Ed), Matls. Sci. Monographs 38 893 (1987).

25. Brook, R.J., Controlled Grain Growth. In Ref. 13, p 331.

4

Processing of Electronic Ceramics

D. Cannell[†] and P. Trigg[*]

† Morgan Matroc Unilator Division, Vauxhall Industrial
Estate, Ruabon, Wrexham, Clwyd, LL14 6HY, UK.

* Filtronic Components Ltd, Royal London Industrial
Estate, Acorn Park, Charlestown, Shipley, West
Yorkshire, BD17 7SW, UK.

1. INTRODUCTION

Electronic ceramics enjoy widespread commercial exploitation in
components such as thermistors, capacitors, transducers, variators, etc.
Their use in these areas has developed for a variety of technical, historical
and economic reasons. Amongst the most prominent are:

1) The novel electronic properties of electronic ceramics. Many of these
materials cannot be prepared by single crystal routes because of a) their
complicated chemical composition or b) their dependence upon their granular
structure for their properties, e.g. grain boundary layer capacitors, ZnO
varistors, $BaTiO_3$ PTC thermistors.

2) The relative ease with which these properties can be tailored to suit
specific applications. This is particularly true in the case of some crystal
structures. The perovskite structure is an example where a wide range of
solid solutions can be accommodated. In this structure cation dopants can be
used to great effect. For example, donors or acceptors can be used to
produce "soft" or "hard" piezoelectric lead zirconate titanate.

3) The economical methods by which components may be fabricated.

The properties of many electronic ceramics permit the construction of
devices which otherwise may not be realised. This is in marked contrast to

structural ceramics where the potential for greater strength at high temperature over an existing metallic component offers a substantial improvement in performance. The problem for the structural ceramist is to attain acceptable levels of reproducibility and reliability of components working at high stresses. This has to be achieved within the broad geometrical and volumetric constraints imposed by the existing design, e.g. a turbine blade in a turbocharger. This problem can be reduced to one of fabricating large volumes of ceramic free of microscopic defects. These difficulties have not inhibited the application of electronic ceramics as, in many cases, their development and application has been dependent upon the availability of a material. Thus, devices have been designed on the basis of what can be achieved with acceptable reproducibility and reliability from an existing material. As a consequence electrical stresses are relatively low, e.g. a multilayer ceramic capacitor designed to work at 50 V is usually specified to be used in a circuit at 5 V but will have a breakdown voltage in excess of 300 V. There are, however, circumstances which require improved electronic ceramic processing to produce more reproducible and reliable components. The first of these is the need to reduce the size of components which inevitably will raise the stress levels at which they must work. This is true for devices made from thin ceramic layers. For larger components there is also the constant desire for tighter tolerances on electrical properties and reproducibility, not only within, but between batches of parts over long time periods. How, then, can modern ceramic processing meeting these challenges?

Perhaps the simplest approach is to start from the basis of a typical ceramic process route which uses solid oxides as its precursors (Figure 1). From here there are two ways in which one can proceed. The first of these is to consider what improvements can be made to the powder processing-fabrication-sintering scheme. The second is to abandon this scheme and use technologies akin to those used in semiconductor fabrication, e.g. sputtering[1], MOCVD[2], etc. This option is one that is being increasingly pursued, especially for producing thin layers; indeed thin film chip capacitors using SiO_2 and Si_3N_4 deposited by CVD are now available[3].

Developments in this field, permitting controlled deposition of complex chemical compositions will lead to greater exploitation of these processes. Although these techniques will not be discussed further here, (see chapter 9) there are lessons arising from them which can be applied to the powder-fabrication-sintering route. The most important of these is the small number

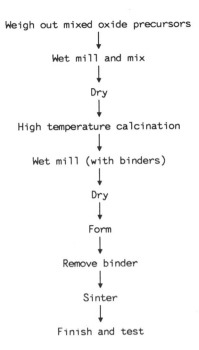

Weigh out mixed oxide precursors

↓

Wet mill and mix

↓

Dry

↓

High temperature calcination

↓

Wet mill (with binders)

↓

Dry

↓

Form

↓

Remove binder

↓

Sinter

↓

Finish and test

Figure 1: Standard mixed oxide processing route.

of process steps involved from the precursor to the final product. Ceramic processing proceeds stepwise, with each new step acting upon the results of the preceding step. If reproducibility and reliability of the end product are to be realised, it is clear that:

a) The process should contain as few steps as possible.

b) The process should contain checks at each step in order to monitor the process and ensure that the product properties are within acceptable limits.

c) Most careful control is needed during the earliest stage of the process as the cumulative effect of variations here is of great significance on final production properties.

Item c) is especially pertinent to ceramic processing as it is well documented that flaws present in unfired compacts persist during sintering and so cannot be removed. Whilst there may be some scope for developing defect-tolerant ceramic compositions, improvements in ceramic processing are likely to be of much greater widespread benefit. In the sections that follow, processing techniques and improvements are discussed (see also chapter 1).

2. POWDER PREPARATION

The first step in ceramic powder processing involves powder preparation. It is vital for the final product that the powder is of the highest quality and therefore initial powder preparation must be carefully controlled. Some of the powder characteristics of particular relevance are shown in Figure 2. In general terms the powder required will have a small primary particle size of the order of 0.5 - 5 µm, a narrow, monomodal size distribution and be free of weakly bonded agglomerates or strongly bonded aggregates. Chemically, each particle should have the same composition as the average for the batch. To ensure uniformity of properties in the final product the same degree of chemical homogeneity or heterogeneity is desirable for each particle.

a) Primary Particles

    Chemical   -   purity
                    -   homogeneity/heterogeneity of dopants
                    -   particle/particle homogeneity
                    -   bulk absorbed species
                    -   surface adsorbed species

    Physical   -   size distribution
                    -   surface area
                    -   shape

b) Agglomerates

    Chemical   -   mixing/segregation of particles in multicomponent system

    Physical   -   size distribution
                    -   shape
                    -   surface area
                    -   internal porosity
                    -   interpaticular bond strength

Figure 2: Powder characteristics.

All of these properties are more easily achieved using liquid rather than solid precursors. The ability to control size and shape of particles by controlling the nucleation and growth is especially attractive as it permits the ceramist to tailor the properties of the powder. Achieving the desired size by growth rather than by reduction also makes the production of very fine powders more practicable. The principal benefits of these powders are shown in Figure 3.

The many and varied chemical preparation routes for the production of electronic ceramic powders are too numerous to discuss in detail here. Typical precursors include chlorides[4], organometallics such as oxalates[5] and alcohol-based complexes[6]. The processes involved include hydrothermal synthesis[7], coprecipitation[8], precipitation of one component followed by coatings of successive dopants, sol-gel preparation[9], polymerisation[10], etc. It will be some time before the benefits of the processes can be assessed. Questions concerning their flexibility for producing different compositions, the quality of results and the problems of cost and scaling up still need to be resolved. What is not in doubt is that chemical methods offer potentially much improved powders compared to those prepared by traditional mixed oxide routes.

	Mixed Oxide	Chemically prepared
Impurity level	Very source dependent	Can be controlled to acceptable level
Particle size	Dependent on ability to reduce pre-existing particle size - usually large	Dependent on ability to control particle growth - can be very small
Particle shape	Dependent upon fracture mechanism of material, method of comminution	Can be controlled by careful chemisty
Chemical homogeneity	Depends upon high temperature reactions with diffusion over particle-size distances	Molecular mixing, low temperature reactions, short diffusion distances
Chemical heterogeneity	Often produced, rarely controlled	Surface doping possible for boundary layer materials

Figure 3: Comparison of mixed oxide and chemically prepared powders

Since the pioneering work on the production of spherical, monisized TiO_2 powders[11], it has been realised that "ideal" powders per se cannot be the complete solution to the problem of producing reliable electronic ceramics. However, they allow the ceramist much greater control at the earliest step in the processing route. It is when the powder is converted into a green compact that the greatest difficulties still remain and where the greatest attention is at present being focussed. In the next sections these powder processing steps are discussed with reference to mixed oxide and chemically prepared powders.

3. MIXING AND MILLING

Mixing and milling are often processes which are carried out simultaneously using one piece of equipment such as a ball mill. The principal aims of the milling process are:

1)   To reduce the primary particle size (if required).

2)   To reduce agglomerates or aggregates in size, ideally to primary particle size.

3)   To achieve the desired particle size distribution.

4)   Having achieved 3) above, to retain the new particle size distribution in a stable state via proper dispersion.

Item (4) is also important to achieve good mixing, the aims of which are:

1)   To ensure that the best degree of mixing possible (a random homogeneous mixture) is achieved.

2)   To ensure that the scale of mixing is sufficiently small to guarantee that any subsequent calcination will produce the correct proportions of new and original phases.

Clearly item (2) is itself dependent upon the milling process as well.

Ball milling is extensively used for mixing and milling electronic ceramics. Care has to be taken to avoid contamination to which materials are very sensitive even at the 0.01 wt% level. The approach is usually to either: a) try and minimise any pick-up by using rubber lined mills and very wear resistant milling media such as zirconia or agate; b) to use media of the same composition as one of the components or the composition of the powders being milled; c) deliberately use media which introduce a predictable amount of material which aids the sintering and/or final electrical properties of the product, e.g. the milling of zirconium tin titanate using steel balls[12]. With judicious choice of shape and size distribution of milling media used, submicron powders can be produced even from mixed oxide material calcined at high temperatures.

Two of the main disadvantages of ball mills are their relative inefficiency and the possibility that they may produce too broad a particle size distribution, promoting abnormal grain growth during sintering. This is largely due to the tumbling action of the media which leads to point contact only, ie the particle milled is that which happens to fall between the contact points. Whilst this can be improved to an extent by using cylindrical media, vibratory energy mills[13] offer the benefits of much faster milling times and narrower size distributions. Such mills use cylindrical media packed in an ordered structure so that their mean displacement within the mill is zero. As the body of the mill vibrates, large amounts of kinetic energy are transferred to the media. These make line contact with each other, ensuring that large particles are crushed first. Due to the fact that these mills do not rotate it is possible to process large volumes of slip using small mills and stirring tanks with recirculating pumps. This offers the possibility of continuous monitoring of the properties of the dispersion and so greater process control.

Another milling technique used for electronic ceramics is jet milling. The material to be milled is made to flow in two opposing jets, the kinetic energy of impact breaking down the particles. This is ideal for materials which cannot tolerate even very low levels of contamination.

All the above techniques rely on high kinetic energy impact for milling and are suited to reducing the ultimate particle size of many solid oxide precursors. With the advent of fine, chemically prepared powders, there is less emphasis on the need for primary particle size reduction but rather deagglomeration. Whilst fine powders may not have strong chemical bonds between them, van der Waals forces can be very large. Ultrasonic milling[14] is increasingly proving to be a useful technique for deagglomerating such powders by using ultrasound to produce cavitation in the milling liquid. As the ultrasound energy is strongly absorbed by common liquids such as water or propan-2-ol, this process is most effectively carried out in small chambers which again gives the possibility of continuous flow and monitoring of the process.

Shear mixing[15] has also been recognised as an effective method of deagglomerating, mixing and dispersing fine powders. This shear may be generated by very high speed stirring of liquid dispersions by forcing very high viscosity powder/polymer compounds through narrow constrictions during, for example, injection moulding or extrusion[16]. The advantage of the latter technique is that it ensures all the powder passes through a high shear zone and does not rely upon good mixing to bring the powder into this zone. With the trend towards finer powders and so a rapidly increasing number of particles per unit volume, the difficulty of ensuring that all the powder receives adequate milling becomes ever greater.

The use of high shear mixing and milling is preferable to those techniques which rely upon statistical probabilities for their effectiveness.

4. DRYING

Drying is used either to remove liquids present during mixing or milling operations or to remove liquid from a compact formed by processes such as tape casting or slip casting. Drying is problematic, especially for fine powders where the surface tension brings particles into intimate contact as the liquid is removed, allowing van der Waals or electrostatic forces to dominate and so form agglomerates. There is renewed interest in casting techniques which avoid the need for liquid removal until after the desired shape has been formed (see chapter 3). However, there are many instances where drying cannot be avoided if powders suitable for subsequent handling are to be produced, e.g. powders which contain pressing aids.

Simple drying of slips in ovens is a rudimentary technique which is not without benefits. If carefully controlled it can produce well packed blocks of material, free from large internal defects, which can be granulated and formed into uniform green compacts. However, due to the relative slowness of this technique, faster and higher volume production routes such as spray drying[17] have found favour.

Spray drying consists of "atomising" a slip into a heated chamber where the liquid is evaporated and the powder collected in a cyclone. This cyclone permits limited control over the achievable granule size distribution in that fine particles may not be collected here but filtered out elsewhere. The two most common types of spray drier use either a nozzle or a spinning disc for generating the liquid droplets. The nozzle type employs a swirl chamber in conjunction with a small carbide aperture. This creates a vortex of droplets which rise toward the top of the drier and the drier granules subsequently fall to be collected near the base. The latter type uses a horizontally spinning disc with holes around its periphery and is mounted in the top of the drier. It produces a radial spray of droplets which dry as they fall against the counter flow of hot gas. Spray driers are capable of large throughputs, of the order of hundreds of kg per hour but the powder they produce may not be ideal. Attempts at using excessive throughput rates require very high chamber temperatures which can degrade not only dispersants but also common organic binders. Hard granules are produced which do not deform adequately or do not bond with their nearest neighbours when pressed. Studies of the morphology of these granules has revealed them often to be hollow spheres or "doughnut" shaped[18]. These internal voids can be a major source of flaws in the green and subsequently sintered material. Much attention is currently focussed[19] on understanding the mechanisms of droplet formation, liquid removal and granule formation in an attempt to produce solid, soft agglomerates which are easily compacted.

An alternative drying technique which neatly avoids the problems of forming hollow particles is that of fluidised bed spray granulation[20]. In this process a "seed" powder is used which is passed through a fluidised bed and coated with a fresh layer of slip on each pass. Using cyclone classifiers, the size and size distribution of the particles can be controlled. To date, this process has not found widespread usage, primarily because of the high capital cost of the equipment.

All of the above methods have the common disadvantage that they remove liquid from the slip and so introduce agglomeration problems caused by surface tension effects. Freeze drying[21] avoids these problems by converting the liquid, first in a solid by freezing and then to a gas by sublimation. This technique has been shown to have remarkable results on the sinterability of fine powder[22].

5. DRY FORMING

Dry forming of electronic ceramic materials is the major method used for the compaction of fine powders prior to densification. The process normally used is dry pressing. In fact, dry pressing is a misnomer as it is necessary for small amounts of pressing additives to be incorporated into the powder to enable the production of handleable green bodies. These additives have different roles in the process. Binders (e.g. polyethylene glycol) are added to endow adequate green strength on the pressed bodies. Lubricants such as zinc stearate ease the release from the die of the pressed body.

Plastercizers may also be added to improve the flexibility of the binder films and so promote plastic deformation of the granules during pressing. Such pressing aids must be able to be decomposed into gases well below the sintering temperature of the ceramic. These gases should then be able to escape through the open porosity of the compacted powder, leaving no residues. Differential thermal analysis (DTA) and thermal gravimetric analysis (TGA) are commonly used to identify the burn off temperatures of ceramics containing such additives. The firing schedules will include a dwell at an appropriate burn off temperature. Such an operation is particularly critical in the firing of ceramics to be used in electronic applications since any residues could dramatically alter the final properties of the sintered body.

The powder to be pressed should have been suitably prepared so that it has uniform flow properties so that it will evenly fill the die (see section 2).

When the die is suitably charged with powder the preferential method of applying the pressure is by a technique known as double ending pressing. This process involves keeping the die in a fixed position while the top and bottom punches move to consolidate the powder. This produces a more evenly compacted body when compared to bodies made by single ended pressing in which the bottom punch is stationary. The reason for this is that because of the particulate nature of the powder, the forces applied are transmitted off perpendicular so that in a single ended pressing the compact will be more compressed at the end nearest the moving punch than at the end furthest away, Figure 4.

Maintaining the uniformity of compaction is vital when manufacturing electronic ceramic as density variations in the green compact can lead to density variations in the sintered body, with a resultant non-uniformity of properties.

Having consolidated the powder by pressing, the green bodies are ejected by holding the die stationary and applying a force to, usually, the bottom punch. When the piece is ejected out of the die elastic strains are relieved with a resultant minute increase in its dimensions. The use of lubricants, both as additions to the powder and as spray onto the die wall, can significantly reduce the pressure required for ejection. It is during the ejection procedure that defects such as lamination and capping appear, Figure 5.

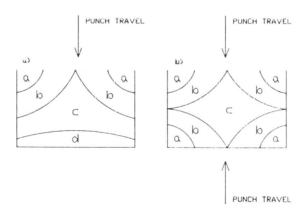

KEY ι LETTERS INDICATE VARIATIONS OF DENSITY
           WITHIN GREEN BODY
       WHERE a>b>c>d>

Figure 4: Density distribution variations between single ended (a) and double ended pressing (b) of a cylindrical compact.

a)

b)

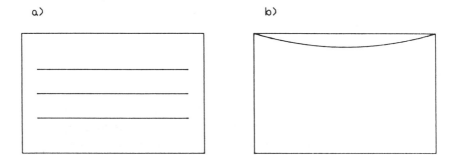

Figure 5: Lamination (a) and capping faults (b) in die pressed cylindrical disks.

These defects usually appear because of die wall friction. The problem is reduced by optimisation of powder flowability and hence uniform die fitting, the use of adequate die pressing additives and suitable lubrication. Such optimisation will lead to the ability to use higher pressures and so obtain a high green density. Maximisation of the green density minimises the shrinkage during firing.

A second method of dry forming of green bodies is isostatic pressing. In this technique the ceramic powder with appropriate strength imparting binders is placed in an airtight, flexible container which in turn is placed inside a sealed chamber filled with liquid. Pressure is then applied to the liquid which consequently applies a uniform pressure over the total surface of the powder filled container.

The powder is thus compacted into a green body of similar shape to the flexible container. There are two basic types of tooling used for isostatic pressing, i.e. wet bag tooling and dry bag tooling. Wet bag toolings are filled with powder sealed and immersed in the pressurising liquid which acts directly on the flexible container, actually wetting it. A dry bag tool is an integral part of the pressure chamber. The liquid is pressurised behind a flexible second lining forcing it onto the tool and the filled container remains dry.

Before the isostatic pressing procedure it may be necessary for the tooling bag containing the powder to be evacuated. Work done on the dry forming of large green bodies (110 mm dia x 24 mm high) from zirconia tin titanate, a dielectric resonator material, has emphasized the necessity of such a step. The pucks were made this size so that the sintered product would be a dielectric resonator at approximately 900 MHz. The dried mixed powder which exhibited good flow characteristics and contained polyethylene glycol as a binder and zinc stearate as a lubricant was critically die-pressed under a low pressure (70 MPa) to form a low density green body. This green body was then placed in a sealed polythene bag for isostatic pressing at 140 MPa.   When the bag was not evacuated prior to sealing the isostatically pressed bodies were observed to be badly cracked.   This was thought to be caused by the release of compressed air from the body when the isostatic pressure was removed.   The cracking problem in the green body disappeared when the bags were evacuated prior to sealing.   The bodies pressed in this way had high green density and were successfully sintered to 95% of theoretical density, Figure 6.

Figure 6: Green and fired bodies for 900 MHz dielectric resonators.

Ceramic powders that are to be isostatically pressed are usually spray-dried. In the spray drying process the powder is dispersed in water to which binders, plasticizers, dispersants, lubricants and pH regulators are added. The spray drying process is intended to yield a free flowing granular powder to ensure uniform tool filling.

An extremely useful attribute of the isostatic pressing process is its ability to be able to form unusually shaped green bodies. For example it has found an application in the mass production of the insulator part of spark plugs. In this instance the dry bag technique is used for mass producing bodies.

6. TAPE CASTING

The advance of microelectronics has led to a significant demand for flat non-conductive circuit carriers. Such substrates require mechanical strength to be able to withstand subsequent processing. Also, because the circuit may generate heat during use, the substrate has to have acceptable thermal conductivity. It should also be unaffected by the chemicals and temperatures used in circuit manufacture.

Other requirements include a high degree of surface smoothness, surface flatness, circuit to substrate adhesion dimensional stability and machinability.

The most common method of producing large quantities of ceramic substrates, mainly aluminium oxide[23], is the doctor blade type casting process first reported by Howatt et al. in 1947[24]. This technique requires the production of a stable slip. It will normally be made by ball milling the powder for several hours in a liquid containing binders, plasticisers and dispersants etc. Basically, the slip is spread over a flat surface and the solvent dried off and the powder and binders form a ceramic tape. In the doctor blading technique the flat surface is in a carrier belt which is fed from a slip chamber. The belt passes under carefully adjusted blades and the cast slip is doctored to a pre-defined level. The layer of slip dries as it is carried through the apparatus. The dried tape is stripped from the belt for further processing.

Such tapes are used in the production of thermistors, piezoelectrics and multilayer capacitors. When the tapes are formed from ceramics such as aluminium oxide and beryllium oxide the sintered bodies can be used as substrates upon which electronic conductors and resistors may be laid down by thick film techniques of screen printing. A number of pieces of green tapes may have various tracks printed on them and then be aligned and laminated together prior to firing, the manufacture multilayer products. An alternative to screen printed tracks is to coat the fired substrate with metal by the thin film technique of sputtering, or chemical vapour deposition. The metal can then be selectively etched off by standard photolithographic techniques to produce an appropriate electronic circuit. It is possible to sputter 2 or 3 different metals, e.g. Nichrome, Nickel and then gold to produce resistors by etching as well as conductive tracks. Such systems are used in microwave integrated circuits (MIC's)[25].

In order to produce substrates with uniform dielectric and mechanical properties great emphasis has to be placed on the preparation and control of the slip and the casting technique.

Slips are prepared by ball-milling the appropriate quantities of powder for a number of hours. It is vital that the viscosity of the slip is closely controlled. Because the viscosity is temperature dependent it should be warmed to and maintained at a temperature close to the normal highest ambient. If the slip has too low a viscosity due to excess liquid when at the operating temperature it can be increased by applying a vacuum. This will cause the liquid to evaporate and also promote de-airing.

The slip should be easily pourable and not exhibit much thixotropy or dilatancy, neither should it have a large yield point. The powder must be stable in suspension, ie there should be minimal settling out. Several additives are required in order to produce a castable slip that will yield a handleable cast tape.

Binders are used to form an adherent layer around the ceramic particles so that when the solvent is removed from the cast tape it has a degree of mechanical strength. Different acrylic polymers are used to this end and usually added at about 3% to 8% by weight. Plasticisers such as polyethylene glycols[26] are added in order to impart flexibility into the dried tape by weight of ceramic.

To produce a slip that does not settle out the addition of dispersing agents is required.   Such agents are surfactants normally bought as a commercial product, e.g. the Dispex[1] range of dispersants.   The dispersant should be added during the initial period of milling.   The dispersing agent also allows increased solids content in the slip so reducing shrinkage and hence cracking during the drying process.

The liquid used in forming the slip will preferably have a low boiling point and a low viscosity.   Usually a non-aqueous solvent will be chosen that will allow the additives to be dissolved but will be inert to the inorganic powder.

Once a stable de-aired slip has been formed it is desirable to pump it through 2 nylon filters with openings of 40 μm and 10 μm to remove agglomerates and coarse particles.

In tape casting there is a spreader or doctor-blade to level off the top surface of the tape.   This blade can be adjusted to allow various thicknesses of tape to be cast.   There are two basic systems used in tape-casting.   The first is batch casting where a fixed carrier (e.g. plate glass) is traversed by a moving doctor-blade.   The slip is allowed onto the substrate and spread over it by moving the blade.   The glass plate and spread slip can then be placed in a drier for the evaporation of the solvents.   When the tape is partially dry it is stripped from the glass and heated to a higher temperature for complete drying.

The other method, more commonly used in industrial applications, is known as the continuous tape method.   Here a carrier film is fed from a spool and enters a vessel containing the slip.   There is an adjustable doctor-blade at the exit side of the vessel where the cast layer of slip is skimmed to a pre-determined height.   The tape passes through a drying chamber to remove the solvent and the dry tape is removed from the film by a sharp displacement of the carrier.   The green tape continues in a flat motion usually through slitters to give lengths of accurate widths.   The tape is then either coiled up onto spools or cut into appropriate lengths for storage.

Casting speeds will vary with drying length, type of solvent and thickness of tape.   When such tapes are to be further processed for use in

---

1 Allied Colloids Ltd, Bradford, UK.

high density circuit applications surface imperfections need to be minimised.
In such cases it may be necessary to perform the doctor-blade process in a
class 10,000 or better clean room.   This will minimise the formation of
surface defects due to foreign matter settling onto the tape during
processing.

The green tape can be stored until required for further processing, ie
stamping, scoring and fixing to produce dense flat sheets of ceramic.

7. SLIP CASTING

Slip casting of ceramics is a technique that has long been used for the
manufacture of traditional ceramics.   The technique, sometimes referred to
as colloidal filtration, is also applied to forming some ceramic bodies used
by the electronics industry.   One of the most well known applications is in
the manufacture of aluminium oxide radomes.   The requirements of a stable
slip for casting have been covered earlier under tape casting.

The advantages of slip casting include its ability to form green bodies
of a complex shape without expensive tooling.   The bodies produced will
almost invariably be thin walled with a uniform thickness; hence its use in
radome manufacture.   It is an inexpensive process when compared with other
ceramic manufacturing techniques since the basic requirements are a ball-
mill and a supply of plaster of Paris.   The following is a brief rundown of
the slip casting process.

A slip is prepared by ball milling the appropriate powders along with
binders, plasticisers, deflocculants, etc. in a solvent or water.   In order
to reproduce the castings it is essential that the slip is characterised by
means of its viscosity, dilatancy, solids content etc.   Such a slip is poured
into a porous mould usually made from plaster of Paris where the liquid part
of the slip will be absorbed by capillary action into the mould leaving a
layer of ceramic and additives formed against the plaster.   Slip can be
periodically added until the desired wall thickness has been formed.   At this
point the excess slip is poured out.   It is possible to cast solid pieces by
leaving the slip in; this technique is called solid casting.   The body is
left in the mould until it is reasonably dry and has sufficient green

strength to be removed. Noticeable shrinking occurs during this stage which facilitates removal of the green body. This body, if the slip has been correctly prepared, will be strong enough to allow trimming of rough edges prior to firing.

It is possible to improve green density and impart higher green strength on a cast body by applying an ultrasonic frequency to the mould during casting. Another way to improve green body characteristics is to apply pressure, e.g. by gas to the slip during casting. This can yield higher densities and minimise shrinkage after casting.

8. ADDITIVE BURNOUT

In the forming processes discussed in this chapter one thing that is common to all of them is that additives such as binders, plasticisers, lubricants, deflocculants etc. are included in the process. It is obvious and necessary that these materials remain in the green bodies formed by these methods in order to impart handlability prior to sintering the inorganic powder.

It is important when choosing the additives to consider what effect they will have on the properties of the sintered body or whether they can be totally removed by oxidation at high temperatures. For instance, when making a piece of zirconium tin titanate, a dielectric resonator material whose densification is promoted by the addition of zinc oxide, it is sensible to consider adding a lubricant such as zinc stearate but not one such as calcium stearate which could have adverse effects on the dielectric losses of the sintered body. Also it would be acceptable to use an ammonium based deflocculant, e.g. Dispex A40, but not a sodium based one, e.g. Dispex N40. In the first instance it will be possible to remove all of the dispersant at a suitably high temperature but in the second sodium ions will not be removed by burning off. The sodium ions would probably increase the losses of the dielectric resonators.

Because most of the additives will be organic polymers it is possible to burn them off at a sufficiently high temperature prior to sintering. It is necessary to establish the temperature at which they oxidise in order to ensure that all carbon residues are removed from the body and so do not have undesired effects on the electrical properties of the ceramic. Determination of the temperatures at which these additives will be burnt off at is relatively straightforward. A small sample of the powder is subjected to the well established techniques of Thermal Gravimetric Analysis (TGA) and/or Differential Thermal Analysis. Details of these techniques can be found in reference 27.

What is not revealed by these analyses is the length of time taken for burn out. This is dependant not only on the quantity of additives but also on the size and geometry of the green body. The time required at burn out temperature has to be determined empirically.

A body is slowly heated to the temperature at which burn out occurs. An estimate of the time required is made and the organics are burnt out for this period. The body is allowed to cool and is then weighed.

This procedure is repeated until a steady weight is recorded indicating that all the organics have been removed. It is considered safe practice to record the time required to reach this steady weight then add 10% to it for burning out in order to ensure total removal of the additives.

It may be necessary to have more than one burn out temperature especially if there are additives with significant differences between the temperatures at which they become gaseous.

A further point to consider when designing a burn off schedule is the heating rate. It is possible that some of the organics may start decomposing at temperatures lower than that determined. If this happens to be the case a fast heating rate could cause the organics to gasify quickly. The pressure resulting from this fast gasification has been known to cause the green body to crack prior to densification.

It is important to re-emphasise that due consideration is paid to burning off the organic additives. If they are not totally removed from the porous body, then at sintering temperatures it is possible that some amount of residue may get trapped and could reduce the performance of some electronic ceramics.

9. FURTHER PROCESSING OF CERAMIC TAPE

There are several ways in which the flexible ceramic tapes produced by the doctor blade techniques may be processed with substrates or multilayer capacitors.   If the end product is to be a substrate the common way of forming the parts is by stamping. The tape will normally have been slitted to a width slightly larger than the part that is to be stamped. This strip is fed into a punch and block die when straight forward rectangular or disk shapes are required. The shape is punched out from the tape into the bottom of the die. The tape is fed through automatically when manufactured under mass production. Obviously more complicated shapes can be made by using more sophisticated punches and dies. The parts are punched oversize in order to allow for up to 17% linear shrinkage, which has to be empirically determined.

An alternative method of production is to score a wide piece of tape prior to firing with shallow grooves. The tape is then fired and the sintered body can have an array of circuits printed on by silk screening. Once these circuits have been fired on, the substrates can be snapped off from each other by hand. Obviously the score should not be so deep that it weakens the sintered material so much that it would cause breakage prior to, or during, circuit deposition.

In larger circuit production outfits CO_2 lasers are used to scribe the substrates after the application of the circuit pattern. Basically the laser burns a line of blind holes around the circuit. When the substrates are broken apart a clean straight edge is normally observed. Lasers can also be used to produce intricate shaped pieces.

In the production of multilayer capacitors (MLC) barium titanate ($BATIO_3$) is by far the most common material used due mainly to its high relative permittivity up to 10,000. A pattern of rectangles of silver palladium (Ag/Pd) ink is silk screen printed onto a length of green $BaTiO_3$ tape. After printing the metallised sheets are dried to remove the solvent from the ink. The ceramic sheets are then stacked one above the other, Figure 7). The stack is laminated at high pressure and slightly elevated temperature. This process forces the air from between the layers and promotes interlayer adhesion. The individual layers should no longer be separable if the procedure is done correctly. If the lamination has not been done at high enough pressures the sheets will delaminate during firing. If

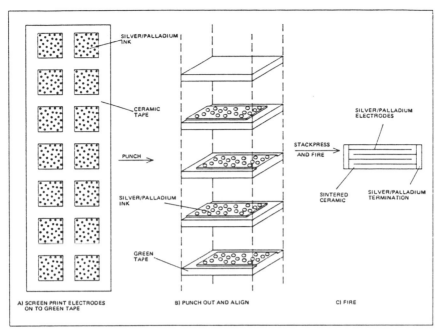

Figure 7: Schematic diagram of multilayer capacitor manufacture.

the lamination pressure is too high the plasticised green sheets will start to flow, causing cracks to appear during firing. The laminated stacks are subsequently punched out into individual capacitors.

Silver/palladium ink is painted onto the ends of the sintered capacitors. These are then heated so that the ink forms silver/palladium terminations for the MLC's.

10. SINTERING

The sintering process converts the green microstructure to the microstructure of the dense component. In this way sintering is the last of the ceramic processing steps where the ceramist has any influence on microstructural development. This influence is limited as the worst inhomogeneities of the compact are usually exaggerated during sintering, e.g. flaws will persist or even grow, large particles may induce abnormal grain growth, etc. For the electronic ceramist the mechanistic links between microstructure and properties are usually much less well understood than is the case for the structural ceramist through, for example, the Griffith equation[28]. Achieving maximum density or a very fine grain size may not be coincident with achieving optimum electrical properties. Generally high densities, small grain size and small pore size are beneficial for producing high quality sintered multilayer devices, realising high electric strengths and avoiding position-dependent properties. Developments in sintering theory have changed the emphasis from studies centred on the atomistic scale to those which are looking at effects on the microstructural and even macroscopic scale[29]. These studies analyse the effect of aggregates or agglomerates on densification rates[30] and calculate the stresses generated by differential sintering[31] between different parts of the compact. This change of emphasis now addresses the problems of sintering using microstructural models which much more closely reflect real microstructures rather than idealised arrangements of spheres[32]. In particular the relationship between the formation of property limiting flaws and sintering is now being investigated (see also Chapter 1). Recent studies have shown that if attempts are made to sinter materials that contain non-densifying inclusions then densification stops completely at well below the maximum density achievable[33]. This is due to stresses which develop in the shrinking matrix around the rigid particle. These increase and oppose the local driving stress for sintering. Further densification in such a system can only begin if the opposing stresses are relieved by creep.

Additives which promote liquid phase formation at sintering temperatures are particularly beneficial. Another approach is to increase the effective sintering stress by applying external pressure using hot pressing or hot isostatic pressing[34]. To achieve high density differential density at the scale of the powder (ie as agglomerates or aggregates) or of the compact (ie as might arise from the forming process) must be avoided. The lessons for ceramic processing are:

1)   Ensure the powder is free from aggregates or agglomerates.

2)   Use a forming technique which minimises density gradients within the compact, e.g. tape casting or slip casting are preferable to die pressing, especially for fine powders.

3)   Employ liquid phases at the sintering temperature to allow relaxation
of stresses generated by differential densification and so delay the end
of densification.

11. FINISHING

Ceramic substrates are becoming increasingly popular for microwave and
hybrid circuit technology.   Initially circuit designs were applied to
substrates with an as-fired surface finish, no economical alternatives being
available.   In recent years the demands placed on the substrate by electronic
engineers such as fine lines and high degrees of uniformity have become more
and more stringent.   The quality of a substrate for electronic applications
is dependent upon four basic parameters.   These are thickness, flatness,
texture and parallelism.

Thickness variation between nominally the same substrates is undesirable
especially when the circuit is being screen printed.   This is because it can
lead to variations in screen clearance and contact, with resultant effects
on the circuit pattern.   Thickness variations of as little as 25 μm (0.001
inch) may be required.

Flatness variations are taken to include camber and waviness in the
substrate.   Waviness is the random ripples and undulations across a surface.
It may be regular, produced by non-flexible grinding wheels, or irregular due
to firing processes.

Camber is a 3-dimensional property involving the combined profiles of
both surfaces including variations in thickness, parallelism and other
defects.   It is usually identified by "go-no-go" slot gauges.

The surface texture or quality is dependent upon homogeneity and density
of the material.   Some porosity in ceramics is unavoidable, however fine or
random.   Isolated pores may occur in the finished plane of a fine polished
substrate.   Such surface imperfections can adversely affect the circuit
subsequently laid down.   The roughness of the surface, usually referred to
as micro inches CLA (centre line average) may be due to poor finishing.   The
method for establishing roughness is the diamond stylus tracing technique.

Here a polished stylus of 2.5 μm (0.001 inch) radius is run over the substrate surface under a load of 1 g. This generates a signal, in a similar manner to a gramophone record stylus, which is proportional to the irregularity of the surface. The roughness value is read from either a meter or a strip chart.

Parallelism is essentially the variation of thickness across a substrate. If the substrate is wedge shaped problems can arise in the screen printing process.

The roughness of 99.5% Al_2O_3 will be relatively fine, of the order of 10 micro inches CLA. That of 96% Al_2O_3 may be as coarse as 20 micro inches CLA. Camber will usually be present sometimes to such an extent it may be observed by the naked eye. Surface defects will almost certainly be present in the form of burrs, pits and ridges varying with the quality of the tape production process.

Improved surface finishes can be achieved by machining the substrate. Machining usually covers the terms grinding, lapping and polishing.

Grinding with an abrasive wheel will remove irregular ripples normally found in as-fired material. Camber may also be reduced when the substrate is precision ground. Unfortunately a ground surface may exhibit a few deep scratches due to a raised piece of abrasive on the wheel. It may even have voids added to by the grinding wheel 'pulling out' grains from the ceramic. Extremely fine diamond wheels can produce uniform surfaces down to 9 micro inches or less.

Lapping involves moving a flat cloth covered tool (lap) against the substrate in the presence of a "fine abrasive medium" either as a layer impregnated in the lap or as a slurry. The pressure between the tool and the surface being lapped will be controlled but may vary from 3.5 to 70 kPa (0.5 to 10 psi) depending upon the substrte, the abrasive and the lubricant. When done successfully, the surface roughness and camber can be significantly reduced. The surface roughness can be closely defined by this technique with minimal variation from substrate to substrate, surface finishes down to 1 micro inch CLA can be produced. This technique is effective for practically any substrate material, shape or size or thickness down to approximately 130 μm (0.005 inch). Lapped substrates offer several advantages including improved screen printing, reduction of rejects and easier assembly.

Polishing is a precisely controlled process which is basically an optimised lapping technique. Usually good quality laps are used, onto whih is sprayed very fine, down to 0.25 μm, diamond powders. Precision polished substrate must have optical grade flatness (2.5 μm or 0.0001 inch), no pullout, no camber and low roughness selected between 10 to 1 micro inches CLA. There must only be the narrowest deviation between pieces and lots. These optical flatness levels are evaluated by observing light interference fringes. Because precision lapped or polished substrates have minimal burrs or localised ripples they are preferred in the production of thin film circuits. This is because such faults cause failures due to local overheating since they may be the weak spots in the film.

A summary of the processes used in the production of substrates can be seen in the flow chart of Figure 8.

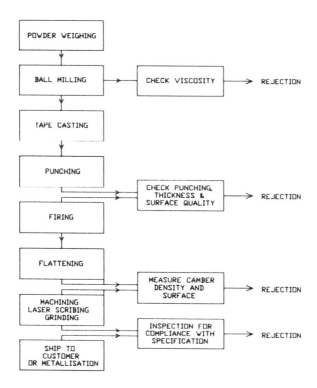

Figure 8: Flow chart of substrate production.

REFERENCES

1. Baumann, R., Rost, T., Stone, B. and Rabson, T., Characterisation of sputtered thin films of barium strontium titanate on silicon substrates, presented at 1st European Conference on Applications of Polar Dielectrics/International Symposium on Applications of Ferroelectrics, Zurich, Switzerland, August 29th-September 1st, 1988.

2. Swartz, S.L., Seifert, D.A. and Noel, G.T., "Characterisation of MOCVD PbTiO$_3$ thin films", ibid.

3. AVX Product Information, "ACCU/F thin film chip capacitor for UHF/VHF".

4. Dow Chemical Company, US Patent 3,725,539, 3rd April, 1973.

5. Clabaugh, W.S., Swiggard, E.M. and Gilchrist, R., Preparation of barium titanate/oxalate tetrahydrate for conversion to barium titanate of high purity. J. Res. Natl. Bur. Stand. 56, 289 (1956).

6. Mazdiyasni, K.S., Dolloff, R.T. and Smith, J.S., Preparation of high purity submicron barium titanate powders. J. Amer. Ceram. Soc., 52, 523 (1969).

7. Ovromenko, N.A., Shvets, L.I., Ovcharenko, F.D. and Kornilovich, B., Hydrothermal synthesis of certain alkaline earth metal titanates. Dopov. Akad. Nauk. Ukr. RSR, Ser B: Geol. Khim. Biol. Nauk., 3 242 (1978).

8. Savoskina, A.I., Limar, T.F. and Kisel, N.G., Neorgan. Mat. 11 (12), 2245 (1975).

9. Berrier, J.C., Sol-gel processing for the synthesis of powders for dielectrics. Powd. Metall. Int. 18 [3] 164 (1986).

10. Pechini, M.P., US Patent 3,330,697 (1967).

11. Barringer, E.A. and Bowen, H.K., Formation, Packing and sintering of monodisperse TiO$_2$ powders. J. Amer. Ceram. Soc., 65, [12], C199 (1982).

12. Thomson - C.S.F., US Patent 4,339,543 (1982).

13. Rose, H.E. and Sullivan, R.M.E., Vibration mills and vibration milling. Constable and Co Ltd., London (1961).

14. Smyth, M.J.B., Niesz, D.E., Haber R.A. and Danforth, S.C., Paper No. 20-W-89 presented at the 91st Annual Meeting of the American Ceramic Society, April 23-27 1989 Indianapolis.

15. ICI European Patent Application No 85308352-5 (1985).

16. Edirisinghe, M.J. and Evans, J.R. G., Compounding ceramic powders prior to injection moulding. In Brit. Ceram. Proc. No 38, Novel Ceramic Fabrication Processes and Applications, R W Davidge (Ed), 67-80, (1986).

17. Masters, K., Spray Drying, 4th Ed. Wiley, New York, (1985).

18. Frey, R.G. and Halloran, J.W., Compaction behaviour of spray dried alumina. J. Amer. Ceram. Soc., 67 (3) 199 (1984).

19. Lukasiewicz, S.J., Spray drying ceramic powders. J. Amer. Ceram. Soc., 72 [4] 617 (1989).

20. Alpine Process Technology Product Information, Fluidised Bed Spray Granulators.

21. Roosen, A. and Hausner, H., In Ceramic Powders, P. Vincenzini (Ed), Elsevier, Amsterdam, 773, (1983).

22. Xue, L.A., Riley, F.L. and Brook, R.J., Tert-butyl alcohol - A medium for freeze-drying: application to barium titanate. Br. Ceram. Trans. J., 85 [2] 47-48 (1986).

23. Mistler, R.E., High strength alumina substrates produced by multiple layer casting techniques. Bull. Amer. Ceram. Soc. 52 [11] 850 (1973).

24. Howatt, G.N., Breckenridge, R.G. and Brownlow, J.M., Fabrication of thin ceramic sheets for capacitors. J. Amer. Ceram. Soc. 30 [8] 237 (1947).

25. Borase, V., Substrates influence thin film performance. Microwaves, p61 (Oct 1982).

26. Hyatt, E.P., Making thin, flat ceramics - A review. Bull. Amer. Ceram. Soc. 65 [4] 637 (1986).

27. Daniels, T., Thermal Analysis. Publ. Kogan Page Ltd, London (1973).

28. Griffith, A.A., Phenomena of rupture and flow in solids. Philos. Trans. R. Soc. Land. A221 [163] 32 (1920).

29. Brook, R.J., State of the art. Stuijts Memorial Lecture 1989. J. Eur. Ceram. Soc. 5 [2] 75-80 (1989).

30. Lange, F.F., Sinterability of agglomerated powders. J. Am. Ceram. Soc., 67 [2] 83 (1984).

31. Hsueh, C.H., Evans, A.G., Cannon, R.M. and Brook, R.J., Viscoelastic stresses and sintering damage in heterogeneous powder compacts. Acta Metall. 34 927 (1986).

32. Zhao, J. and Harmer, M.P., Effect of pore distribution on microstructure development: II, First and second generation pores. J. Amer. Ceram. Soc., 71 [7] 530 (1988).

33. Tuan, W.H., Gilbart, E. and Brook, R.J., Sintering of heterogeneous ceramic compacts, Part I Al_2O_3 - Al_2O_3. J. Mat. Sci. 24 1062 (1989).

34. Haertling, G.H., Improved hot pressed ceramics within the $(PbLa)(ZrTi)O_3$ system. J. Amer. Ceram. Soc., 54, 303 (1971).

5

Processing of Ceramic Composites

R. Rice

W. R. Grace and Co., 7379 Route 32, Columbia, Maryland
21044, USA.

1. INTRODUCTION

Ceramic composites are attracting increasing attention because of the broader diversity, and especially improvement, in properties they can frequently provide. Properties and, hence, the benefits of composites typically depend upon the achievement of a specific range of microstructures. It is in fact, in part, this design and tailoring of microstructures in current composites which distinguishes them from many earlier ceramics which were accidental composites, that is, had a composite character as a result of the choice of raw materials in empirically developed processing. The selection of raw materials and processing parameters in these earlier composites was not predicated upon their design as composites. Most modern ceramic composites, of course, go well beyond traditional ceramics in composition, microstructure, and properties.

Design or tailoring of composite microstructures to achieve improved or new properties presents processing challenges. As with many other high-tech ceramics, limited or near zero porosity is commonly desired, as is a substantial degree of homogeneity. However, homogeneous dispersion of the phases and low porosity is much more difficult to achieve in many composite systems because of intrinsic and extrinsic factors limiting mixing and densification. The broad compositional diversity of composites is a serious challenge because of the attendant opportunities for reaction between constituents, especially as temperatures are raised to overcome densification limitations. The combination of the fineness of most dispersed phases and

123

the high volume fractions often desired commonly make achieving a high density and degree of homogeneity a challenge. These challenges of ceramic composite processing have resulted in a significant shift in the emphasis of processing technologies in comparison to other ceramics.

This paper reviews the processing of both ceramic particulate and fiber composites, the latter including use of continuous and short fibers, which in turn includes both chopped fibers and whiskers. This review will be divided into two main parts, namely, powder based methods of processing and non-powder based methods. Powder based methods discussed include sintering, hot pressing, HIPing and reaction processing. Non-powder based methods discussed are polymer pyrolysis, chemical vapor deposition and melt processing. In all cases, the concept of applying these processing techniques will be discussed first for particulate composites, then for fiber composites. Generally, each processing method will be introduced by a description of it, with more description for less familiar methods, followed by a review of its actual use with composites. Following coverage of the various process methods, there will be a discussion of future trends and a summary. However, since many ceramists are unfamiliar with the specifics of microstructure/mechanisms of toughening sought, or the processing to achieve these, they are reviewed first. Fiber handling and processing of fiber composites are emphasized, since they are expected to be the least familiar to ceramists.

Finally, some constraints need to be noted. It is impossible to give fully comprehensive referencing in such a review. Instead, key or representative references are generally cited. Also it is impossible to give all processing details. Further, many processing details have not yet been understood, or optimized, and some processing specifics are not available because of government restrictions. Further, in order to keep the topic tractable, it has been restricted to all ceramic composites. Thus metal-ceramic (e.g. cermets) and polymer-ceramic composites, each extensive topics themselves, are not treated. Also, the emphasis is on composites for mechanical performance, in part because many composites for non-mechanical purposes involve metallic or polymeric constituents[1,2].

2. OVERVIEW OF COMPOSITE MECHANISMS/MICROSTRUCTURES AND PROCESSING

2.1 Mechanisms/Microstructures

Ceramic composite toughening mechanisms are briefly reviewed pertinent to processing; the reader is referred to other reviews[3-5] for details. Mechanisms pertinent to particulate, as well as to some fiber, especially short fiber composites, are transformation toughening, microcracking, crack deflection, and crack branching.   Transformation toughening currently is essentially exclusively derived from appropriate dispersion of metastable tetragonal ZrO_2 particles in an appropriate matrix, mainly cubic zirconia, alumina, or mullite. Both experience and modelling show that a very uniform spacial and size distribution of the ZrO_2 particles is desired for optimum results, Figure 1. Figure 1a shows how the level of toughening varies as a function of particle size distribution, after Evan's[5].   Figure 1b illustrates the deleterious effects of agglomeration of the toughening phase. Not only may such agglomerates act as a flaw, they often result in the surrounding matrix area being devoid of toughening, thus emphasizing the need for homogeneous mixing. While the optimum particle size depends on several factors, such as the stiffness of the matrix and the volume fraction of ZrO_2 (which is also typically related to the extent it is partially stabilized), typical ZrO_2 particle sizes desired are in the 0.5-1 µm range.

Microcracking is also typically viewed as requiring a uniform spacial and size distribution of particles. The particles chosen are typically those that undergo a phase transformation (i.e. again essentially ZrO_2 to date) or have a substantial thermal expansion mismatch with the matrix.   The concept is to develop a sufficient population of microcracks in the vicinity of the crack tip such that the individual microcracks are well below the size of a failure causing crack and their density is high enough to provide toughening, but not so high as to provide ready crack propagation due to easy linkage of a series of microcracks. An important concept in maintaining good strength levels with microcracking is to have the microcracks be induced due to the combination of the mismatch strains between the particle and the matrix and the high stress concentrations in the vicinity of the tip of a stressed crack, as opposed to having microcracks that preceded the formation or loading of the main crack.   The particle diameter, d, leading to the development of a microcrack in the absence of an applied stress can be estimated by the following relation:

A. Effects of Particle Size

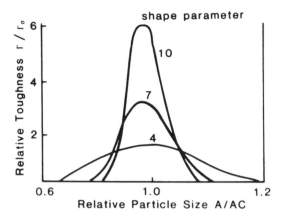

B. Effects of Particle Spacing

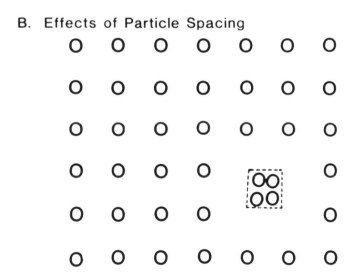

Figure 1: Schematic of transformation toughening as a function of particle size and distribution (see text).

$$d \approx \frac{9\gamma}{E(\Delta\epsilon)^2}$$

1

where γ = pertinent fracture energy (e.g. of a single grain boundary for an intergranular microcrack), E = the Young's modulus of the matrix and Δ = the particle-matrix mismatch strain.   Particles below this size generally will not result in pre-existing microcracks, but can result in their generation above a stress level inversely related to particle size.   While modelling of microcracking has been more difficult, it is again important to have a narrower particle size distribution, Figure 2.

One of the major challenges and uncertainties in such microcrack development is that many of the particles of interest have anisotropic expansions, e.g. Al_2TiO_5 or BN. The combination of different expansion coefficients and misorientations leads to a significant range over which microcracking can occur, but is difficult to predict precisely.   Another challenge is presented by non-equiaxial particles, e.g. platelets such as graphite or BN, often used in some composites.

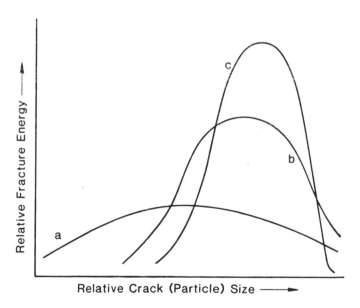

Figure 2: Schematic of the effect of the size distribution of microcracks (and, hence, of the particles causing them) on fracture toughness. Note the benefits of a narrower distribution (i.e. curve c vs curves a and b). After Evans, ref. 5.

Crack deflection models show it depends substantially on particulate geometry, Figure 3, with rod shaped particles being most desired. This mechanism, therefore, clearly, overlaps with that of short fibers. Effects of variation in particle dimensions and of spacial distribution are not precisely known. Agglomeration, of particles or short fibers is clearly detrimental and thus one of the particular challenges of processing whisker composites is to prevent or eliminate agglomerates.

Crack branching has not been modelled, but is commonly a result of either microcracking, crack deflection, or both. Overall, particle size and distribution requirements to cause crack branching are similar to those for microcracking and crack deflection. For branching to be effective in controlling strength, it must occur on a scale equal to, or more likely substantially less than, the typical failure causing flaw size, thus indicating the need for fine particles of limited spacing. However, to limit thermal shock damage where one is often concerned with controlling crack propagation on a much larger scale, particle size and spacing requirements can be substantially relaxed.

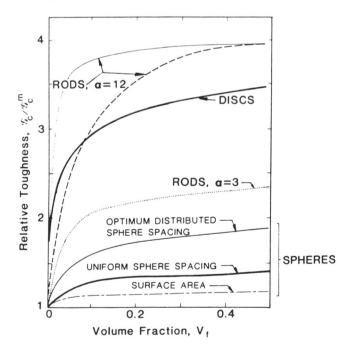

Figure 3: Plot of effects of particle, size, shape, volume fracture, and distribution of toughening due to crack deflection after Evans and Faber, ref. 5.

Some of these mechanisms will naturally combine as noted above. However, significant benefit may be obtained by purposely designing the composite and, hence, the processing to accentuate combination possibilities. This may well introduce processing challenges such as obtaining two particulate phases , e.g. one for microcracking and one for transformation toughening, differing in composition, size and/or shape. Further, some combinations may in fact be enhanced by introducing a controlled heterogeneity to extend microcracking possibilities, Figure 4. The concept is to introduce clusters of higher density microcracks to obtain a net increase in the overall density of microcracks and, hence, in the resultant fracture energy, while at the same time limiting the ease of crack propagation via microcrack linking by having the clusters of higher density microcracks sufficiently separated that crack linkage between clusters is limited[2,3].

Toughening in fiber composites can involve the above mechanisms for particulate composites, especially microcracking, crack deflection, and crack branching. Load transfer can also be important, however, it appears to be less essential for ceramic composites than most other composites[4]. The one factor that has now been quite well established as central to the

DUAL MICROCRACKING COMPOSITE

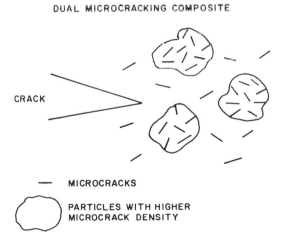

CRACK

— MICROCRACKS

PARTICLES WITH HIGHER
MICROCRACK DENSITY

Figure 4: Schematic of a dual microcracking composite (see text).

toughness of ceramic fiber composites is fiber pullout. Further, it is now well accepted, and with reasonable direct demonstration[4], that toughness due to fiber pullout in ceramic composites is typically achieved by having little or no chemical bonding between the fibers and matrix.

While fiber composites with matrices of very low Young's modulus such as plaster or cement, may utilize very low volume fractions, such as a few percent, of fibers, composites of high-tech interest typically require relatively high volume fractions of fibers, e.g. 20-60 vol%. It is also quite often important to control fiber orientations or architecture as discussed later. Use of smaller diameter fibers is often important. Large fibers such as CVD filaments or wires having dimensions of 100-200 μm are more prone to generate microcracks due to expansion differences with the matrix, as per equation 1. Because such microcracks are on the scale of the fiber, large fibers can seriously limit strengths. On the other hand, having fibers too small may present problems both in terms of the practicality of handling as well as from possible limited surface reaction. For example, a 0.1 μm depth of chemical interaction between the fiber and matrix may not be serious for a 10 μm diameter fiber, but could be rather serious for a 0.5 μm diameter fiber. Fibers most extensively used in ceramic fiber composites today are typically in the range of 5-10 μm in diameter. Table 1 lists properties of typical ceramic fibers and whiskers.

While the emphasis, both in this chapter and in the ceramic community, has been on ceramic composites for mechanical performance, there is also considerable opportunity and interest for ceramic composites for non-mechanical, especially electrical, performance[1,2]. While some concepts and limited amounts of experimental work have been done on more complex composites requiring various degrees of contiguity of one or more of the phases, the majority of work has been on simple particulate composites. Here, the challenge is first to retain appropriate chemical integrity of both phases as is typically a requirement for mechanical composites, as well as to be cognizant of many of the basic requirements for mechanical composites. Thus, particulate composites for other purposes, such as electrical/ electronic applications involving two phases having significant differences in thermal expansion, typically require that microcracking must be addressed. Usually, it must be avoided, but it is utilized in some composites, e.g. for humidity sensors[2].

Material	Diameter μm	Density g/cm³	Thermal expasion ×10⁻⁶ °C	Young's Modulus GPa	Tensile Strength GPa	Max. Use Temp. °C	Source
Fibers							
$85\%Al_2O_3 - 15\%SiO_2$	10→15	3.2	9	250	2.5	≤1200	Sumitomo
$96-97\%Al_2O_3 - 3-4\%SiO_2$	3	3.4	9	300	2.0	≤1200	ICI
$\alpha-Al_2O_3$	20	3.9	9	380	1.4	≤1200	DuPont
$62\%Al_2O_3 - 14\%B_2O_3 - 24\%SiO_2$	8→12	2.7	3.5	150	1.7	1100→1400	Nextel 312
$Al_2O_3 - SiO_2$	10→12	2.7	5	210→250	1.4→2.0	1200→1400	Nextel 440
SiC	10→15	2.6	5	175→210	3.0	1200→1500	Nippon Carbon
SiO_2	10	2.2	0.5	70	0.6		Quartz Prod Corp
$Al_2O_3 - ZrO_2$	21		10	380	2.1	1300	DuPont
Carbon	5→10	1.7→2.0	-0.4→-1.8	200→700	1.4→1.5	>2000ᵃ	Various
Filaments							
SiC	140	3.2	5	430	3.5	≤1500	AVCO
B	100-140	2.3	5	430	3.5	≤1300	AVCO
Whiskers							
Si_3N_4	0.2→10	3.2	3	280→350	1.4→3.5	1400→1800	Tataho
SiC	0.05→0.2	3.2	5	450→630	1.4→90	1500→1900	Tataho, AVCO

Table 1: Properties of ceramic fibers and whiskers at room temperature.

a: in inert environments; <1000°C in air.

2.2 Composite Processing

Processing of ceramic particulate composites typically has overall similarity, and frequently detailed similarity, to more traditional ceramic processing. The most generic difference and greatest challenge is obtaining homogeneous mixing of the dispersed phase; whilst obtaining good densification can also be a problem. Specifics of processing such composites, especially densification, will be the subject of the subsequent two sections. Whisker (and some short fiber) composites present even greater challenges for homogeneous mixing. Most whiskers also contain a substantial amount of debris, mostly particulates (e.g. from broken whiskers, catalysts for whisker growth and growth media). While reducing or removing such debris is in part a factor in whisker preparation (a subject beyond the scope of this review), it cannot be neglected in processing. Whisker debris or inhomogeneities, separately or collectively, can limit strengths by acting as failure sources.

More significant differences occur for processing continuous fiber composites from that of normal ceramic processing; i.e. fiber orientation/architecture, and achieving these in the size and body shape desired. While the great majority of processing studies on ceramic fiber composites, to date, have been on unidirectional fiber composites, in practice, these have limited uses. Some work on different fiber architectures has been done, but there needs to be a great deal more, since this will be required for most applications, as indicated by use of other composites, e.g. polymer matrix and carbon-carbon. Filament winding, wherein the angle of the filaments relative to the resultant component and relative to different layers is controlled to achieve a good balance of properties, is likely to be important. Similarly, use of tapes of uniaxially aligned fibers or various cloth weaves, and laminating these with controlled orientation between the layers, is important. The type of weave or the type of fibers within each layer may be varied to balance different processing and property opportunities. Felted mats of chopped or short fibers may also be used. There are also expected to be specialized applications of multidimensional weaves, braids, etc.

An important factor in achieving any of the above fiber architectures with fine fibers is the use of fiber tows. Typically, these are strands of a few thousand fibers with some random intertwining. Handling of fiber tows thus becomes a major factor in fiber composite processing. Many fibers also

come with a polymer coating ("sizing"), which must first be removed, usually
by solvent or oxidation. The most common approach for introducing the matrix
is to put fiber tows (as well as larger, individual filaments) through a bath
which is the source of the matrix or its precursor, and then lay this up
directly by filament winding or other related techniques, Figure 5A. Most
commonly, the bath is a slurry or slip of matrix particles, but it may also
be, for example, a sol or pre-ceramic polymer. Typically, the wet filaments
are wound onto a drum analogous to a filament winding operation. The green
composite tape is subsequently removed from the drum for further processing,
e.g. drying and hot pressing. (One might also dry the matrix precursor onto
the fibers or into the tows for later operation, but this has generally not
yet been done with ceramic composites, presumably due to lack of adequate
adherence of most dried matrix precursors to the fibers.) A typical
challenge in such bath coating operations is to get a uniform coating on the
fibers. This usually requires some technique of opening or spreading the
tow, which can be limited by fiber intertwining. Another challenge is to get
the distribution of matrix precursor between tows similar to that within the
tows. Cloth may also be infiltrated then laminated, but may often present
greater challenges in controlling the quantity and uniformity of matrix (or
precursor) infiltration.

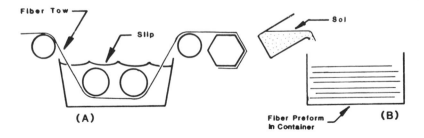

Figure 5: Schematic of continuous fiber ceramics composite processing.

An alternate technique is to make a preform, usually from a lay-up of cloth, or mats made from slurries by extracting the fluid from liquid-chopped fiber (or whisker) mixtures by vacuum forming or paper-making techniques. Although mats of chopped fibers are currently commonly limited to 5-15 vol% of fibers, work is underway to further increase the resultant volume fractions of fibers. Preforms are subsequently infiltrated with a fluid source of the matrix precursor, (see Figure 5b), sometimes using pressure. However, this method can present various challenges, e.g. the preform may act as a filter if a suspension of particles is used as the matrix source. For reference, it should also be noted that in the area of carbon/carbon composites, where the technology is considerably developed, a variety of pressure infiltration techniques have been developed, e.g. infiltration using cold and hot isostatic pressing to impregnate preforms with appropriate carbon producing polymers.

3. POWDER BASED METHODS

3.1 Sintering

Sintering is the most widely practised method of processing ceramics, especially for medium- and high-tech applications. While evaporation-condensation, and surface diffusion can aid in sintering, providing bonding and neck formation, only bulk diffusion can lead to actual densification in pure systems. This is the only mechanism that can lead to center-to-center motion of adjacent particles. Alternatively, additives or impurities that give a liquid phase may result in densification through dissolution and related liquid transport mechanisms, but may compromise properties.

The requirements of center-to-center motion for shrinkage and, hence, densification in the absence of a liquid phase presents a basic challenge, and probably basic limitations, to sintering of many ceramic composites. In order to maintain adequate phase distinction, mixing of the phases by interdiffusion (or liquid dissolution) must be limited, but such diffusion limitations are also likely to limit densification. Consider first sintering of a matrix around an isolated second phase particle whose size is similar to that of the particle size of the matrix material. Densification would presumably be slower relative to the matrix by itself due to the fact that the isolated particle will not shrink, only the matrix around it shrinks. Since changes in shape of the isolated particle due to surface diffusion may aid densification by accommodating geometrical fitting of it into the matrix, inadequate surface diffusion of the second phase particles under matrix sintering conditions can limit densification temperatures (e.g. as for some non-oxide particles dispersed in oxide matrices).

This problem of shrinkage of the matrix around an isolated dispersed particle becomes more serious as one or more dimensions of the dispersed particles becomes large in comparison with the matrix particles.   This results from the fact that shrinkage of the matrix must be three dimensional, introducing the requirement for significant local shrinkage anisotropy around such particles with larger dimensions.   This anisotropy is commonly accommodated either by cracking or lack of densification. (This is similar to dense agglomerates inhibiting sintering of a single phase material.) Figure 6a shows schematically the incompatibility of shrinkage in the axial direction of a fiber. Since the whiskers or fibers do not shrink, but the matrix does, this results in either inhibited shrinkage of the matrix i.e. pores (P), cracking (C) or both, as illustrated.

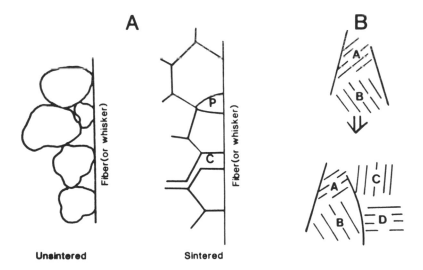

Figure 6: Incompatible shrinkage between the matrix and dispersed whiskers or short fibers.

A second fundamental problem with sintering composites also arises from shrinkage, but on a multi-particle scale. This becomes of increasing importance as the volume fraction or number of particles increases, resulting in increasing interaction between the dispersed particles. This problem is most serious, and most easily observed, when the dispersed phase has one or two dimensions larger than the matrix particle size, e.g. for fibers or flakes. Figure 6b illustrates incompatible shrinkage between two fibers, i.e. on a larger scale than 6a. Since there is considerable difficulty in obtaining homogeneous matrix packing between fibers (regions A and B), there will commonly be inhomogeneous shrinkage between fibers resulting in bending (exaggerated) of one or both fibers in order to accommodate the shrinkage motion. Fiber bending then counteracts the shrinkage, yielding regions of different density, e.g. regions A-D.   Note that the problem of non-uniform shrinkage between dispersed particles is greatly accentuated by inhomogeneous packing of matrix particles between them.

3.1.1 Sintering Of Particulate Composites

A clever case of liquid phase sintering of a composite for electrical applications is that of the sintering of $BaTiO_3$ with $NaNbO_3$[6]. The latter causes not only liquid phase sintering, but also a $NaNbO_3$ grain boundary phase which provides the benefit of maintaining a high capacitance over a larger voltage range, Figure 7.   Another important example for electrical/electronic purposes is the development of a magneto-electric material by investigators at Philips[7]. Sintering particles of $BaTiO_3$, which is piezoelectric, and a magnetostrictive ferrite results in a magneto electric material having about 20 times the sensitivity of the best known natural magneto-electric material.

Sintering composites for enhanced mechanical properties has a long history. Many traditional sintered ceramics are composites, but they were not originally recognized and designed as such. Pioneering studies of model particulate composites systems were carried out by various investigators such as Hasselman, Fulrath and colleagues[8-10], and Binns[11]. The earliest of the modern composites that offer high strength and toughness are those that result in a fine (e.g. 0.4-2 μm) grain body of all, or nearly all, tetragonal ZrO_2 phase. These materials are sintered from extremely fine (100 Å) powders of ZrO_2 that are chemically prepared, allowing stabilizers to be added via chemical solution for greater homogeneity. A common precursor is $ZrOCl_2$[12]. Such bodies are typically partially stabilized with (3-6 wt%) Y_2O_3 (as opposed to MgO or CaO used in other partially stabilized ZrO_2 bodies),

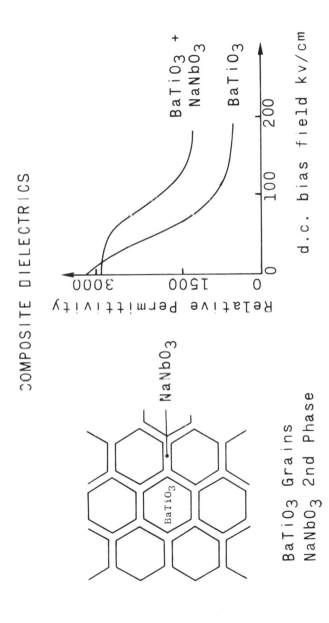

Figure 7: Schematic of a ceramic composite for electrical, as opposed to mechanical performance, see text.

presumably because of the much higher phase stability of this system. Powders are typically sintered to near theoretical density at temperatures of 1300-1400°C, Table 2. These systems which were apparently first developed by Nauman[13-15] and colleagues at Union Carbide and subsequently (apparently independently) by Gupta[16,17] of Westinghouse, give typical strengths of ≥700 MPa and toughnesses of ≈7 MPam$^{\frac{1}{2}}$ (Table 2). Strengths were commonly limited by processing defects such as voids, agglomerates and foreign particles[13,18,19], shown in Figure 8. A) shows failure initiation from an isolated large void (at 22°C, 700 MPa), and B) from an agglomerate near the specimen edge (at 500 MPa). Note insert showing some areas of separation between agglomerate and the rest of the body. While these specimens represent processing from several years ago, and substantial improvements have occurred in the interim, they illustrate the processing challenge.

Although there has been further development of such fine grain PSZ materials by a variety of laboratories[20-26], they are not yet available on any significant commercial scale. Some of the practical challenges to commercial development have been the cost of chemically prepared powders as well as the combination of the cost and challenge of handling and densifying fine powders. Removal and/or avoidance of agglomeration as well as removal of chemically adsorbed species, e.g. residual chlorine, left from the original chemical preparation of many of these powders and consolidation, are also difficulties to overcome. Thus, for example, serious lamination and related problems were experienced by Reith et al[12], in cold pressing. Sedimentation to remove agglomerates, washing to remove residual Cl, and colloidal suspension and pressure filtering have been used to attack the above problems on a laboratory scale[14,15]. Hot pressing or HIPing (Table 2) have also been successfully used to eliminate pores and agglomerates as sources of failure as discussed later, as is work on superplastic forming.

General trends have been to use lower Y_2O_3 levels, e.g. 5 wt% vs. the 6-8 wt% used by Naumann and colleagues. These newer materials appear to exhibit greater stability against high temperature grain growth and resultant degradation. There has also been investigation of hydrothermal preparation of the powders[24,25] (as well as a report of hydrothermal reaction sintering of unstabilized ZrO_2[26]). More recently, promising results have also been reported for similarly processed, e.g. $ZrOCl_2$ derived powder, but higher temperature sintered material with approximately 12 m/o CeO_2[22] (Table 2). Note also, that use of CeO_2 can give good strengths with substantially larger grain sizes (e.g. 40 µm, Table 2)[21].

Stabilizer	Fabrication method(s) and conditions[a]	Grain size μm	Flexure strength MPa	K_{IC} MPam$^{\frac{1}{2}}$	Investigator
6-8 wt% Y_2O_3	DP 70 MPa → 60% rel density PC 0.7-1.4 MPa → 45% rel den. S 1300°C / 3-8 hrs S 1450°C / 2-5 hrs	0.4	600-840	9	Nauman, Reed, Ruth, Scott, (13-15)
Low Y_2O_3 levels 2-3 mol% Y_2O_3	S 1300-1400°C/2hrs DP 40 MPa HIP 1400°C/0.5 hrs IP 300 MPa 100-150 MPa	<0.3 <1	700 1300-1700	6-9 6-11	Gupta et al (16,17) Tsukuma and Shimada (20)
18 mol% CeO_2	S 1450°C-1550°C / 15-20 hrs DP 40 MPa IP 170 MPa HT 1550-1650°C / ≤ 236 hrs	2.6		4-7	Coyle (21)
12 mol% CeO_2	S 1400-1600°C/2hrs DP 40 MPa IP 300 MPa	0.5-2.5	700-800	8-12	Tsukuma (22)
2-3 mol% Y_2O_3 (+ 20 wt% Al_2O_3)	S 1400°C / 2 hrs HIP 1500°C/0.5 hrs DP 40 MPa IP 300 MPa 100 MPa		2400	17	Tsukuma et al (23)
3-5 wt% CaO	S 1700°C (4 hrs) - 1900°C (1 hr) DP, IP 200 MPa HT 1300°C (30 hrs) - 1500°C (3 hrs)	50	400-650	4-7	Garvie et al (36)

Table 2: PSZ processing and room temperature mechanical properties.

a: DP = die pressed, PC = pressure cast, IP = isopressed, S = sintered, HP = hot pressed, HT = heat treated, HIP = hot isopressed.

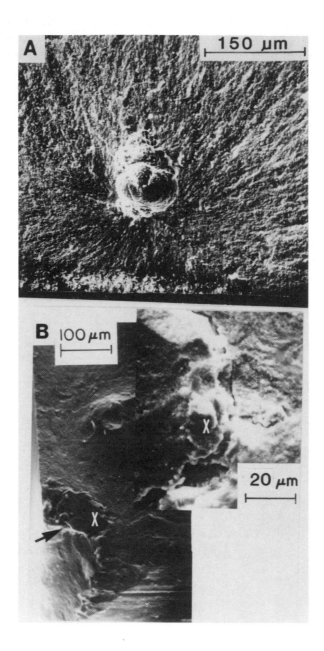

Figure 8: Samples of processing defects acting as fracture origins in fine grain, sintered PSZ (6 wt% Y_2O_3).   See text.

The second type of partially stabilized ZrO_2, precipitation toughened ZrO_2, is sintered from more typical ceramic powders, since sintering temperature and resultant grain sizes are not so highly constrained. While these materials in their more developed form appeared after the preceding fine grained PSZ materials, their development started in about the same period. Development of commercial 2.8 wt% MgO PSZ by King and Yavorsky[27-29] that proved quite successful for metal extrusion dies[30,31] and subsequently nozzles for continuous casting of Cu[32] was the starting point. The excellent performance and subsequent reproducibility problems of these materials led Garvie and colleagues[33-38] to study the mechanisms of these materials and subsequently control the process and resultant material and performance.

The typical procedure to prepare these precipitation toughened PSZ materials is to sinter them to as high a density as is practical and heat treat them to achieve a single phase cubic solid solution, with these operations often being done simultaneously, e.g. 1-4 hours at 1700-1900°C for ZrO_2 with about 3.6 mol% CaO[36]. The resultant solid solution body is then heat treated in the two phase region, e.g. during cooling in the 1300-1500°C range for ZrO_2 + 3.6 wt% CaO, to develop a tetragonal precipitate phase. Obtaining a solid solution and appropriate heat treatment to control the precipitate structure is the key to processing such bodies. Figure 9 illustrates some typical processing effects. Note that most of the development of these bodies has been with MgO or CaO partial stabilization and very little with Y_2O_3 stabilization. This is because of the relatively lower solid solution temperature for ZrO_2 with MgO or CaO versus Y_2O_3, making achievement of solid solution conditions reasonably practical with the former two, but not the latter. Temperatures are on the order of 200-400°C, less for the MgO system. Figure 10 shows the differences in the precipitate structures for the most common stabilizers. Recent work has focused more on use of MgO for partial stabilization[39,40]. Bodies of both MgO and CaO partially stabilized ZrO_2 made by this process are commercially available.

The third type of zirconia-toughened body is that made by addition of ZrO_2 to some chemically different matrix, the most prominent of which is Al_2O_3. Extensive use of Al_2O_3 as a matrix is predicated on extensive available Al_2O_3 technology, its moderate cost and high Young's modulus (toughening is proportional to the matrix Young's modulus). However, several other matrices have been investigated, especially mullite, as discussed below.

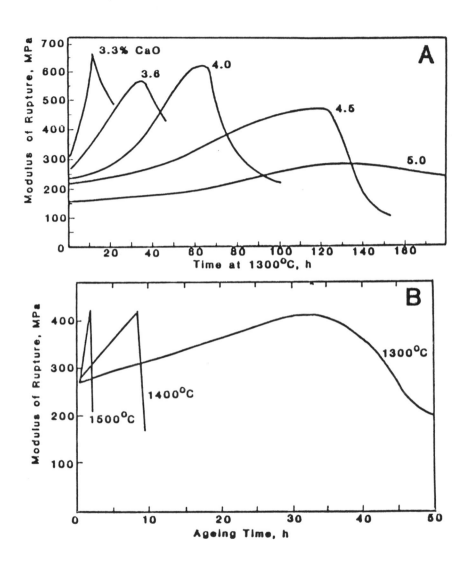

Figure 9: Outline of heat treating parameters for optimum precipitation toughening of CaO-PSZ.

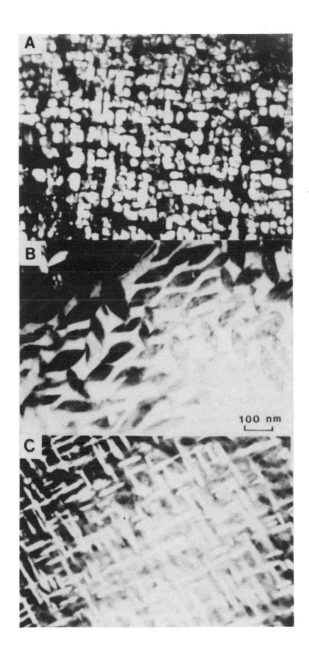

Figure 10: Different precipitate shapes (light objects in A and C, dark objects in B) in PSZ as a function of stabilizer. A) 4 wt% CaO - cuboidal, B) 2.8 wt% MgO - lenticular, C) 8 wt% Y_2O_3 - rod. Note common scale.

The basic requirement is to obtain a sufficiently homogeneous mixture of matrix and ZrO_2 powders which will allow good densification while maintaining a relatively narrow ZrO_2 particle size distribution with a mean on the order of 0.5 μm, see Figure 1. The exact particle size that one must stay below depends upon the volume fraction of ZrO_2 used, which in turn depends upon the amount of stabilizer added to the ZrO_2. It is now generally recognized that the optimum ZrO_2 content is about 10 vol% when no stabilizer is used (e.g. Figure 12 below). As the stabilizer (usually Y_2O_3) content is increased above zero, larger volume fractions of ZrO_2 can be added.

One of the original developers of the Al_2O_3-ZrO_2 system was Clausen[41,42] who first demonstrated toughening, but not strengthening, Figure 11, (actually using hot pressing, Table 3). The upper plot shows the fracture toughness as a function of ZrO_2 content (no stabilizer added to the ZrO_2). The lower plot shows the corresponding room temperature fracture strength for the same materials. Note that no strengthening whatsoever was obtained and, in fact, strengths were, in general, degraded. His failure to initially obtain strengthening along with toughening is attributed to agglomeration of either the ZrO_2 or Al_2O_3 or both[43]. Either type of agglomerate (as well as voids) can act as failure origins. Dworak[44] was also (apparently independently) developing Al_2O_3-ZrO_2 bodies in the same time frame. Again, he did not demonstrate toughening, but did achieve reasonable sintered strengths (but higher ones, by hot pressing - Table 3). His development led to a commercially available sintered Al_2O_3-ZrO_2 cutting tool. Becher[45] was possibly the first to demonstrate both strengthening and toughening (Figure 12) in the Al_2O_3-ZrO_2 system. He avoided agglomeration problems by making powders via gelling a mixture of an Al_2O_3 sol and a ZrO_2 sol (as well as hot pressing his bodies, Table 3) giving excellent intermixing. The inserts in Figure 12 show the extremely uniform microstructure of his specimens; micrographs A and B are, respectively, of a fracture and an as-sintered surface (bar is 1 μm). The third major Al_2O_3-ZrO_2 development was the demonstration, originally by Lange[46,47], that the addition of a stabilizer to the ZrO_2 allowed a greater volume percent to be added with increased toughening resulting.

Two general avenues for further development of Al_2O_3-ZrO_2 have been followed. The first has been to develop other methods of preparing powders of uniformly mixed Al_2O_3 and ZrO_2. Good results have been reported for CVD prepared powders, made by oxidizing (with H_2O) mixtures of $AlCl_3$ and $ZrCl_4$. These have been sintered to respectable (and hot pressed to still higher)

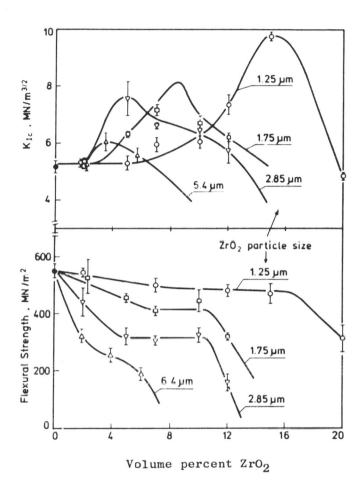

Figure 11: Summary of Claussen's original Al_2O_3-ZrO_2 strength and fracture toughness results.   See text.

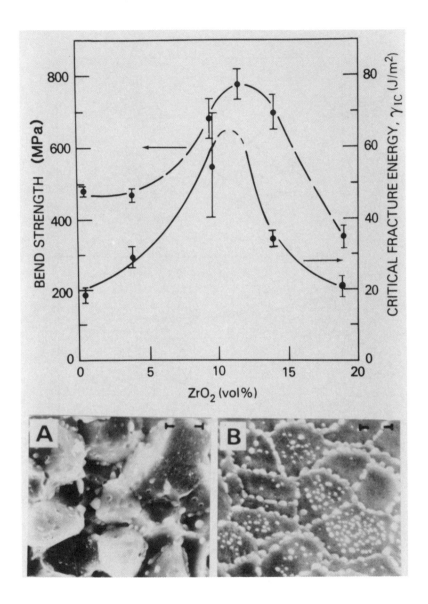

Figure 12: Becher's room temperature strength and fracture toughness data for Al_2O_3-ZrO_2 as a function of ZrO_2 content (unstabilized).

Materials		Fabrication method(s) and conditions[a]	K_{IC} MPa·m$^{\frac{1}{2}}$	Flexure strength MPa	Investigator
Al$_2$O$_3$	ZrO$_2$				
Alcoa A16 1.25-6.4 μm 2-20 vol%		HP 1400-1500°C / 30 min 40 MPa	10	≤550	Claussen et al (41,42)
5 μm	1μm; 2.5-20 vol% + 30 wt% TiC-TiN	IP-S 1600°C / 60 min HP 1750°C / 3 min		500 750	Dworak (44)
sol	sol; 0.5-30 vol% 6-80 vol% sub-μm	HP 1550°C (180 min) - 1650°C (15 min) HP 1500-1600°C / 20 min	6.5	80	Becher (45)
Linde B	0,2,7.5mol% Y$_2$O$_3$ added as nitrate		8	1100	Lange (46,47)
CVD powders by co-oxidation of AlCl$_3$ + ZrCl$_4$ => 10-20 wt% ZrO$_2$		S 1550-1600°C / 60 min HIP 1450°C / 1 min; 98 MPa	4-6.5 4.5	650	Hori et al (48-50)
submicron	Toyosoda (2 mol% Y$_2$O$_3$); 60 wt%	DP 40 MPa; S 1400-1450°C/120min IP 300 MPa; HIP 1500°C/30 min 100 MPa		2200	Tsukuma et al (55)
Alcoa A16 SG	30 vol% ZrO$_2$ (+ 2 mol% Y$_2$O$_3$)	Colloidal filtration (50-55% rel density); S 1600°C/120 min		900	Aksay et al (56)
	10-15 vol% ZrO$_2$	Slip casting; S <1600°C		850	Willfinger & Cannon (57)

Table 3: Al$_2$O$_3$ - ZrO$_2$ processing and room temperature mechanical properties.

a: DP = die pressed, IP = isopressed, S = sirtered, HP = hot pressed.

strengths, see Table 3[48-50]. Spray pyrolysis has also been investigated[51,52], and hydrothermal preparation of powders is also being explored[53].

The second general avenue to improved Al_2O_3-ZrO_2 processing has been to achieve better densification, especially reduction of isolated voids which are common sources of failure[43,54]. A very successful approach has been to hot press or HIP the materials (Table 3), or hot form them as discussed later. Another approach has been to improve powder processing, especially to reduce or eliminate agglomerates, e.g. by colloidal techniques[56,57]. Recently, successful preparation of bodies with surface layers of higher Al_2O_3-ZrO_2 content using tape lamination has also been reported[58].

A derivative of the above Al_2O_3-ZrO_2 processing is the (somewhat belated) recognition that ZrO_2 is a very effective grain growth inhibitor[59-63]. Thus, for example, Sato et al[63], report that sintering Al_2O_3 with 5-10 vol% ZrO_2 to 1% porosity at 1500°C (for 1 hr.) reduces the Al_2O_3 grain size to 2.5 μm vs. 5 μm without the ZrO_2. Thus, part of the improvement in strength as a result of adding ZrO_2 is the refinement and homogenization of the Al_2O_3 microstructure, as well as aiding densification[58-61]. This in turn has led to the use of low levels (e.g. 1-3 vol%) of ZrO_2 for such benefits.

Several other matrices have been investigated for ZrO_2 toughened composites, most extensively mullite ($3Al_2O_3 \cdot 2SiO_2$). While mullite has a lower Young's modulus (hence, reduced toughening), it offers lower thermal expansion, high creep resistance, and lower cost (as well as opportunities for reaction sintering[64-66], Table 4 and discussed below). Although mullite can often be difficult to sinter, Prohaska et al[67] notes that ZrO_2 can retard grain growth and promote densification of mullite. Moya and Osendi[68,69] and Yuan et al[70,71] have shown some strengthening and toughening in mullite-ZrO_2 (Table 4), but this has generally been modest.

A variety of other matrices have been investigated for ZrO_2, with beta alumina[74-77] being the most extensive of these. While sintering has been fairly successful (Table 4) both reaction processing and hot pressing have been investigated as discussed below. Significant benefits have again been observed in limiting grain growth, hence, reducing sintering temperatures, e.g. by 150°C[74], and homogenizing beta alumina microstructures, which was a major factor in improved strengths. Other matrices investigated include

Mullite matrices

Materials / Preparation	Fabrication method(s) and conditions[a]	K_{-C}[b] $MPam^{\frac{1}{2}}$	Flexure strength[c] MPa	Investigator
$3Al_2O_3$ + $2ZrSiO_4$ → M + ZrO_2 $3\mu m$ $5\mu m$	HP 1450°C/≤40 min; 25 MPa			DiRupo et al (64)
$3Al_2O_3$ + $2ZrSiO_4$ → M + ZrO_2 $7m^2/g$ $1m^2/g$ attritor milled 6 hrs; spray dried	IP 600 MPa; S 1400°C / 60 min R 1550°C / 120 min	4.5	400	Claussen, Jahn (65)
$3Al_2O_3$ + $2ZrSiO_4$ → M + ZrO_2 $1\mu m$				Anseau et al (66)
Fused mullite + 10-25 vol% ZrO_2 (<44 μm); attritor milled 8 hrs	IP 630 MPa; S 1610°C / 180 min			Prochazka et al (67)
Submicron mullite + 10-20 vol% ZrO_2; attritor milled 1 hr	IP 200 MPa; S 1570°C / 150 min	3.2	290	Moya, Osendi (68,69)
$(3+x)Al_2O_3$ + $2ZrSiO_4$ → M + ZrO_2 + xAl_2O_3				Cambier et al (70)
Submicron fused mullite + 10-20 vol% ZrO_2; 1 μm	DP 250 MPa → 55-60% rel density S 1610°C / 360 min.	2.6	130	Yuan et al (71,72)
Rapid solidified mullite + 5wt% ZrO_2 powder	HP 1040°C HT 1200-1600°C / 1 min			McPherson (73)

(continued)

M = mullite ($3Al_2O_3 \cdot 2SiO_2$)

Materials / Preparation	Fabrication method(s) and conditions[a]	K_{IC}[b] $MPam^{\frac{1}{2}}$	Flexure strength[c] MPa	Investigator
Beta aluminas				
βAl_2O_3 prepared by 3 different methods, with 15 vol% ZrO_2 (+ 3 mol% Y_2O_3)	IP ≥ 200 MPa; S 1480°C / 60 min	4.5	350	Lange et al (74)
$Al_2O_3 + Na_4ZrO_4 \rightarrow \beta Al_2O_3 + ZrO_2$	R 1250°C			Viswanathen et al (75) Binner and Stevens (76)
βAl_2O_3 + 15 vol% ZrO_2 (+ 2.2 mol% Y_2O_3)	Slip cast; S 1535°C	4-5	350–410	Green and Metcalf (77)

Table 4: Other oxide-ZrO_2 processing and room temperature mechanical properties.

a: DP = die pressed, IP = isopressed, S = sintered, R = reaction sintered, HP = hot pressed
b: For reference, typical K_{IC}'s of good mullite and beta-alumina are both 2-3 $MPam^{\frac{1}{2}}$
c: For reference, typical flexural strengths of good mullite and beta-alumina are 200-350 MPa and 200 MPa, respectively

$MgAl_2O_4$[78], MgO[78], fosterite[80], zircon[81], Cr_2O_3[82], cordierite[83], glasses[84,85] (sol-gel prepared in addition to melt prepared systems discussed later), dental porcelain[86], Si_3N_4[87], and SiC[88].   Again, various other (e.g. hot pressing) or additional (e.g. reaction) processing methods have been used.   Clearly, some matrices are limited by reactions. While $MgAl_2O_4$ did not present compatibility problems (i.e. MgO stabilization of the ZrO_2), Si_3N_4 and SiC matrices can clearly present ZrO_2 reduction-stabilization problems.   However, it is interesting to note that Yamaguchi[82] notes almost no sintering of Cr_2O_3 + 30 wt% ZrO_2 in air, but achieves 98% of theoretical density by sintering in carbon powder at 1500°C.

Consider next composites utilizing dispersed phases of mainly hard carbides such as SiC, B_4C, and especially TiC, mainly in Al_2O_3 matrices.   In contrast to the zirconia-toughened composites discussed above, there is less compatibility between the diffusion and related sintering characteristics of these dispersants and alumina and, hence, they are much less sinterable. This has generally limited sintering of these composites to low volume fractions of dispersants, greater porosity, or both.

Much of the processing data on Al_2O_3+TiC is in the extensive patent literature on such compositions widely used for cutting tools and wear surfaces.   This shows sintering temperatures are commonly in the 1600-1850°C range using normal powders (i.e. 0.5-1 µm) and 20-40 vol% TiC.   However, because of more difficult densification and the desire to reduce processing defects, such as isolated voids and void clusters, much of the processing of these materials is by hot pressing or sinter/HIPing (discussed later), especially for commercial products.   In the case of sintering prior to HIPing, sintering temperatures are often lower (e.g. 1600°C).   Densification aids, most commonly one or more of nearly a dozen common oxides (which are known, or probable, sources of liquid phases) are used (often even with hot pressing and HIPing).   Typical Al_2O_3-TiC composites have 1-2 µm TiC grains with the Al_2O_3 grain size similar but often somewhat smaller, thus the TiC acts as an effective inhibitor of Al_2O_3 grain growth.   Room temperature strengths are typically 500-800 MPa with K_{IC}'s of 4-5 $MPam^{\frac{1}{2}}$.   Such strengths are still often limited by pores and larger Al_2O_3 grains (e.g. due to local TiC deficiency[89]).

The difficulties of sintering Al_2O_3-TiC are illustrated by the work of Hojo et al[90].   They report achieving <85% of theoretical density at 1700°C, even using an Ar instead of a vacuum atmosphere (to suppress vaporization).

Use of MgO (a common densification aid) gave densities of 95% of theoretical at 1600°C. Recently improved sintering results have been reported. Hoch[91] reports sintering Al_2O_3 + 30% TiC to full density at 1350°C in H_2 using alkoxide derived mixtures of Al_2O_3 + TiO_2 (reacted with C introduced by CVD using methane on the particles to make TiC in-situ). TiC grains of 0.5 μm in a finer grain Al_2O_3 matrix were reported but no properties are given. Recently Borom and Lee[92] have reported sintering of Al_2O_3 + 30 wt% TiC to >97% of theoretical density using typical high purity powders (no additives) by rapid (>200°C/min) heating to 1950°C where it was held for <2 minutes.

Although work on most other composites with dispersed carbide particles has again been by hot pressing, some sintering has been achieved. Again, increasing particle content typically decreases density. However, improved densification can be obtained by using a liquid phase, e.g. porosities of only a few percent in Al_2O_3 with up to 25 vol% SiC by Berneburg[93]. Radford reports sintering of Al_2O_3-B_4C[94,95]. Again, densities achievable at normal Al_2O_3 sintering temperatures dropped off significantly with a few percent B_4C additions. Reasonable sintering kinetics required temperatures >1700°C.

Several useful composites have also been made using C or BN (primarily in their hexagonal forms). Refractories consisting of oxides (mainly Al_2O_3) and C (mainly graphite flakes), widely used for steel casting, are made by refractory processing techniques, often as very large pieces (i.e. 1m long and many cm in diameter) and require little densification[96-98]. The high thermal shock resistance required for their application can only be tested using a thermite bomb[98].

Some sinterability of Al_2O_3 has been achieved with up to 15 vol% BN (i.e. well below the ≥30 vol% often desired) by using Al_2O_3 with a substantial amount of glass phase to provide liquid phase sintering[93]. While such liquid phase sintering has not been fully optimized, various limitations are still expected in the volume fraction of dispersant that can be sintered to reasonable densities. The flake-like character of the BN is felt to limit densification. Thus, most BN composites (where high densities have been sought) have again been hot pressed (or HIPed), as discussed later.

Despite potentially greater challenges in sintering, some work on all non-oxide composites has been carried out. Prochazka and Coblenz[99] sintered β-SiC with B_4C using carbon (0.1-1%) as an additive. Sintering submicron powder compacts with green densities of ≥50% at temperatures of

2080°C, they achieved porosities as low as 3% with 11 wt% B_4C, i.e. better than the 5 and 7% porosities obtained using only 6 wt% and 1 wt% B_4C, respectively. Higher levels of B_4C again increased porosity, to 5, 7, 17, and 23%, respectively, at 22, 25, 35, and 50 wt% B_4C. Tanaka et al[100] sintered Si_3N_4 (0.5 μm) with 5-60 wt% SiC particles (1 μm) using 15 wt% addition of a 70/30 mixture of Al_2O_3/Y_2O_3 as a sintering aid and temperatures of 1600-1850°C. They observed that the SiC retarded both pore and grain growth during sintering, but porosities increased from 3% at 0 wt% SiC to 10% at 60 wt% SiC. (Subsequent HIPing at 2000°C under 100 MPa N_2 reduced all porosities to <5%.) While fracture toughness was about constant, (3.5-4 $MPam^{\frac{1}{2}}$), strengths were of the order of 300-600 MPa and increased with increasing SiC content (after HIPing), which was attributed to the finer microstructures. More recently, McMurtry et al[101] sintered α-SiC (submicron) with 16 vol% TiB_2 to 98-99% of theoretical density at temperatures over 2000°C. Again, the second phase particles (final TiB_2 particle sizes 2 μm) limited the matrix (SiC) grain growth. This refined grain size was probably a factor in the (30%) higher strength of the composite (500 MPa) vs. the SiC alone (350 MPa); however, the even larger (90%) increase in K_{IC} to nearly 9 $MPam^{\frac{1}{2}}$ is also a factor.

3.1.2 Sintering Fiber Composites

Almost all processing of whisker composites has been by hot pressing[102-107], e.g. Table 5, discussed later. Nonetheless, progress is being made in sintering whisker (primarily SiC) composites. Tamari et al[108] have recently reported sintering of Si_3N_4 plus 10 or 20 wt% SiC whiskers at 2000°C with, respectively, 20 or 30 mol% Y_2O_3 + La_2O_3 sintering aid at 1MPa N_2 pressure (Table 6). Presumably, the good densification is due to the higher levels of sintering aids and probable resultant liquid phase and possible resultant HIPing action due to the N_2 over pressure. Note, however, the decrease in density associated with increasing whisker contents (Table 6). More recently, Tiegs and Becher[109] (Table 5) have shown that Al_2O_3 + 16 vol% SiC whiskers can be sintered to 95% of theoretical density at 1700-1800°C using 0.5% MgO + 2% Y_2O_3 (for liquid phase sintering) to give reasonable mechanical properties. However, higher levels of whisker additions limited achievable densities, e.g. to 80% of theoretical with 20 vol% SiC whiskers.

Only one significant attempt is known to sinter ceramic fiber composites. In addition to the much greater constraint of sintering due to

Materials	Fabrication methods and conditions[a]	K_{IC} MPam$^{\frac{1}{2}}$	Flexure strength MPa	Investigator
Al$_2$O$_3$ + 20-30 vol% SiC whiskers (0.7^3 μm diam, 30 μm long) ultrasonically dispersed.	HP 1850-1900°C, 40-60 MPa	9	800	Becher and Wei (102-104)
Mullite + 20 vol% SiC whiskers	HP 1600°C, 70 MPa	4.6	440	Wei and Becher (103)
Mullite + 20-30wt% SiC whiskers	HP 1600-1700°C / 60 min, 43 MPa	3.5	390	Samanta and Musikant (105)
ZrO$_2$ (+ 3 mol% Y$_2$O$_3$) + 30 vol% SiC whiskers; 0.06 μm diam, 10-40 μm long, >95% SiC.	HP 1450°C / 10 min		1200	Claussen et al (106)
Si$_3$N$_4$ (+ 5 wt% MgO) + 30 vol% SiC whiskers; <98%	CP 6.9 MPa + 45-48% rel density HP 1750°C	7	700	Shalek et al (107)
Si$_3$N$_4$ (20-30 mol% Y$_2$O$_3$ + La$_2$O$_3$) + 10-20 wt% SiC whiskers	Green body 53-55% rel density; S 1800-2000°C		560-600	Tamari et al (108)
Al$_2$O$_3$ + 10 vol% SiC whiskers (0.6 μm diam, 25 μm long) + MgO + Y$_2$O$_3$	CP 140 MPa S 1700-1800°C IP 280-470 MPa	7	330	Tiegs and Becher (109)

Table 5: SiC whisker composite processing and room temperature mechanical properties.

a: CP = cold pressed, IP = isopressed, S = sintered, HP = hot pressed

Whisker content (%)	% Theoretical green density	% Theoretical sintered density[a]
0	56.5	
10	55.3	99.8
20	53.2	86.5
30	50.3	77.1

Table 6: Effect of SiC whisker content on percent of theoretical density[b]

a: Sintered at 2000°C with 20 mol% Y_2O_3 + La_2O_3 for 60 min with 1 MPa N_2
b: Data of Tamori et al[108]

the continuous fiber nature, there is also the serious limitation on sintering temperatures in order to avoid either fiber-matrix reaction or strong bonding, either of which would greatly reduce, if not totally eliminate, any toughening. Coyle, working with Gugut and Jamet[110], made a composite having unidirectionally aligned SiC based (Nicalon) fibers in an Al_2O_3 matrix. Al_2O_3 particles (0.2-0.3 μm diameter) were made into a slip of about 50 wt% concentration, then introduced into the fiber preform by pressure casting (0.6-1.0 MPa argon pressure on top and vacuum on the bottom) to give a green compact that was sintered at 1235°C for 1.5 hours (using a 5-10°C/min heating rate). The resultant bodies had about 25 vol% fibers and 30 vol% (25-28%) open porosity and room temperature tensile strengths of 300-450 MPa with failure strains of 1-2%. While these are encouraging results, the small size and simple shape of the cast specimens (2 x 2 x 65 mm) and uniaxial nature of these composites probably minimized sintering limitations.

3.2 Hot Pressing, HIPing, Hot Forming

3.2.1 Particulate Composites

Hot pressing has been used for a substantial amount of particulate composite processing, e.g. Tables 2-5. Despite some reduction by the typical graphite hot pressing environment, limited hot pressing of PSZ has been used. For example, Toray has produced very high strength (e.g. >1 GPa) PSZ via hot pressing. HIPing, which poses less reduction problems, has also been very successfully applied to PSZ, e.g. by Tsukuma and colleagues[20, 23] see Table 2. Most Al_2O_3-ZrO_2 composites are hot pressed; it was the major, or only, densification method in the key studies of Clausen[41,42], Dworak[44], Becher[45], and Lange[46,47]. Note, however, that stresses believed to result from gradients in reduction (as indicated by gradients in darkness of the material) may have been the cause of delamination and cracking in large zirconia toughened alumina (ZTA) plates, and thus may limit scaling up of hot pressing to large bodies[43].

While some of the other ZrO_2 toughened composites using oxide matrices show good sinterability, e.g. β-aluminas as discussed earlier, many of these have been hot pressed, often in connection with reaction processing, discussed later. Fabrication with non-oxide matrices, e.g. Si_3N_4[87] or SiC[88] again requires hot pressing, e.g. at 2000°C for SiC[88]. The high temperatures and reducing conditions of such pressing can cause reactions with the ZrO_2, such as the formation of oxynitride with Si_3N_4, and resultant oxidation problems[111]. However some systems, such as TiB_2-ZrO_2, show good promise despite the requirements of hot pressing at 2000°C for 3 hours[112].

Hot pressing has been used even more extensively for composites containing non-oxide dispersants. Thus, while there has been some reasonably successful sintering of Al_2O_3-TiC composites most production has been by hot pressing, commonly around 1600°C[113]. Clearly, hot pressing is most practical for simple, flat shapes, such as cutting tools and some wear parts. The practicality of hot pressing is also greater for composites that require processing under non-oxidizing environments. Such environments reduce or remove two of the important advantages that sintering normally has over hot pressing; namely, the use of lower-cost heating facilities for the lower temperatures required for most air-fired materials, and the high throughput available for air-firing, e.g. tunnel kilns.

Composites of Al_2O_3 (or BeO, or MgO) with a wide range of SiC contents are commercially hot pressed (by Ceradyne, Inc.) for use as controlled microwave absorbers in travelling wave tubes. These materials typically show strengthening and toughening due to the SiC, with maxima at ≈50% SiC[114,115]. Other oxide-carbide composites have received limited study, e.g. hot pressing of Al_2O_3 - 30 vol% WC to give 97-98% of theoretical density at 1600-1700°C[116]. Some fabrication of Al_2O_3 + diamond composites has also been undertaken by high pressure (6 GPa) hot pressing of 1 μm powders at 1300°C for 1 hr[117]. Optimum toughnesses of 7 MPam$^{\frac{1}{2}}$ were obtained by subsequent heat treatment at 1100-1300°C to convert much of the diamond to graphite insitu.   Note that the diamond serves as a good grain growth inhibitor, grain sizes were 30 μm at 0 vol% and 3 μm at 15 vol% diamond.

Except for some limited sintering of Al_2O_3-BN particulate composites noted earlier, they have typically been made by hot pressing.   Earlier composites used large (e.g. 100 μm) diameter BN flakes[118,119] similar to the large graphite flakes used in the refractories described earlier. However, more recent development has concentrated on the use of much finer (2-5 μm dia.) BN flakes.   A SiO_2-BN hot pressed composite has also been developed[120] and sold commercially.   It offers good thermal shock resistance along with a low dielectric constant and reasonable thermal conductivity. Considerable development has been carried out, first on Al_2O_3-BN[114,121,122] and then on mullite-BN[122].   Typically, 30 60 vol% BN was added by ball milling followed by hot pressing, e.g. at 1750-1900°C for Al_2O_3-BN.   Figure 13a & b show at low and high magnifications respectively, fracture surfaces of an Al_2O_3 - 30 vol% BN composite made by mixing the two ingredients and then hot pressing. (This should be contrasted with a similar composite made by reaction processing and shown in Figure 19.)   Strengths as high as 350 MPa have been achieved with good thermal shock resistance.   There has also been some promising reaction processing of mullite-BN as discussed later.

While hot pressing of such oxide matrix composites produces greater densification, it is limited in the densities achievable at higher levels of non-oxide additions despite the general ease with which oxides may be hot pressed.   Thus, for example, Wahi[123-124] reports about 0.75% porosity in Al_2O_3 with 5-10 wt% TiC and about 1.5% with 20-40 wt% TiC (the hot pressed bodies were apparently made using 0.5 to 1 μm powder particles).   Nakahira et al[125] also obtained near zero porosity by hot pressing Al_2O_3 with fine (2 μm) SiC at 1800°C, but found greater porosity with larger (8 μm) SiC particles, see Figure 14.   Similarly, while Sato et al[63] showed sinterability of Al_2O_3 + 5 or 10 vol% ZrO_2 to 1% porosity at 1500°C (1 hr) (with significant reduction in Al_2O_3 grain sizes due to ZrO_2, as discussed earlier), even with hot pressing (1500°C, 2 GPA, 30 min.) porosities increased by 5%, when 10 vol% additions of TiN, TiC, SiC, or B_4C were made to the ZTA.

A wider variety of all non-oxide composites has been made by hot pressing (Table 7) than by sintering. While some of these composites have

Figure 13: Al_2O_3 – 30 vol% BN composite microstructure produced by mixing and hot pressing.

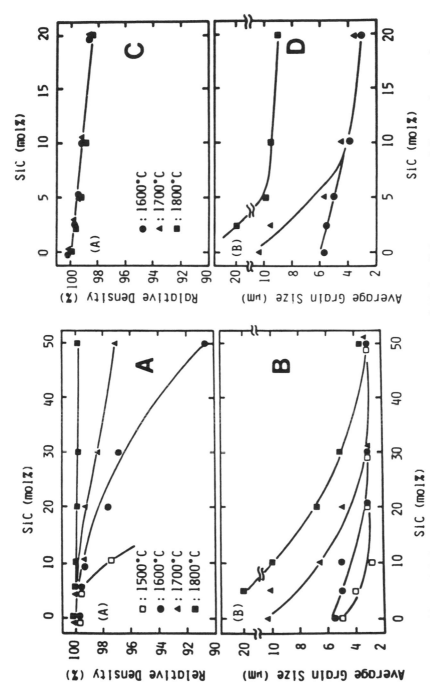

Figure 14: Effect of SiC particle size and content on sintering of Al_2O_3 – SiO_2 composites. SiC particle size 2 μm in A) and B); 8 μm in C) and D). Data after Nakahira et al (ref 125), published by permission of Yogyo-Kyokai.

Composite	Hot pressing conditions	K_{IC}[a] MPa	Flexure strength[a] $MPam^{\frac{1}{2}}$	Investigator
Si$_3$N$_4$ (+ 3 wt% MgO) + 10-40 vol% SiC (5, 9 or 32 μm)	1750°C 120 min 28 MPa	8-5	600-300	Lange (126)
Si$_3$N$_4$ (+ 5 wt% CeO$_2$) + 20-50 vol% TiC (400 mesh)	1750°C 30 min 35 MPa	4.5-6.5	600-500	Mahet et al (127)
Si$_3$N$_4$ (0.7 μm) + 0-15 vol% diamond (0.5 μm)	1600°C 30 min 6000 MPa	2-7		Noma and Sawaoka (128)
SiC (+ 1wt% Al + 1wt% C; 0.7μm) + 0-24.6 vol% TiC (1μm)	2000°C 70 min	4-6	500-800	Wei and Becher (129)
SiC (+ 1 wt% B$_4$C + 0.5 wt% C) + 0-15 vol% TiB$_2$	2000°C 45 min 35 MPa	3-4.5	380-500	Janney (130)
B$_4$C (+ 5% Al$_2$O$_3$; <10μm) + 50 vol% graphite (2μm)	2000°C		150	Rice et al (114)
Si$_3$N$_4$ (+ 6% CeO$_2$; 0.5μm) + 5-50% BN (1 μm)	1750°C		650-400	Mazdiyasni and Ruh (131)
AlN (+ 5 wt% Y$_2$O$_3$) + 5-30 wt% BN (10-15 μm)	1800-2000°C 60-15 min 11.5 MPa		300-200	Mazdiyasni et al (132)
SiC + 0-25 vol% BN	2000°C 34.5 min		300-450	Valentine et al (133)

Table 7: Non-oxide particulate composite processing and room temperature mechanical properties.

a: Properties in decreasing order, e.g. 600-300 MPa, indicates that the property generally decreased with increasing additive content, and vice versa.

increased strengths over the matrix itself, some have lower strengths, which often reflects, at least in part, the nature of the additive. Thus, in a few cases strengths decreased with larger additive particle sizes, e.g. for the 9 and 32 μm SiC particles added to Si_3N_4 by Lange[126]. Also, use of weaker second phase particles, such as graphite or BN (hexagonal), typically decrease strengths, but give large increases in thermal shock resistance[114]. Noma and Sawaoka[128] hot pressed Si_3N_4 + diamond (with no additives) at high pressure (Table 7) in a similar manner to their Al_2O_3-diamond work noted earlier. Subsequent vacuum heat treatment at 1300°C to convert a significant fraction of the diamond to graphite increased toughness. Some composite strengths, whether higher or lower than the matrix alone, are also probably limited by non-optimized processing. Thus, for example, Janney[134], in a more detailed study of SiC-TiC composite processing, showed that both the type of carbon addition (carbon black or phenolic resin) and the extent of ball milling can significantly effect mechanical properties. However, his observation of opposite trends for strength and Weibull modulus with process parameters clearly shows the need for much more process study.

Several types of particulate composites have also been HIPped. These include PSZ, ZTA, and Al_2O_3-TiC (see Tables 2-4). While most of these have been HIPped after sintering, some have been HIPped after hot pressing. Typically, HIPing results in somewhat higher and more uniform strengths. HIPing is generally carried out under non-oxidizing conditions, nevertheless conditions are usually less reducing than for hot pressing and, hence, there is less concern regarding the use of phases that are subject to reduction, such as ZrO_2. Sintering (or hot pressing) to closed porosity allows HIPing without any canning and therefore is much more practical. This allows processing on an industrial scale as shown by its wide use for WC based cutting tools. Such HIPing of bodies sintered to closed porosity greatly reduces isolated pores and pore clusters which are often a major factor in lower strength failures. Fabrication of large numbers of small components in a single run is the most economical method of using HIPing.

Some study of particulate composites having two dispersed phases has been made, again mainly by hot pressing or sintering/ HIPing. The two main types have been: 1) substitution of a second hard material (e.g. TiN, SiC or B_4C) for some of the TiC in Al_2O_3-TiC, and 2) substitution of ZrO_2 for some of the Al_2O_3 in various Al_2O_3-carbide or nitride composites or addition of some carbide or nitride materials to oxide - ZrO_2 composites (e.g. see

Sato et al[63]). Considerable development was undertaken on composites of HfB$_2$ and especially ZrB$_2$ with SiC + C by hot pressing relatively coarse (10 μm) powders at 1800° at high pressure (840 MPa)[135]. These composites have been produced commercially. More recently, composites of TaN + ZrB$_2$, with either ZrN or WC have been hot pressed (2100°C for 30 min. with 18 MPa in N$_2$) for wear studies[136].

Turning to hot working, Rice et al[137] have demonstrated press forging (i.e. slow compressive hot deformation) of PSZ crystals similar to press forging single phase ceramic crystals[138]. Kellett and Lange[139] have recently reported press forging of fine grain Al$_2$O$_3$-ZrO$_2$ at 1400-1500°C and 100 MPa obtaining strains of >80% without tearing. This is not unexpected in view of the extensive creep in fine grain PSZ at ≤1200°C[140] and the results of press forging studies in Al$_2$O$_3$[138]. Japanese investigators have recently reported over 100% tensile (superplastic) elongation of fine grain PSZ at <1200°C[141].

3.2.2 Whisker Composites

Almost all development of whisker composites has been via hot pressing (Table 5) with Becher and colleagues of Oak Ridge National Laboratory being leaders in such developments[101-103,142]. Their work has been with SiC (mostly ARCO) whiskers in an Al$_2$O$_3$ matrix, made via hot pressing at 1750-1900°C (usually 1850°C). While achievement of homogeneous mixing was not fully realized at significant whisker volume fractions (e.g. ≥20%)[102], promising results and progress have been achieved. Use of 0.2 μm average particle size and less agglomerated Al$_2$O$_3$ powder versus a more agglomerated one with 0.3 μm average particle size[101,102], and ultrasonic dispersion[102] have improved uniformity. The resultant bodies have anisotropic properties due to the partial planar whisker orientation developed during hot pressing. Room temperature fracture toughness for crack propagation in the plane of hot pressing is about 20% greater than for pure Al$_2$O$_3$ (≈4.6 MPam$^{\frac{1}{2}}$)[101], but is about double that for crack propagation perpendicular to the plane of hot pressing with 20-30% vol% whiskers[101-103]. Corresponding strengths of 300-450 MPa are the same, or less than, pure Al$_2$O$_3$, and about 650 MPa, with 5-10 vol% whiskers, and especially 850 MPa respectively with 20-30 vol% and 40-60 vol% are above those reliably achieved for pure Al$_2$O$_3$. Evidence of crack-deflection and limited whisker pull-out is observed, Figure 15. The lower degree of strengthening versus toughening generally found with whisker composites probably reflects effects of whisker debris and inhomogeneous distribution.

Figure 15: Fracture microstructure of hot pressed Al_2O_3 – SiC whisker composite tested at room temperature. Note clear imprints of whiskers on the surface showing crack deflection, whisker pull-out, or both occurring along the crack surfaces.

Whether the maxima or plateau of toughness at about 20-30 vol%, and strength, at approx. 40 vol% whiskers[104], is real, or an artifact, e.g. of increasing homogeneity problems as whisker content increases, is not yet known. Recently Homeny et al,[143] obtained similar mechanical properties with Al_2O_3 + 30 vol% SiC whiskers to those achieved by the Oak Ridge investigation.

Use of SiC whiskers in other matrices (mostly oxide) is also being explored, with mullite[104,106] again being of interest. Samanta and Musikant[105] hot pressed such bodies at 1600-1700°C for 1 hr. with generally better results at the higher temperatures. Room temperature strengths had a definite maxima of nearly 400 MPa at 30 vol% whiskers, over twice the strength achievable with no whiskers (approx. 170 MPa). Wei and Becher[103] have hot pressed similar bodies with 20 vol% SiC whiskers, to give strengths of about 440 MPa and a K_{IC} of 4.6 MPa$^{\frac{1}{2}}$, again about twice the value obtained without whiskers (2.2 MPam$^{\frac{1}{2}}$). Since the best pure mullite bodies have strengths of 250-350 MPa and toughnesses of 2-2.5 MPa$^{\frac{1}{2}}$, greater progress has again been in improving toughness than in strength. Less relative improvement in strength may again reflect effects of whisker debris and inhomogeneous distribution.   Panda and Seydel[144] have recently made $MgAl_2O_4$-SiC composites by mixing 30 vol% whiskers (Versar) in a solution of $Mg(NO_3)_2$ and $Al(OH)_3$. After drying (350°C) and calcining (970°C), specimens were press-forged (basically hot pressing of a preform that was smaller than the die diameter so some lateral flow occurs) in graphite dies (in argon) at 1680°C. Flexure strengths (3-point, 13 mm span) averaged 400+ MPa to about 900°C, where they began to decrease to approx. 200 MPa at 1300°C.   In contrast the matrix only displays strengths of approx. 300 MPa but with only a slight decrease over this temperature range.   Finally, Gadkarce and Chung[145] have recently reported processing of whisker composites using glass and crystallisable glass matrices similar to processing of continuous fiber composites (discussed later).

Shalek et al[107] reported hot pressing of SiC (LANL) whiskers in Si_3N_4 matrices (using 5 wt% MgO densification aid).   Room temperature flexure strengths for materials hot pressed at 1600°C were constant at about 400 MPa for whisker contents of 0-20 vol%, then decreasing to about 250 MPa at 40 vol% whiskers.   Hot pressing at 1750 or 1850°C gave strengths of about 650 MPa with no whiskers, decreasing to about 500 MPa at 30 vol% whiskers (the maximum addition for these hot pressing conditions).   Young's moduli were scattered between 300 and 370 GPa for the various hot pressing conditions to about 30 vol% whiskers, but were much lower (240 GPa) at 40 vol%. Fracture toughness (Chevron notch test) of samples hot pressed at 1600°C increased from ≈5.5 to ≈8 $MPam^{\frac{1}{2}}$ over the range of 0-40 vol% whiskers, while specimens hot pressed at 1750 or 1850°C increased from ≈7 to ≈10 $MPam^{\frac{1}{2}}$ over the range of 0-30 vol% whiskers. Most of the increase in the latter case occurred for the initial level of whisker addition (10 vol%). The drops in strength and Young's modulus at higher whisker volume fractions are attributed to whisker debris and inhomogeneous distribution.   More recently, Buljan et al[146] have reported successful toughening and strengthening, at least at higher (e.g. 30 vol%) SiC whisker additions to Si_3N_4 hot pressed with 6 wt% Y_2O_3 + 1.5 wt% Al_2O_3 at 1800°C.   Whether their improved results are due to differences in whiskers or processing used is not clear.

Composites consisting of SiC whiskers dispersed in ZrO_2 toughened matrices have also been investigated for possible dual toughening. Claussen et al[106] hot pressed fine particles of ZrO_2 (+3 mol% Y_2O_3) with 30 vol% SiC whiskers at 1450°C for 10 min.   Room temperature fracture toughness was reported to be doubled (from 6 to 12 $MPam^{\frac{1}{2}}$) as were strengths at 1000°C (200 vs. 400 MPa), but not at room temperature. Similarly, Becher and Tieges[147] have shown that the toughening due to the incorporation of both tetragonal ZrO_2 particles (0-20 vol%) and SiC whiskers (0-20 vol%) in a mullite matrix by hot pressing (1450°C, for 60 min) is additive, at least up to 20 vol% SiC whiskers (the limits of investigation).

Limited work on HIPing whisker composites has begun. Thus, Becher and Tieges[109] HIPed (at 1600°C) some of the Al_2O_3-SiC whisker composites they sintered, but such efforts are limited by the limited levels of whiskers in composites that can be sintered to closed porosity for HIPing. However, as discussed later, these restrictions do not necessarily apply to reaction processed matrices.

3.2.3 Fiber Composites

The great majority of ceramic fiber composites have been prepared by hot pressing. This reflects the much greater ease of densification by hot pressing versus sintering with a broader range of fibers and fiber architectures. Thus, in addition to uniaxial fiber composites, there has also been considerable hot pressing of 2-D lay-ups of cloth and other orientations of uniaxial tapes. This application of hot pressing to continuous fiber composites is driven not only by the easier densification, but also by the need to keep temperatures sufficiently low to minimize reaction and bonding with the fibers. Only fine SiC-based (Nicalon) or graphite fibers, or large CVD-SiC filaments have been successfully used, since they are the only ones sufficiently non-reactive and sufficiently unaffected by hot pressing conditions to produce reasonable composites. Even with these fibers, especially the fine SiC-based fibers, one is often limited in hot pressing temperatures in order to avoid degradation or reaction of the fibers, Table 8.

There have been a variety of earlier efforts to make fiber reinforced ceramics, mainly using CVD filaments and especially refractory metal wires, usually in polycrystalline matrices, as discussed in earlier reviews[1,2,148-150]. However, much of the original impetus for the current interest in ceramic fiber composites stems from work by Sampbell, Phillips, and colleagues[151-156], and Levitt[157] using carbon fibers hot pressed in silicate glass-based matrices (Table 8). The significant toughening and strengthening they observed with sufficient (e.g. >30 vol%) continuous uniaxial (but not chopped) fibers ultimately motivated other researchers in this area.

Matrix	Fiber	Fabrication method and conditions	E GPa	Flexure strength MPa	K_{IC} MPa·m$^{\frac{1}{2}}$	Investigator
Soda lime, lithium alumino silicate, & borosilicate glasses	C (40 vol% uniaxial)	HP 1200-1250°C 34 MPa		680		Sambell et al (151-156)
$Li_2O.Al_2O_3.8SiO_2$	C	HP 1375°C 5 hrs 7 MPa		840		Levitt (157)
Al_2O_3 (dry mixed with fibers, then mixed as slurry with acetone)	C (18vol% chopped, 1-5 mm long) E = 20-40 GPa; σ = 690-900 MPa; E = 200 GPa σ = 3 GPa	HP 1600°C (20 hrs) - 1750°C (5 hrs)				Yoshikawa, Asaeda (158)
Al_2O_3 Mullite	C (40-50vol% uniaxial) E = 192GPa; σ = 2.26GPa	HP 1700°C 10 hrs porosity = 5-7%	50 50	300-500[a] 650[a]	3-10[a] 7[a]	Yasuda and Schlicting (159)
7740 Borosilicate glass	SiC (35-65 vol% 140 μm diam CVD uniaxial)	HP 1150°C 20 hrs 6.5 MPa		650-850	19	Prewo et al (166-168)
	SiC (40 vol%; 10 μm diameter, pyrolysed)	HP 1200°C 60 hrs 14 MPa		300	11.5	
ZrO_2 - SiO_2	SiC (50-70 vol%)	HP 1250°C (25 hrs) - 1325°C (45 hrs) 34.5 MPa		100-350[b]	10[b]	Lewis and colleagues (176)
ZrO_2 - TiO_2	SiC (50-60 vol%)	HP 1175°C (60 hrs) - 1250°C (25 hrs) 34.5 MPa		370-670[b]	12-20[b]	

Table 8: Ceramic fiber composite processing and room temperature mechanical properties.

a: Fibers CVD-coated with SiC;   b: Best results generally obtained with CVD-coated fibers (eg with BN).

Yoshikawa and Asaeda[158] were amongst the earlier investigators to follow up on the above continuous fiber composite developments but, instead, utilized non-glass-based matrices. Carbon fibers of low density (approx. 1.7 g/cc) and a very low Young's modulus (approx. 30 GPa), or carbon fibers (Young's modulus approx. 200 GPa) with strengths proportional to their Young's moduli, were used in chopped form (1-5 mm length). These were hot pressed in an Al_2O_3 matrix in graphite dies (under nitrogen, which resulted in conversion of some of the Al_2O_3 to AlON) at temperatures of 1600-1750°C, with pressures of 100-350 MPa for times of 5-20 minutes. While the room temperature flexural strength of the composites decreased with increasing fiber content, apparently due to microcracking (indicated by audible sounds emitted during cooling), the thermal shock resistance was increased. Specimens with 14% or 18% fiber content retained 3-4 times as much strength as all other materials tested after quenching from a temperature of 800°C, and also showed the best maintenance of strength with multiple quenching cycles (from 400°C).

Yasuda and Schlichting[159] made Al_2O_3 and mullite composites reinforced with carbon fibers (having a Young's modulus of 192 GPa and tensile strengths of 2.3 GPa, in tows of 10,000 fibers each), either chopped, or unidirectionally aligned. The fibers were infiltrated with either an Al_2O_3 sol, or a mixture of Al_2O_3 and SiO_2 sol, then hydrated in air and hot pressed in graphite dies at 1700°C for 10 minutes. Chopped fiber results were not particularly encouraging; the best room temperature flexural strengths achieved with mullite and 20% chopped fibers were about 250 MPa. Thus, most work was on unidirectional continuous fiber composites. Strength, Young's modulus, and fracture toughness of composites with the Al_2O_3 matrix all initially decreased with increasing fiber content in contrast to a more general increase of these properties with the mullite matrix. SiC coatings (applied by CVD) on the fibers gave higher strengths and toughnesses (e.g. 600-800 MPa and 6-8 MPam$^{\frac{1}{2}}$ at 50 vol% fiber), especially for the Al_2O_3 matrix.

Fitzer and Schlichting[160] have reviewed much of the work of Yasuda and Schlichting and related studies along with presenting their own work. They showed the strength of SiO_2-10% TiO_2 (alkoxide derived) matrix composites hot pressed at 1600°C to be constant at 70 MPa with 0 to 50 vol% unidirectional SiO_2 fibers, but fracture toughness increasing from 2 to 5.5 MPam$^{\frac{1}{2}}$. Similarly, hot pressing of Al_2O_3 or mullite matrices (again alkoxide derived) at about 1700°C with unidirectional Al_2O_3 (DuPont FP) fibers gave strengths, respectively, of 240 and 110 MPa, again essentially independent of fiber content.   The fracture toughness of the all Al_2O_3 composite was reported to increase from 4 to 12 $MPam^{\frac{1}{2}}$ from 0 to 60% fibers.   In contrast to the above, use of SiC coated unidirectional C fibers gave peak strengths of 500, 350, and 300 MPa, respectively, for SiO_2, mullite, and Al_2O_3 matrices.   Finally, they briefly review work showing strengths of 600 MPa and 900 MPa, respectively, for unidirectional CVD-SiC filaments (46 vol%) and CVD-B filaments (30.5 vol%) in a PbO glass matrix.

Two investigations of metal wire or ceramic monofilament reinforcement of ceramics have been carried out since the earlier work cited above.   The first was by Brennan et al[161] who investigated wire reinforcing of hot pressed Si_3N_4.   Poor results using tungsten wires due to formation of a brittle silicide reaction product led to somewhat more promising experiments using (unidirectional) tantalum wires (approx. 25 vol %, 0.63 mm or 1.27 mm diameter).   Hot pressing (using 5% MgO or 10% Y_2O_3 additions with the Si_3N_4) at 1750°C gave flexural strengths of about 560 MPa or 700 MPa at room temperature, respectively, i.e. about 20% below those of the two matrix compositions alone.   Tests of the true room temperature tensile strengths proved very disappointing, approx. 170 MPa, with failure apparently initiating from the Ta fibers.   On the other hand, flexural strengths of the composite were about 50% greater than those of the matrix alone at 1300°C and room temperature.   Charpy impact strength was increased about 4-fold before the onset of damage.

Shetty et al[162] have more recently hot pressed large SiC (AVCO) filaments in Si_3N_4 (+ 8 wt% Y_2O_3 + 4 wt% Al_2O_3).   Composites with up to 44 vol% of SiC filaments were fabricated by infiltrating the matrix via a slurry and hot pressing in graphite dies at pressures of 27 MPa and temperatures to 1750°C.   The room temperature flexure strengths of the composites were about one half of the strength of the matrix alone (930 MPa) and independent of filament volume fraction over the range investigated (10-44 vol%) consistent with the brittle failure of the composites, which resulted from degradation of the filaments and their strong bonding to the matrix.

The first of the most active groups in recent ceramic fiber composite development is that of Prewo and colleagues at UTRC who initially duplicated the earlier Harwell carbon fiber glass-based matrix work.   However, they subsequently pursued two major changes in fibers used.   First, using glasses whose thermal expansion matched that of SiC, they successfully used CVD-SiC filaments[163] (Table 8).   The other major fiber change was to explore use of the SiC-based fibers from polymer pyrolysis based on Yajima's[164,165] pioneering developments. These fibers also proved to be quite successful in appropriate silicate-based glass matrices[166,167] (Table 8). Improvements in properties were demonstrated by crystallizing some of the matrices (as also demonstrated by Sampbell and colleagues with carbon fibers). An important discovery from this crystallization work by Prewo and colleagues[166,167] was that TiO_2 nucleating agents for crystallizing the matrix glasses attacked the Nicalon fibers, while ZrO_2 nucleating agents did not. They further found that Nb_2O_5 added for nucleation tended to preferentially form NbC coatings on the fibers with beneficial effects.

Three developmental trends have occurred in fiber composites based on glass-based matrices. First has been a broadening of groups investigating such composites[166,167], with further variations in compositions and processing. Second, there is a growing understanding and focus on the role of the fiber-matrix interface. Thus, it is now recognized that little, and probably no, chemical bonding can exist at this interface if good toughness, i.e. non-catastrophic failure, is desired. As a specific case, it is now recognized that successful use of Nicalon (SiC based) fibers is typically accompanied by the formation of a carbon rich layer on the fiber surface during processing[167] (the source of the NbC coating noted above). More aspects of fiber-matrix development will be noted below. Third, the fabrication of a wider range of components shapes has begun to be addressed, with hot pressing and other hot forming methods being utilized. In Figure 16 can be seen A) a hot pressed fabric reinforced cup, B) a hot pressed hybrid air foil, C) a matrix transfer moulded cylinder, D) an injection molded igniter shape, and D) an injection moulded cylinder.

Another major development has been pursuit of non-glass-based matrices. This author and colleagues[171-176] began exploring such matrices somewhat later than, but unaware of, efforts at UTRC. While some had apparently thought that hot pressing may not be applicable to composites other than glass-based matrices because the latter would provide much greater plastic flow and less damage to the fibers, no unequivocal disadvantage has been demonstrated to hot pressing ceramic fiber composites with non-glass based matrices. The choice of such matrices has been driven by both the need to limit reactivity with the fibers as well as to attempt to achieve matrices with more modest Young's moduli so that there could be some possible load transfer between the fibers and the matrix. A variety of composites with polycrystalline oxide matrices have now been hot pressed with SiC based

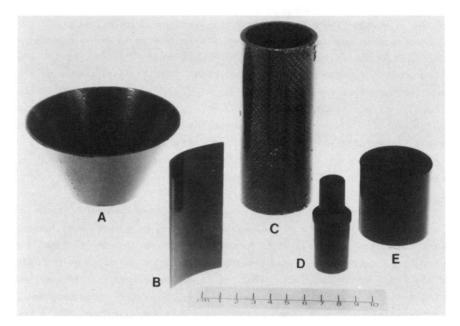

Figure 16: Shapes made of continuous fiber composite utilizing glass-based matrices.   Photos courtesy of Drs Karl Prewo and John Brennan, United Technology Research Centre.

(Nicalon) fibers[176].   The most successful matrices subsequently reported by Lewis et al have been ZrO_2-SiO_2 and ZrO_2-TiO_2[173-175] (e.g. Table 8). An important factor in obtaining the highest strength and toughness in many of these composites has been the use of CVD fiber coatings, initially BN[172-176,177] (Figure 17) and later multilayer coatings[173,177].   In Figure 17, load deflection curves are shown for two ceramic fiber composites which are identical, except that one was fabricated using fibres with a thin (0.1 μm) coating of BN.   Note that the composite without the BN coating on the fibres has a much lower strength and total catastrophic failure in comparison with the much higher strength and very non-catastrophic failure of the composite with coated fibres.   Note also that the brittle composite shows smooth, i.e. low strength, fiber fracture (lower insert), whereas the higher strength, tougher composite shows high strength failure in the fiber as is indicated by the presence of fracture mirrors on the fibers (upper insert).

Almost all of the composites described in the references above have been hot pressed from bodies in which the tows were infiltrated with a slurry of the powder to produce the matrix.   Some have also been processed with sol-

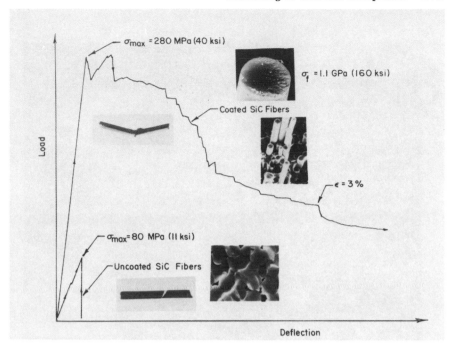

Figure 17: Microstructures of ceramic fiber composites illustrating fiber-matrix inhomogeneities.

gel matrices, the work illustrating one of the important challenges that must be observed in fabricating composites. For example, in some composites utilizing BN coated fibers poor results were obtained which have been tentatively attributed to substantial oxidation of the coating by products of the pyrolysis of the gel-derived matrix. Similarly, control of the atmosphere during hot pressing can be important, e.g. use of a N_2 vs. a vacuum atmosphere to retard degradation of fibers (which can also be limited by fiber coatings). One of the key challenges in making fiber composites is achieving homogeneous distribution of the fibers. Figure 18, an early, fairly inhomogeneous composite which still had reasonable strength (300 MPa) and good non-catastrophic failure, illustrates a major problem which is obtaining similar distribution of matrix within fiber tows as between them. A) and B) show, respectively, lower and higher magnifications of a polished section. Substantial progress has since been made in achieving uniformity.

Only limited HIPing of continuous fiber composites has been done, in part because the available canning techniques using metal or glass for encapsulation are cumbersome. HIPing is also potentially somewhat limited for such composites because of architectural issues. Thus, if one has high

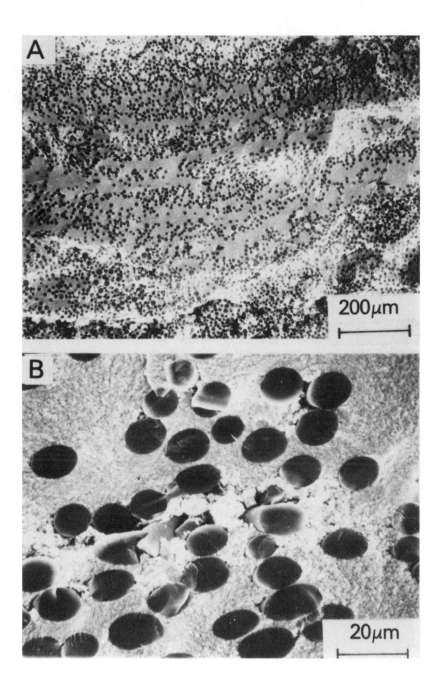

Figure 18: Effects of BN fiber coating on SiC ceramic fiber composite strength and fracture, see text.   Data after reference 172.

stresses parallel with the direction of fibers, damaging fiber buckling may result. Hot forming of fiber composites in the normal sense of deforming the composite is greatly limited by their very fibrous nature, except for some possibilities with chopped fiber composites. However, the hot forming demonstrated by UTRC[166,168], wherein glass is extruded into fiber preforms, may hold significant promise. This may make it feasible to form various long parts such as rods, tubes, beams, plates, etc., processes analogous to extrusion or pultrusion (but at slow rates). Some possibilities may also exist for doing this with very fine grain polycrystalline bodies.

3.3 Reaction Processing

A variety of reactions may be used to produce composites. Many reactions are used in preparing ceramics ranging from powder preparation (e.g. decomposition of salts, gels, polymers, and other organometallics) to direct production of a product, i.e. CVD or polymer pyrolysis as discussed in subsequent sections. However, what is of interest here are reactions that produce composite powders for consolidation into composites, or more commonly reactions involving a powder compact to yield a composite product in conjunction with heating, possibly with pressure, such as in hot pressing, or HIPing.

Clearly, there are various roots to reaction technology, with reaction sintering (or bonding) of Si_3N_4 or SiC from Si or SiC + C compacts, respectively, being a major component. Processing of composites such as Si_3N_4 bonded SiC and SiC bonded B_4C made by infiltrating compacts of SiC + Si and B_4C + C, respectively, with N_2 or NH_3 gas, or molten Si at high temperatures are important examples of industrial application of this technology for refractory, wear, and armor uses. More recently, reaction formed Si_3N_4 or SiC matrices have been investigated for fiber and whisker composites. While these entail Si at high temperatures where it is very reactive and, hence, may also attack fibers or whiskers, it has the advantage of "sintering" a matrix without shrinkage and the attendant limitations discussed earlier. A recent demonstration of combining reaction and other processing is the work of Lundberg et al[178]. They incorporated up to 30 vol% SiC whiskers in a Si_3N_4 matrix by reaction sintering. While they observed decreases in green and nitrided densities with increasing whisker content, they had added 6% Y_2O_3 and 2% Al_2O_3 to the Si powder such that after nitriding they could sinter or HIP. Only the latter was successful in achieving high densities (via glass canning). Strength and toughness decreases by approx. 1/3 at the highest whisker content, probably due in part to whisker debris and possibly too strong a whisker-matrix bond.

Of potentially much broader applicability are a variety of other reactions for producing composites. Some important examples of reaction processing of composites have already noted for ZrO_2 toughened mullite and beta alumina (Table 4). Thus, for example, investigators have used the reaction:

$$2ZrSiO_4 + 3Al_2O_3 \rightarrow 3Al_2O_3.2SiO_2 + 2ZrO_2 \qquad 2$$

to form mullite-ZrO_2 composites, yielding some of the best mechanical properties. An advantage of this is its use of $ZrSiO_4$, the most common (and a low cost) natural source of ZrO_2. More complex ZrO_2-SiO_2-Al_2O_3 bodies involving CaO[180] and TiO_2[181] have also been investigated. However, not all reaction processing has been successful. Thus, Yangyun and Brook[179] were unsuccessful in reaction sintering dense composites of fosterite (Mg_2SiO_4) + ZrO_2 from MgO and $ZrSiO_4$ because the rate of reaction was faster than the rate of pore removal. Only about 60% of theoretical density was achieved by sintering at 1275°C for 200 min. Samples fired under more extreme conditions, 1420°C for 2 hrs., contained extensive monoclinic ZrO_2. However, useful bodies were produced by hot pressing (1325°C for 15 min. at 20 MPa). Similarly, Binner and Stevens[76] had less success using sodium zirconate to form beta Al_2O_3 + ZrO_2 than simple addition of ZrO_2 powders (Table 4).

The potential for reaction processing may go well beyond use of lower-cost raw materials and possible mixing advantages noted above. Sintering (or hot pressing or HIPing) reactant compacts, then reacting them may provide opportunities to control composite microstructures by controlling nucleation and growth of the new phases formed by the reaction, i.e. to provide some, possibly all, of the opportunities for microstructural control offered by crystallization of glasses. Thus, for example, Coblenz[182] has demonstrated advantageous processing of mullite-BN composites by reactions, e.g.;

$$B_2O_3+2Si_3N_4+9Al_2O_3 \rightarrow 3(3Al_2O_3.2SiO_2)+8BN \qquad 3$$

Besides using cheaper raw materials (than mullite and BN), production of the BN in-situ after densification avoids its inhibiting densification while taking advantage of the B_2O_3 based liquid phase for densification.   This produces excellent microstructures (Figure 19) and good properties. Note the greater homogeneity of the microstructure relative to specimens made by mixing BN particles with the matrix phase as a powder, followed by densification (Figure 14).

An intriguing possible extension of the above type of reaction processing is for in-situ development of whiskers. While earlier work at in-situ whisker formation[183] was unsuccessful, recent work by Hori et al[184] to form Al_2O_3 platelets/whiskers in-situ in an Al_2O_3-TiO_2 compact is promising. They made very fine powders of intimately mixed Al_2O_3+TiO_2 by oxidation of $AlCl_3$ + $TiCl_4$ for good sinterability below the temperatures for forming Al_2TiO_5 (1300°C). Use of small (0.85%) additions of Na compounds for aiding densification and in-situ formation of Al_2O_3 platelets/whiskers improved fracture toughness from <3 $MPam^{\frac{1}{2}}$ with no platelets/whiskers to over 6 $MPam^{\frac{1}{2}}$ at 35 vol% platelets/whiskers.

Another class of reactions that has attracted considerable recent interest are those often referred to by terms such as self-propagating high temperature synthesis, which is referred to by such acronyms as SPHTS, SHTS, and SPS (the latter acronym may be more appropriate, since it emphasizes the unique propagating character of these reactions). The substantial exothermic character of these reactions results in a high temperature reaction front that actually sweeps through a compact of the reactants once the reaction is ignited at some point. While such reactions can be used to produce single phase compounds, e.g. Ti + C to yield TiC and Ti + B yield to TiB_2, they can also be used to produce composites directly. The classical thermite reaction Fe_2O_3 + 2Al → 2Fe + Al_2O_3 is an example of a metal ceramic composite product. However, there are a variety of reactions that can yield all ceramic composite products[185], e.g.;

$$10Al + 3TiO_2 + 3B_2O_3 \rightarrow 5Al_2O_3 + 3TiB_2 \qquad\qquad 4$$

$$4Al + 3TiO_2 + 3C \rightarrow 2\ Al_2O_3 + 3TiC \qquad\qquad 5$$

There are, however, serious challenges to be met in such direct reaction processing. These include intrinsic problems due to the products inherently having smaller molar volumes than the reactants, e.g. by 20%, with the discrepancy tending to increase as the enthalpy (and, hence, ease of ignition and speed of reaction) increases[186]. This intrinsic volume change means

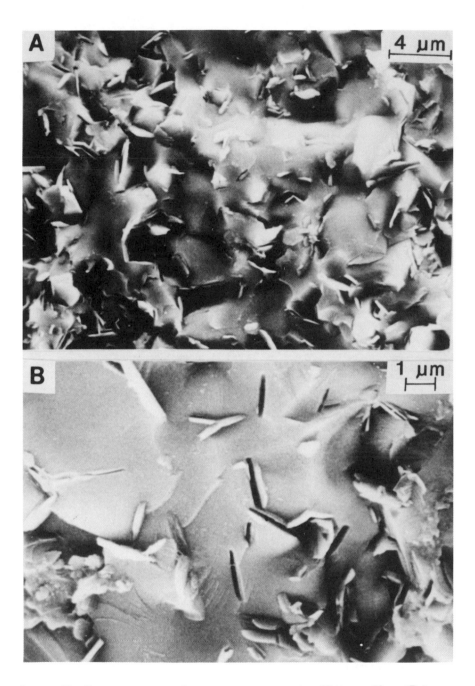

Figure 19: Microstructure of reaction processed mullite - 30 vol% boron nitride composite.

that even in a fully dense compact of reactants there would be porosity generation due to the higher density of the resultant products, unless there is some driving force for densification. Such a driving force is typically lacking unless it is applied externally, such as by hot pressing[187] (or hot rolling[188]). Further, there are significant extrinsic problems due to the high temperature nature of the reactions and the speed with which these high temperatures are reached. These typically result in serious out-gassing problems from the powders that present a very significant porosity problem. While such processing has shown some progress, there are still substantial problems in achieving high and uniform densification.

There are a variety of microstructural issues and possible opportunities in such SPS type reactions. Thus, both particle size (and probably morphology) and especially porosity affects the propagation rates[189]. It has also been proposed that reaction microstructures can effect the course, and possibly the final products, of the reaction[182]. The latter would be a reflection of the broader suggestion of SPS reactions yielding different phases because of the very transient nature of the reactions. Figure 20 is a schematic of the potential effects of using coated particles for SPS and related reactions. While mixtures of individual particles (1) may lead to

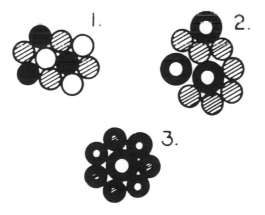

Figure 20: Schematic of potential effects of using coated particles for SPS and related reactions. See text.

one set of reaction products, coating of one reactant on particles of another (2) or reactant, or on both reactants (3) may change the resultant products. For example, mixing of Ti + B + C particles will normally lead to predominantly Ti-B-C products and little or no TiC or B_4C. However, reaction of Ti particles coated with C and B particles coated with C may result in more of these products.

Another interesting possibility suggested by these reactions is that some of them appear to produce whiskers in-situ. Besides the possibilities of directly producing a product body, such reactions may be used to prepare powders. This, in turn, would introduce another microstructural issue and opportunity in processing, that is, the size of the composite particles and how this effects densification and resultant composite performance.

Another type of reaction processed composite is that under development by Lanxide Corporation[190]. They have reported "growth" of Al_2O_3 by reaction of molten Al at the interface between the Al_2O_3 product (fed by capillary action along microscopic channels in the Al_2O_3 product) and a gaseous source of O_2 (e.g. air). Since this is a growth process it can "grow" around particulates, or fibers to form composites. The extent to which such composites can be varied in composition and microstructure is not yet fully known, but appears to be quite broad. The process can also apparently be applied to other matrix materials such as AlN, TiN, and ZrN.

Composites formed from cementitious derived matrices are potentially applicable to the fabrication of fiber composites. While portland cement might be considered as a matrix for some applications, in general more refractory cements are expected to be of interest. These are likely to include calcium aluminate, phosphate bonded, and possibly chloride based cements. Work in this area is apparently only in its early stages. The only known work on composites applicable to high-tech applications is that of Washburn and colleagues at Accurex who investigated $AlPO_4$ as a matrix for Nicalon (SiC based) fibers or sapphire filaments[176]. Working from a chemically derived aluminum phosphate matrix they obtained modest but useful properties (e.g. 140 MPa) by firing. In order to significantly reduce the porosity and improve the resultant properties, they subsequently hot pressed some of these composites which did result in considerable improvement in some strengths, e.g. to 300MPa.

4. NON-POWDER BASED METHODS

4.1 Polymer Pyrolysis

Polymer pyrolysis[165], i.e. the decomposition of polymer to produce ceramics, other than prototype pyrolysis of polymers to produce carbon matrices (as well as fibers) for carbon-carbon composites, offers potential advantages of significant importance in processing composites. First, in principle, one forms a composite exactly as with a normal polymer and then converts the polymer matrix in-situ to a ceramic matrix. This takes advantage of the broad technology for making polymer based as well as carbon-carbon composites. Second, the conversion of the pre-ceramic polymer to a ceramic matrix can typically be done at temperatures on the order of 1000°C or less, hence providing much broader compatibility with available ceramic fibers. Third, there is potential synergism between the polymer pyrolysis process and the composite character, i.e. the fibers control shrinkage cracking of the pyrolysing polymer matrix. Figure 21 A) and B) show lower and higher magnification views of cracking in the matrix perpendicular to the fibers. The density and length of the cracks are inversely related to the fiber's spacing. C) and D) show lower and higher magnification views of a transverse section of this composite. Note longitudinal cracking in this composite made using a SiC producing polymer for the matrix with no filler.

In principle, polymer pyrolysis consists of taking an appropriate polymer that contains the atoms of the desired ceramic product in a form that can be pyrolysed without excessive polymer backbone losses (i.e. ideally the only losses are of hydrogen) to yield the desired ceramic product[165]. Several polymers are now known, and additional ones are being developed, that yield basically Si_3N_4 and/or SiC. Polymers are also being developed to yield products which are basically BN or B_4C. (The term "basically" is used since typically a pure stoichiometric compound cannot be produced by this process.) Such polymers, in addition to or in combination with various carbon producing polymers, are all of interest for producing ceramic composites because of the high temperature and other capabilities of these materials.

Several factors strongly encourage the use of polymer pyrolysis for fiber- as opposed to particulate-composites. These include their cost, shrinkage, and low temperature processing. Thus lower temperatures, as noted above, allow much wider use with fibers, which are commonly much more limited

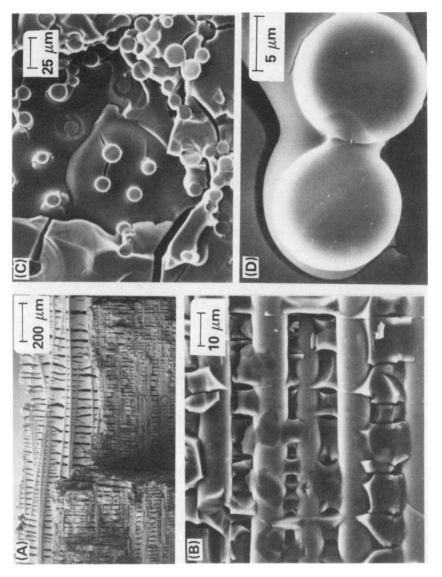

Figure 21: Shrinkage cracking in a matrix derived from polymer pyrolysis; see text.

in their temperature capabilities than most particulates of interest. The large density change, e.g. typically from about 1 g/cc for the polymer to over 3 g/cc for the theoretical density of resultant products such as SiC and Si_3N_4, means that one either has large shrinkages, substantial porosity, and/or microcracking. With a high density of particulates and a small composite body of simple shape, it may be feasible to achieve useful properties. However, in general, it is expected that the limited strengthening and toughening achievable in particulate composites will not produce attractive composites via polymer pyrolysis. On the other hand, the strengthening and especially toughening available in fiber composites can yield quite useful properties despite the presence of porosity and microcracking from polymer pyrolysis.

Sol-gel processing and resultant gel pyrolysis can be considered to be similar in some respects to that of polymer pyrolysis. However, the greater weight losses and associated density changes of gel pyrolysis versus polymer pyrolysis impose a severe limitation, as may the oxidizing character of some pyrolysis products. Further, pyrolysis of gels frequently results in a more particulate type matrix as opposed to a coherent matrix. Thus gel-derived matrices were discussed primarily under powder-based methods, because they usually require (and are favourable to) sintering, e.g. by hot pressing.

The limited experiments on forming composites via polymer pyrolysis illustrate well the points made above. Thus, dispersion of about 15 vol% fine (e.g. approximately 2 μm) graphite particles in a mainly SiC producing polymer resulted in substantially less shrinkage, cracking, and improved mechanical properties, (approximately 30 MPa flexural strength) upon pyrolysis[191]. Although the composite was clearly not optimized, it appears that the strengths would be limited to 2-4 times that already achieved and that this would only be achievable in relatively limited size parts.

Two initial studies have shown substantial promise in making ceramic fiber composites via polymer pyrolysis. Both utilized the SiC based Nicalon fibers. Jamet and colleagues[192] made composites by infiltrating fiber tows with a SiC producing polymer using a solvent plus a ceramic filler and then moulded the pre-pregged uniaxial aligned tows to produce a composite with very promising strength (about 330 MPa) and stress-strain behaviour. Fiber coatings (by CVD) have also been demonstrated to be beneficial to the performance of fiber composites made by polymer pyrolysis[192]. Chi and colleagues[193] achieved similar properties using polymer infiltration of uniaxial fiber preforms.   As many as 4-6 impregnations were used with diminishing returns for each impregnation, i.e. similar to processing of carbon/carbon composites.

Substantial further opportunity is seen for polymer pyrolysis with a number of potentially important processing developments. First, it has good potential for being of significant utility in making whisker and chopped fiber composites, quite possibly allowing the use of injection moulding and other effective forming procedures. It may also be feasible in some cases to make composites by copyrolysis, that is to pyrolyse both the fiber and the matrix at the same time and, hence, greatly reduce or avoid the shrinkage problems noted above when only the matrix is pyrolysed with an already formed fiber. However, this is likely to be rather limited because of the extensive possibilities for chemical interaction of the fiber and matrix during the copyrolysis.

4.2 Chemical Vapor Deposition/Chemical Vapor Infiltration

Chemical vapor deposition (CVD) is an important example of one of the less used but developing techniques of processing ceramics which is likely to see significant increase in use. Its potential arises both from the intrinsic advantages of CVD as well as potential synergism between CVD and its use for making ceramic composites, as will be discussed below.

While readers are referred to extensive literature on chemical vapor deposition[194] for more detailed background, (see also Chapter 9) it is useful to briefly outline the process. It basically consists of introducing one or more gasses into a chamber such that one and possibly more of the resultant reaction products would be a solid under the reaction conditions utilized. Typically, these conditions are at reduced pressures, e.g. a few to several torr pressure, and elevated temperatures, ranging from a few hundred to nearly 2000°C. Examples of some reactions of interest are shown in Table 9.

The major use of CVD is, of course, to produce powders which are used widely for paint pigments, as well as for ceramic processing. There have also recently been some successful uses of chemical vapor deposition to produce composite powders for processing of ceramic composites, e.g. Al_2O_3-TiO_2 and especially Al_2O_3-ZrO_2, as discussed earlier[48-50,184].

Material	Temperature / °C	Reactants
Si_3N_4	1100 - 1300	$SiCl_4 + NH_3$
BN	800 - 1100	$B_4Cl_3 + NH_3$
SiO_2	800 - 1100	$SiCl_4 + CO_2 + H_2O$
Al_2O_3	800 - 1200	$AlCl_3 + CO_2 + H_2O$
Si	800 - 1100	$SiCl_4$ or $SiHCl_3$
AlN	1000 - 1200	$AlCl_3 + NH_3$

Table 9: CVD/CVI materials and parameters.

Basically, temperatures for chemical vapor deposition can be divided into two regimes. Those done below 1000°C typically utilize organometallic precursors, whereas those conducted above about 1000°C typically utilize inorganic precursors. These two categories also correspond with basic cost factors. Generally, organometallic precursors are substantially more expensive than the inorganic precursors, but are still within the cost range for a variety of applications. Use of low cost precursors contributes to one of the important potential advantages of CVD, namely its overall low cost. Thus, for example, methyltrichlorolsilane, which can be used as the sole precursor for SiC, costs about $1/lb. Its cost is low, despite good purity, because of its other industrial uses. Since the theoretical yield of SiC from this precursor is about 1/3 lb and, in practice, one should be able to obtain about 1/5 lb or more, the process clearly has the potential for being low cost from the raw materials standpoint. Further, since the cost of facilities for CVD are relatively modest and the process has the potential of being a net shape process, its potential for reasonable cost is further enhanced. Deposition rates are, of course, an important factor in the resultant cost and practicality of the process. These vary widely from on the order of microns per hour to microns per second. High deposition rates, however, are frequently not practical because they commonly lead to poor microstructural control, such as large grains and/or large growth cones, and may also lead to porosity entrapment due to highly irregular growth fronts. Thus, the low deposition rates often quoted for CVD appear to reflect more the limitation of understanding how to fully control the process, and less its potential. An important problem that frequently can arise in CVD (as well as other deposition processes) is residual stresses. Such stresses, which can be extreme and appear to arise from variations in stoichiometry and deposition conditions, have been major limitations in the use of CVD to date. CVD offers important potential for processing of particulate composites due to the fact that, in principle, one can get extremely homogeneous mixing of reactant species in the gas phase.   This should then lead to homogeneous codeposition of two or more phases to form a resultant composite directly from the deposition process. Certainly, one of the easiest and possibly the only practical method of making particulate composites is to have both phases deposited out from reactions that occur in similar temperature and pressure ranges to reduce the limitations of using CVD for making conventional ceramics.   Such codeposition may give significant enhanced microstructural nucleation due to its two phase character, thus limiting grain and growth cone sizes, and possibly residual stresses.

To date, there has been only modest investigation of direct chemical deposition of particulate composites. Earlier efforts, e.g. on SiC-C[195] and ZrC-C[196] were directed more at microstructural control and CVD variations than on clear composite concepts. This author and colleagues conducted some preliminary trials of SiC with each of several metals added in the CVD reaction, e.g. Ti, W, and B. More recent work at Oak Ridge National Laboratories[197] has been on SiC with some added phases and studies in Japan[198] have been on Si_3N_4-TiN.

Another possibility for using CVD for forming particulate (and possibly whisker or fiber) composites that has received little or no exploration is to feed the particulate phase onto the substrate while the matrix is being simultaneously chemically vapor deposited. A key issue, of course, is getting the particulates (whiskers or chopped fibers) to "stick in place" (not an issue for continuous fibers). This might be possible by using electrostatic techniques to attract and hold the particulates (whiskers or chopped fibers) on the deposition surface. However, another issue also arises, and that is whether or not there would be adequate deposition underneath the particles (whiskers or fibers) to bond it to the matrix, since circulation and temperature gradients may not be sufficient to cause complete deposition over the volume between the particle and the surface on which it is held.

The most extensive utilization of chemical vapor deposition for ceramic composites has been for making continuous fiber composites. In this case, the typical practice is to form a fiber preform and infiltrate this by CVD to form the matrix.   Because of this infiltration aspect, the process is often referred to as chemical vapor infiltration (CVI).   The preforms may be formed by a variety of processes such as cloth lay-up, filament winding, vacuum forming, or various weaving and braiding processes.

CVI offers the same basic synergisms for making fiber composites as for particulate composites, since the fibers limit grain and growth cone sizes by providing nucleation cites and actual physical limitations to such growth. Such inherent microstructural control should allow shorter deposition times via use of higher deposition rates and by deposition occurring throughout the composite simultaneously.   Thus, even if one has limited rates of deposition, since this is commonly occurring on each fiber simultaneously, the overall rate of building up a composite is greatly enhanced.   The fiber structure also limits the potential of residual stresses through enhanced nucleation and especially by load transfer and toughening via the fibers.   An important challenge and limitation of CVI is the fact that deposition on the fibers typically leads to substantial residual porosity as a result of the buildup on the individual fibers closing off gas paths between them leaving the interstices unfilled as illustrated in Figure 22.   A) schematically illustrates chemical vapour deposition (D) occurring on individual fibers (F) so the matrix builds up approximately concentrically on them until the growing matrix section on the fibers intersect, leaving a pore in the interstices (I) between the original fibers.   B) is a micrograph showing a CVI composite microstructure using 10 μm diameter fibres, i.e. illustrating deposition on the fibers.

After hot pressing, CVI has been the most extensively used method of forming fiber composites, but almost exclusively for continuous fiber composites.   Clearly, extensive work on carbon-carbon composites and limited work on BN-BN composites[200] by CVD/CVI have served as a model for such fabrication of other ceramic fiber composites.   CVI has been used for fabricating probably more complex shapes and the largest composites.   For example, utilizing tubes of braided oxide fibers with a SiC matrix deposited on them at temperatures probably on the order of 1100°C, Figure 23.   (Picture B shows; a carbon-carbon disk (Concorde) brake, Al_2O_3 - SiC heat exchanger tube section, Al_2O_3 - SiC thermocouple well, and C - SiC rocket nozzle and exit cone.)

The great majority of recent CVI fiber composites have been with SiC matrices, mainly by four organizations.   The most extensive and successful

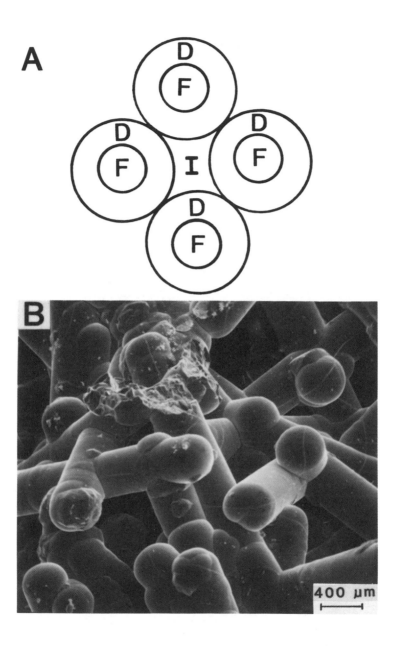

Figure 22: Porosity in composites made by chemical vapor infiltration; see text. (Micrograph courtesy of Drs A. Caputo of ORNL and W. Lackey of Georgia Tech.)

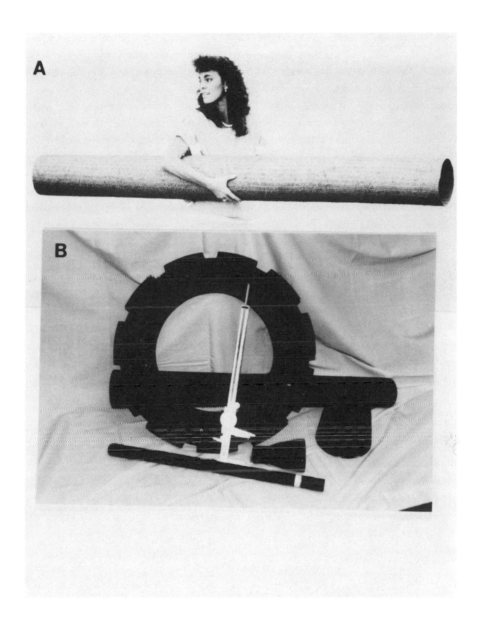

Figure 23: Examples of ceramic fiber composites made by chemical vapor infiltration (CVI); see text.   A) courtesy of R. Fisher of Amercom; B) courtesy of J. Warren of RCI.

work has been by investigators at SEP using C or SiC (Nicalon) fibers[201-204]. They have apparently carried out the SiC infiltration at maximum temperatures of about 1000-1200°C in several stages to control microstructure better, especially porosity levels. While values as low as about 5% porosity has been apparently achieved, typical values of 15-20% appear to be much more common. Both uniaxial and cross-plied tapes, as well as cloth lay-ups, have been successfully infiltrated. Room temperature flexural strengths in the 350-700 MPa range have been achieved with good toughnesses and the non-catastrophic failure expected of such fiber composites. Potentially even more significant is the good strength retention after 100-200 hrs. at 1200°C (although it dropped off more rapidly with longer exposure). Such retention, though not as extensive as desired, is better than other ceramic fiber composites, possibly due to the deposition protecting the fibers.

Fisher and colleagues[205,206] of Amercom have investigated CVI of SiC into SiC (Nicalon) fibers and oxide (Nextel) fibers, the latter especially in braided tube form. Strengths thus far have been at or below 150-200 MPa, but some impressive size parts have been made (see Figure 23A). Warren and colleagues[207] of RCI have also infiltrated SiC into both SiC (Nicalon) and C fibers (often in cloth form) achieving room temperature strengths at 300-500 MPa, again with good fracture toughness and the non-catastrophic failure expected of fiber composites.

More recently, investigators at ORNL[208-211] have reported similar strengths by CVI of SiC into unidirectional SiC (Nicalon) or C fibers. Fitzer and Gadow[212] obtained higher strengths (700 MPa) using CVI of SiC with C fibers versus over 900 MPa with SiC (CVD) filaments. However, the composites were respectively nearly, and fully brittle (i.e. catastrophic) in their failure mode.

Colmet et al[213] have infiltrated Al_2O_3 (via $AlCl_3$-H_2-CO_2) into various Al_2O_3 fibers (DuPont, FP, ICI, and Nextel AB-312). Low deposition pressures and temperatures (950-1000°C) gave porosities as low as 12% with 50 vol% fibers. Strengths were typically 100-200 MPa at room temperature but started decreasing rapidly at ≤1000°C, and were down to 50 MPa when tested at 1400°C.

Another use for CVD and processing of ceramic fiber or whisker composites is to coat the fibers or whiskers. Thus, for example, Rice and colleagues[172-177] have demonstrated significant benefits of coating ceramic fibers with BN by CVD with resultant significant improvement in many cases in both the strength and toughness of using such coated fibers in the resultant composite, as discussed earlier, Figure 17. Clearly, such coating of fibers and whiskers can, in many cases, also be done by other methods, such as other vapor deposition techniques, as well as liquid chemical routes. However, overall CVD is the most general and versatile method.

4.3 Melt Processing

The high melting temperatures of most ceramics pose serious challenges to melt processing of ceramic composites. The frequently very high liquid to solid shrinkages of many ceramic materials on solidification (typically 10-25 vol%)[214] can also be an important limitation. Despite this, there are at least three ways in which some selected ceramic composites may be processed using melting techniques, aided by advances in melting technology. Melt processing of composites commonly draws upon the uniform solutions achieved in the melt phase and subsequent solidification of eutectic structures or precipitation in the solid state on or after solidification. One of the most obvious applications of melt processing is to produce powders for processing of composites, as discussed by Rice and colleagues[137,215,216]. Not surprisingly, the most extensively investigated melt-derived powders are $Al_2O_3-ZrO_2$[215-220] and PSZ[213-214]. The eutectic structure of $Al_2O_3-ZrO_2$ melts providing a significantly different and much more homogeneous mixing of the two ingredients than does simply mixing of Al_2O_3 and ZrO_2 powders process via the conventional approach discussed in earlier sections. Figure 24 illustrates this, A) showing a polycrystalline body consolidated from melt-derived particles. Note the fracture roughness on both the particle and sub-particle scale (the latter due to the eutectic structure). B) shows more detail of the eutectic microstructure.

The challenge is clearly to obtain melt-derived powder particles sufficiently small for good sinterability, but larger than the precipitate or lamellae dimensions. Since the latter are typically less than 1μm for ZrO_2 toughening this should be feasible, and initial studies support this. A possible practical aid in this balance between particle size and phase dimensions is atomization of a molten stream. Splatting molten particles increases cooling rates, giving finer structures than solidification of bulk ingots and aids grinding to desired particle sizes. Whether this technique will be required remains to be determined. This author and colleagues[137,215,216]

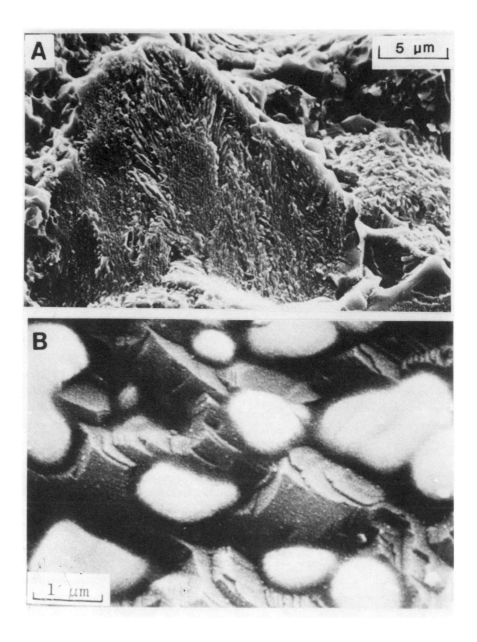

Figure 24: Microstructure of Al_2O_3 – ZrO_2 composite obtained using eutectic melt processing; see text.

have previously hot pressed melt-derived Al_2O_3-ZrO_2 abrasive grits into a solid body achieving respectable strengths (over 500 MPa). These strengths were, however, believed to have been limited by several percent porosity in the resultant body due to the relatively coarse nature of the particles that were hot pressed. Further, the material was significantly reduced, due to the original casting of the abrasive material in a graphite mould, as well as the graphite environment in subsequent hot pressing. Krohn et al[220] similarly hot pressed melt-derived Al_2O_3-ZrO_2 eutectic powder at 1450°C (for 4 min.) and obtained strengths >800 MPa and fracture toughnesses of 15 MPam$^{\frac{1}{2}}$.

This author has also proposed that similar processing of PSZ powders, especially those partially stabilized with Y_2O_3, may offer significant advantages for developing precipitation toughened PSZ for two reasons. First, the achievement of solid solution by normal processing requires high temperatures which results in significant grain growth. Thus, grain sizes of at least 50-100 µm are typical in precipitation toughened PSZ. This substantial grain growth is often accompanied by accumulation of impurities as well as some excess stabilizer at the grain boundaries which appears to be a factor in limiting strengths[221]. Second, temperatures needed for solid solution are quite high and difficult to achieve. This is especially true for PSZ systems other than those using CaO or MgO as stabilizers. Thus, the solid solution temperatures for systems utilizing Y_2O_3 which offer greater potential are well beyond the achievement of most processing facilities.

The second method of melt processing composites is to infiltrate compacts consisting of the second phase. This may be the most limited melt approach in that it is most likely to be successful mainly with particulate composites. However, infiltration of B_4C + C compacts with molten Si to make B_4C + SiC composites is one practical example. Similarly, infiltration of compacts of carbon fibers and powder with molten Si has been used by Hillig and colleagues[222] to make C-SiC composites. Hillig[223] has since been exploring other possible methods of making ceramic composites by molten matrix infiltration using model systems.

The third method of melt processing of composites is solidification of the composite from the melt. The extensive work on fusion-cast refractories, typically using multiphase compositions, with many of these involving eutectic structures, some for very refractory non-oxide systems[224] are a guide to some of the possibilities. More recently, Rice and colleagues[137] have obtained strengths of about 600 MPa in limited studies of PSZ materials, and about 500 MPa for Al_2O_3-ZrO_2 eutectics solidified from the melt. Krohn et al[220] report strengths of about 400 MPa and fracture toughnesses of 5 $MPam^{\frac{1}{2}}$ for their solidified eutectic specimens. The challenge is to keep grain sizes sufficiently fine and control the amount and size and especially location of solidification pores.

A major method of controlling solidification porosity is directional solidification. Such solidification of eutectics results in unidirectional lamellar or rod "reinforced" composites, wherein the thickness of the lamellae or rod diameters are inversely proportional to the solidification rates, as discussed in various reviews[225,226]. Also important are the crystallographic relation of the two phases and the frequency and nature of growth variations. From a practical standpoint, the limitation of such directionally solidified bodies to cylinders has been a basic limitation. However, use of crystal growth shaping techniques[227] may offer important opportunities for some practical shaping.

As expected, directional solidification has been explored most for oxide systems[225-236] with ZrO_2 containing, especially Al_2O_3-ZrO_2, compositions being very common. However, non-oxide systems have also been studied[237-239]. Most studies have focused on growth-orientation microstructure relations, but some have indicated potentially interesting mechanical properties. Thus, for example, Hulse and Batt[229] reported work of fracture values for Al_2O_3-ZrO_2 (Y_2O_3) eutectic compositions at least twice those measured for commercial polycrystalline Al_2O_3. Similarly, strengths of 200-700 MPa were reported for various ZrO_2 containing eutectics (commonly having lamellar spacings of a few microns) with solidification rates of a few cm/hr to tens of cm/hr, generally with limited, or no decrease in these strengths to temperatures of at least 1600°C. Fracture toughnesses of 4-8 $MPam^{\frac{1}{2}}$ have been reported[235] with the higher values being obtained with decreasing yttria content. A ZrC-ZrB_2 eutectic was reported to be grown pore and crack free between 0.4 and 5.4 cm/hr. Studies of these bodies showed a maximum in fracture toughness of nearly 5.5 $MPa^{\frac{1}{2}}$, as well as of hardness and wear resistance at a lamellar spacing of 1.85 µm (2.9 cm/hr growth rate)[239]. Another example of directionally solidified oxide eutectics is $NaNbO_3$-$BaTiO_3$, which is reported to have a promising electro-optic effect[236].

Another approach to melt processing of composites is preparation of a glass, then crystallizing it. While this introduces the large subject of so-called "glass-ceramics" (a poor and contradictory term) which is beyond the scope of this review, a few very relevant examples are in order. Thus, some investigators have made ZrO_2 containing glass that allow tetragonal ZrO_2 precipitates to be formed (e.g. references 240-243). Few properties have been reported, but Mussler and Shafer[242] report fracture toughness increasing by $\approx 100\%$ with up to 12.5% ZrO_2 obtained in a cordierite-based glass body made by melting at 1650°C, then sintering at 860-1100°C. More recently, Leatherman[243] reports room temperature fracture toughness increases of approx. 50% in a $Li_2O-SiO_2-ZrO_2$ glass with approx. 12 vol% fine tetragonal ZrO_2 precipitates. Particularly promising is the directional solidifications of a $CaO-P_2O_5$ glass ($CaO/P_2O_5 = 0.94$) at 20 µm/min. at 560°C, by Abe et al[244]. This resulted in room temperature strengths of up to 600 MPa and non-catastrophic failure, see Figure 25.

There are clearly other ramifications and extensions of the above melt processing. A direct extension is broader investigation of rapidly quenched materials, e.g. in the $Al_2O_3-SiO_2-ZrO_2$ system[245,246], while investigations of other systems giving fibrous structures on solidification[247] and melting

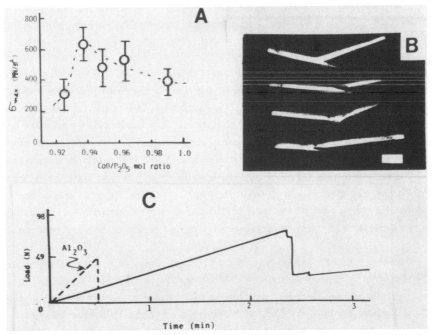

Figure 25: Summary of results of directionally solidified calcium phosphate glass. Courtesy of Abe et al (ref 244), published with permission of the American Ceramic Society.

and solidification of composite surfaces are others. While porosity, cracking, and shape are issues, some possibilities for solving these challenges have been explored with some success[215].

5. SUMMARY, NEEDS AND TRENDS

Since the field of ceramic composites, in the modern context of designing composites for specific performance improvement, is relatively new, much of the effort, to date, has been in the form of preliminary investigations. The purpose of such initial development is, of course, to sort out major aspects of composite performance and processing capabilities to indicate the best level of performance to be expected, to determine size and shape capabilities of various processes and their practicality, and to identify key needs which will allow understanding and improvement in the performance of composites.

The purpose of this review has been to summarize this wealth of information. From the overall performance standpoint, we have particulate composites which may increase strength, fracture toughness or both, or alternatively provide good improvement in thermal shock resistance although at the expense of some reduction in strength and possibly fracture toughness. However, in all cases, these composites fail catastrophically. Whisker composites, to date, have typically performed in a similar fashion to particulate composites. Thus far, continuous fiber composites have been the only ceramic materials clearly demonstrating significant non-catastrophic failure (in the absence of plastic deformation). A key question for whisker and other short fiber composites is whether they can be made to exhibit significant non-catastrophic failure, e.g. by increasing the level, or the degree of orientation of whisker additions, or both (controlled orientation may be necessary in part, to achieve much higher levels of whisker loadings).

Another basic issue is that of the frequent, high temperature embrittlement of continuous ceramic fiber composites and the extent to which this may be controlled by processing, for example the possibility that embrittlement may be less fast or as serious in composites made by chemical vapor infiltration as opposed to other processes. These latter questions heavily focus around the issue of the mechanism of toughening in these

composites and in particular, the issue of fiber pull-out. This, in turn
leads to the issue of fiber-matrix interfaces. As discussed earlier for
continuous composites, this interface is increasingly being recognized as a
critical and complex one. The mechanisms by which a carbon-rich surface
forms on Nicalon fibers is not yet fully understood and, hence, how this can
be more appropriately controlled and how, in turn, this can guide better use
of other fibers in other composites is not yet fully determined. However,
the results on using fiber coatings are clearly promising and an important
step in this direction, but require substantial further work. It is also
beginning to be evident that the surface chemistry of whiskers is important
to their performance in composites with some limited evidence that whiskers
with higher carbon contents near the surface may again be preferred.
Clearly, both this as well as the analogy with continuous fibers would
strongly suggest investigation of whisker coatings. Some preliminary studies
are underway by this author and other investigators' laboratories, but no
significant results are known to date.

A major trend shown in this review is the shift away from (pressureless)
sintering as the dominant method of processing to other methods ranging from
hot densification to a variety of chemically-based methods ranging from
reaction processes to polymer pyrolysis and chemical vapor deposition, as
well as utilization of other methods, such as melt processing. While a
dispersed phase in a composite typically provides grain growth control for
the matrix and, in some cases, may even aid densification (e.g. with ZrO_2
toughened composites), the performance of all composites is still limited by
processing defects. Further, particulate composites and also probably
whisker and short fiber composites (at least as long as they exhibit
catastrophic failure) tend to fail from processing defects. While these
composites typically exhibit improved mechanical performance, the statistics
of mechanical failure and, hence, the uniformity of processing must be
adequately controlled to take full advantage of these improvements. The
single most common method of processing composites, to date, has been by hot
pressing. Clearly, sintering improvements will continue to come and, hence,
expand the application of sintering, with sintering to closed porosity
followed by HIPing being a definite trend. This route is substantially more
challenging with whisker and short fiber composites, and especially
continuous fiber composites, so that hot pressing is likely to remain a
common mode of processing. For glass-based matrices (and possibly very fine
grain polycrystalline matrices) the concept of hot injection of the matrix
into a preform appears quite promising. A challenging and exciting extension

of this could be the extent to which such processing of continuous fiber composites could be made by processes such as extrusion or pultrusion to make rods, tubes, beams, etc.

The challenges of processing continuous fiber composites by powder-based methods, especially sintering, have led to the investigation of a variety of other processing techniques. These offer a range of capabilities in terms of composite character, constituent materials, and composite sizes and shapes. After hot pressing, the next most common method of processing continuous fiber composites has been by CVD/CVI. While this method typically cannot reduce porosity below the 10-20% level, it has potential for being cost effective and clearly can produce substantial sizes and considerable variation in shape. It may also provide some advantages in terms of avoiding oxidation embrittlement at high temperatures due to the deposition tending to encircle the individual fibers and, hence, potentially giving them greater protection. This protection may also result, in part, from the porous nature providing alternate paths for microcracking, i.e. between the pores rather than through the deposition to the fibers allowing environmental attack of the fiber matrix interfaces. Clearly, much more needs to be known about how to achieve uniformity and control porosity gradients for this type of processing. Nonetheless, this technique appears to have significant potential for future processing of continuous fiber composites.

Polymer pyrolysis also appears to hold considerable promise for processing continuous fiber composites. However, the very limited availability of even a few polymer precursors, mostly in the early stages of their development, leaves a great many unknown factors. These include much understanding of the size and shape of composites that can be practically achieved by this route because of very limited information on shrinkage and gas evolution details during pyrolysis. Similarly, developing composites with cementitious based matrices and by reaction processing appear promising. However, the levels of porosity, their uniformity, fiber attack by matrix constituents, etc., are all unknown. Finally, melt and related processing, e.g. by directional solidification, have promise within the substantial constraints of size and shape (mainly approximately prismatic bodies), to which this can be applied.

Clearly, beyond the overall generic issues of processing, there are many specifics of each processing method that need to be filled in. These were beyond the scope of this paper, both because of length as well as the basic level of knowledge in this area. Considerable chemical sophistication has already been brought to bear in processing of particulate composites, but more development is required. Trade-offs as well as possible combinations with other methods, in particular melt derived powders and possibly some direct melt processes, are clearly one of many areas for further exploration and development. Similarly, considerable work is now underway in developing more processing details for whisker composites. Thus, work appears to be getting underway to improve the definition of issues such as the relevance of matrix powder particle size distribution, whisker aspect ratios, and particularly, the interrelationship of particle and whisker dimensions. Also, as noted earlier, the important issue of controlling interfacial chemistry has begun to be fairly extensively explored in continuous fiber composites and is now beginning to be looked at for whisker composites.

Major issues remaining in the development of composites are finding suitable applications and achieving the cost performance necessary for these. Clearly, both are directly coupled to processing issues. A fairly encouraging start has been made with whisker composites wherein there is already a whisker toughened cutting tool on the market. This is a relatively high value added application which is an excellent place to introduce high-technology materials. This follows on the use of other composite materials, i.e. particulate composites, for cutting tools and special wear components. However, for the field to significantly further develop, it will require other high value added applications as well as a spectrum of progressively less high value added applications that can be addressed as both material costs and processing costs are reduced by further development.

Most of the field of ceramic composites is still in an early state of development. However, the very encouraging progress to date, combined with the increasing recognition of the potential advantages of ceramics, is expected to provide substantial and growing stimuli for further development. Such development will, of course, be challenging to not only meet these goals, but also to meet the competition from other materials which will range from improving polymeric based composites, developing metal matrix composites, as well as continued improvement of more traditional ceramics, as well as ceramic coatings (mainly on metal substrates). However, while there will be competition between ceramic composites and other more traditional ceramics, there will also be synergism, for example, from the observations of processing improvements obtained from microstructural refinement resulting from low levels of second phase additions. Continued development of ceramic composites and the interaction of this with other improvements in ceramic technology are expected to play major roles in broadening the use of ceramics.

ACKNOWLEDGEMENT

The extensive aid and patience of Mrs. Ruth Ann desJardins in assembling the many pieces that ultimately made up this review, as well as the aid of Mrs. Linda G. Talbott in assembling the references, is greatly appreciated.

REFERENCES

1. Rice, R.W., A Material Opportunity: Ceramic Composites. Chemtech 13 230-239 (1983).

2. Newnham, R.E., Composite Electroceramics. Chemtech 732-739 (Dec 1986).

3. Rice, R.W., Mechanisms of Toughening in Ceramic Matrix Composites. Ceram. Eng. and Sci. Proc. 2 [7-8] 661-701 (1981).

4. Rice, R.W., Ceramic Matrix Composite Toughening Mechanisms: An Update. Ceram. Eng. and Sci. Proc. 6 [7-8], 589-607 (1985).

5. Evans, A.G. (Ed), Fracture in Ceramic Materials: Toughening Mechanisms, Machining Damage, Shock; Noyes Publications (1984).

6. Payne, D.A., Ph.D. Thesis, Pennsylvania State University, (1973).

7. Bommgaard, J. and Born, R.A.J., A Sintered Magnetoelectric Composite Material $BaTiO_3$-Ni(Co,Mn) Fe_2O_4. J. Mat. Sci. 13 1538-1548 (1978).

8. Fulrath, R.M., In Ceramics for Advanced Technologies; Hove, J.E. and Riley, W.C. (Eds), Wiley (1965).

9. Hasselman, D.P.H. and Fulrath, R.M., Micromechanical Stress Concentrations in Two-Phase Brittle-Matrix Ceramic Composites. J. Am. Ceram. Soc. 50 [8], 399-404 (1967).

10. Stett, M.A. and Fulrath, R.M., Mechanical Properties and Fracture Behavior of Chemically Bonded Composites. J. Amer. Ceram. Soc. 53 [1], 5-13 (1970).

11. Binns, D.B., Some Physical Properties of Two-Phase Crystal-Glass Solids. Sci. of Ceram. 1 316-334 (1962).

12. Union Carbide Corporation, Finely-Divided Metal Oxides and Sintered Objects Therefrom. UK Patent 1,345,631, May 10 (1971).

13. Rice, R.W. and McDonough, W.J., Ambient Strength and Fracture of ZrO_2. Mechanical Behavior of Materials IV, 394-403, The Society of Materials Science, Japan (1972).

14. Rieth, P.H., Reed, J.S. and Naumann, A.W., Fabrication and Flexural Strength of Ultrafine-Grained Yttria-Stabilized Zirconia. Amer. Ceram. Soc. Bull. 55 [8], 713-728 (1976).

15. Scott, C.D. and Reed, J.S., Effect of Laundering and Milling on the Sintering Behavior of Stabilized ZrO_2 Powders. Amer. Ceram. Soc. Bull. 58 [6], 587-590 (1979).

16. Gupta, T.K., Bechtold, J.H., Kuznicki, R.C., Cadoff, L.H. and Rossing, B.R., Stabilization of Tetragonal Phase in Polycrystalline Zirconia. J. Mat. Sci. 12, 2421-2426 (1977).

17. Gupta, T.K., Lange, F.F. and Bechtold, J.H., Effect of Stress-Induced Phase Transformation on the Properties of Polycrystalline Zirconia Containing Metastable Tetragonal Phase. J. Mat. Sci. 13 1464-1470 (1978).

18. Rice, R.W., In Processing of Crystalline Ceramics; Palmour III, H., Davis, R.F. and Hare, T.M. (Eds), Plenum Press (1978).

19. Lewis, III, D., Huynh, T-C.T. and Reed, J.S., Processing Defects in Partially Stabilized Zirconia. Amer. Ceram. Soc. Bull. 2, 244-245 (1980).

20. Tsukuma, K. and Shimada, M., Hot Isostatic Pressing of Y_2O_3-Partially Stabilized Zirconia. Amer. Ceram. Soc. Bull. 64 [2], 310-313 (1985).

21. Coyle, T.W., Coblenz, W.S. and Bender, B.A., Transformation Toughening in Large Grain Size CeO_2 Doped ZrO_2 Polycrystals. To be published.

22. Tsukuma, K., Mechanical Properties and Thermal Stability of CeO_2 Containing Tetragonal Zirconia Polycrystals. Amer. Ceram. Soc. Bull. 65 [10], 1386-1389 (1986).

23. Tsukuma, K., Ueda, K. and Shimada, M., Strength and Fracture Toughness of Isostatically Hot-Pressed Composites of Al_2O_3 and Y_2O_3-Partially-Stabilized ZrO_2. J. Amer. Ceram. Soc. 68 [1] C4-5 (1985).

24. Tani, E., Yoshimura, M. and Somiya, S., Hydrothermal Preparation of Ultrafine Monoclinic ZrO_2 Powder. J. Amer. Ceram. Soc. 64 C18 (1981).

25. Tani, E., Yoshimura, M. and Somiya, S., Formation of Ultrafine Tetragonal ZrO_2 Powder Under Hydrothermal Conditions. J. Amer. Ceram. Soc. 66 [1], 11-14 (1983).

26. Yoshimura, M. and Somiya, S., Fabrication of Dense, Nonstabilized ZrO_2 Ceramics by Hydrothermal Reaction Sintering. Amer. Ceram. Bull. 59 [2], 246 (1980).

27. King, A.G. and Yavorsky, P.J., Stress Relief Mechanisms in Magnesia- and Yttria-Stabilized Zirconia. J. Amer. Ceram. Soc. 51 [1], 38-42 (1968).

28. Johns, H.L. and King, A.G., Zirconia Tailored for Thermal Shock Resistance. Ceramic Age 29-31 (May 1970).

29. Yavorsky, P.J., Properties and High Temperature Alications of Zirconium Oxide. Ceramic Age 64-69 (June 1962).

30. Gulati, S.T., Helfinstine, J.D. and Davis, A.D., Determination of Some Useful Properties of Partially Stabilized Zirconia and the Alication of Extrusion Dies. Amer. Ceram. Soc. Bull. 59 [2] 211-219 (1980).

31. Jaeger, R.E. and Nickell, R.E., Thermal Shock Resistant Zirconia Nozzles for Continuous Core Casting. Mats. Sci. Res. vol 5; Kriegel, W.W. and Palmour III, H. (Eds), Plenum Press (1971).

32. Swain, M.V., Garvie, R.C., Hannink, R.H.J., Hughan, R. and Marmack, M., Material Development and Evaluation of Partially Stabilized Zirconia for Extrusion Die Alications. Proc. Brit. Ceram. Soc. 32 343-353 (March 1982).

33. Garvie, R.C., Hannink, R.H. and Pascoe, R.T., Ceramic Steel. Nature 258 763-764 (1975).

34. Garvie, R.C. and Nicholson, P.S., Structure and Thermo-mechanical Properties of Partially Stabilized Zirconia in the CaO-ZrO$_2$ System. J. Amer. Ceram. Soc. 55 [3] 152-157 (1972).

35. Garvie, R.C., Hannink, R.H.J. and Pascoe, R.T., Calcia Stabilized Zirconia. Australian Patent 503,775, October 13 (1975).

36. Garvie, R.C., Hughan, R.R. and Pascoe, R.T., Strengthening of Lime-Stabilized Zirconia by Post Sintering Heat Treatments. Mat. Sci. Res. 11 263-274 (1978).

37. Garvie, R.C., Hannink, R.H.J. and Urbani, C., Fracture Mechanics Study of a Transformation Toughened Zirconia Alloy in the CaO-ZrO$_2$ System. Ceram. Int'l. 6 [1] 19-24 (1980).

38. Hannink, R.H.J. and Swain, M.V., Magnesia-Partially Stabilised Zirconia: The Influence of Heat Treatment on Thermomechanical Properties. J. Australian Ceram. Soc. 18 [3] 53-62 (1982).

39. Hannink, R.H.J. and Garvie, R.C., Sub-Eutectoid Aged Mg-PSZ Alloy with Enhanced Thermal Up-Shock Resistance. J. Mat. Sci. 17 2637-2643 (1982).

40. Hannink, R.H.J., Microstructural Development of Sub-Eutectoid Aged MgO-ZrO$_2$ Alloys. J. Mat. Sci. 18 457-470 (1983).

41. Claussen, N., Fracture Toughness of Al2O$_3$ with an Unstabilized ZrO$_2$ Dispersed Phase. J. Amer. Ceram. Soc. 59 [1-2] 49-54 (1976).

42. Claussen, N., Steeb, J. and Pabst, R.F., Effect of Induced Microcracking on the Fracture Toughness of Ceramics. Amer. Ceram. Soc. Bull. 56 [6] 559-562 (1977).

43. Rice, R.W., Ceramic Composites - Processing Challenges. Ceram. Eng. and Sci. Proc. 2 [7-8] 493-508 (1981).

44. Dworak, U. and Olapinski, H., Sintered Ceramic Material of Improved Ductility. U.S. Patent 4,218,253, October 2 (1978); assigned to Feldmuhle Aktiengesellschaft.

45. Becher, P.F., Transient Thermal Stress Behavior in ZrO$_2$- Toughened Al$_2$O$_3$. J. Amer. Ceram. Soc. 64 [1] 37-39 (1981).

46. Lange, F.F., Transformation Toughening Part 4: Fabrication, Fracture Toughness and Strength of Al$_2$O$_3$-ZrO$_2$ Composites. J. Mat. Sci. 17 247-254 (1982).

47. Lange, F.F., Al2O$_3$/ZrO$_2$ Ceramic. U.S. Patent 4,316,964, July 14 (1980); assigned to Rockwell International Corporation.

48. Hori, S., Yoshimura, M., Somiya, S. and Takahashi, R., Al$_2$O$_3$-ZrO$_2$ Ceramics Prepared from CVD Powders. In Science and Technology of Zirconia II, Advances in Ceramics, vol. 12; Claussen N. and Heuer A. (Eds), 794-805, Amer. Ceram. Soc. (1984).

49. Hori, S., Kurita, R., Kaji, H., Yoshimura, M. and Somiya, S., ZrO$_2$-Toughened Al$_2$O$_3$; CVD Powder Preparation, Sintering and Mechanical Properties. J. Mat. Sci. 4 [4], 413-416 (1985).

50. Hori, S., Yoshimura, M. and Somiya, S., Strength Toughness Relations in Sintered and Isostatically Hot-Pressed ZrO$_2$-Toughened Al$_2$O$_3$. J. Amer. Ceram. Soc. 69 [3] 169-172 (1986).

51. Sproson, D.W. and Messing, G.L., Preparation of Alumina- Zirconia Powders by Evaporative Decomposition of Solutions. J. Amer. Ceram. Soc. 67 C92 (1984).

52. Kagawa, M., Kikuchi, M. and Syono, Y., Stability of Ultrafine Tetragonal ZrO$_2$ Coprecipitated with Al$_2$O$_3$ by the Spray-ICP Technique. J. Amer. Ceram. Soc. 66 [11] 751 (1983).

53. Somiya, S., Yoshimura, M. and Kikugawa, S., Preparation of Zirconia-Alumina Fine Powders by Hydrothermal Oxidation of Zr-Al Alloys. Mat. Sci. Rec. 17 155-166 (1984).

54. Lange, F.F., Processing-Related Fracture Origins: I, Observations in Sintered and Isostatically Hot-Pressed Al$_2$O$_3$/ZrO$_2$ Composites. J. Amer. Ceram. Soc. 66 [6] 396-398 (1983).

55. Tsukuma, K., Ueda, K., Matsushita, K. and Shimada, M., High-Temperature Strength and Fracture Toughness of Y$_2$O$_3$-Partially-Stabilized ZrO$_2$/Al$_2$O$_3$ Composites. J. Amer. Ceram. Soc. 68 [2] C56-58 (1985).

56. Aksay, I.A., Lange, F.F. and Davis, B.I., Uniformity of Al$_2$O$_3$-ZrO$_2$ Composites by Colloidal Filtration. J. Amer. Ceram. Soc. 66 C190 (1983).

57. Wilfinger, K. and Cannon, W.R., Processing of Transformation-Toughened Alumina. Proceedings of the 13th Automotive Materials Conference, published by the Amer. Ceram. Soc. (1986).

58. Boch, P., Chartier, T. and Huttepain, M., Tape Casting of Al$_2$O$_3$/ZrO$_2$ Laminated Composites, J. Amer. Ceram. Soc. 69 [8] C191-192 (1986).

59. Lange, F.F. and Hirlinger, M.M., Hindrance of Grain Growth in Al$_2$O$_3$ by ZrO$_2$ Inclusions. J. Amer. Ceram. Soc. 67 [3] 164-168 (1984).

60. Hori, S., Kurita, R., Yoshimura, M. and Somiya, S., Supressed Grain Growth in Final-Stage Sintering of Al$_2$O$_3$ with Dispersed ZrO$_2$ Particles. J. Mat. Sci. Lett 4 1067-1070 (1985).

61. Hori, S., Kurita, R., Yoshimura, M. and Somiya, S., Influence of Small ZrO$_2$ Additions on the Microstructure and Mechanical Properties of Al$_2$O$_3$. Science and Technology of Zirconia III, Advances in Ceramics Vol. 24, 423-430 (1988).

62. Kibbel, B. and Heuer, A.H., Exaggerated Grain Growth in ZrO$_2$-Toughened Al$_2$O$_3$. J. Amer. Ceram. Soc. 69 [3] 231-236 (1986).

63. Sato, T., Shiratori, A. and Shimada, M., Sintering and Fracture behavior of Composites Based on Alumina-Zirconia(Yttria)-Nonoxides. J. de Physique 47, C1-733 (1986).

64. DiRupo, E., Gilbart, E., Carruthers, T.G. and Brook, R.J., Reaction Hot-Pressing of Zircon-Alumina Mixtures. J. Mat. Sci. 14 705-711 (1979).

65. Claussen, N. and Jahn, J., Mechanical Properties of Sintered, In Situ-Reacted Mullite-Zirconia Composites. J. Amer. Ceram. Soc. 63 [3-4] 228-229 (1980).

66. Anseau, M.R., Leblud, C. and Cambier, F., Reaction Sintering [RS] of Mixed Zircon-Based Powders as a Route for Producing Ceramics Containing Zirconia with Enhanced Mechanical Properties. J. Mat. Sci. Lett 2, 366-370 (1983).

67. Prochazka, S., Wallace, J.S. and Claussen, N., Microstructure of Sintered Mullite-Zirconia Composites. J. Amer. Ceram. Soc. 66 C125-126 (1983).

68. Moya, J.S. and Osendi, M.I., Effect of ZrO_2 (ss) in Mullite on the Sintering and Mechanical Properties of Mullite/ZrO_2 Composites. J. Mat. Sci. Lett 2 599-601 (1983).

69. Moya, J.S. and Osendi, M.I., Microstructure and Mechanical Properties of Mullite/ZrO_2 Composites. J. Mat. Sci. 19 2909-2914 (1984).

70. Cambier, F., Baudin DeLaLastra, C., Pilate, P. and Leriche, A., Formation of Microstructural Defects in Mullite- Zirconia and Mullite-Alumina-Zirconia Composites Obtained by Reaction-Singering of Mixed Powders. Brit. Ceram. Trans. and J. 83 [6] 196-200 (1984).

71. Yuan, Q-M., Tan, J-Q. and Jin, Z-G., Preparation and Properties of Zirconia Toughened Mullite Ceramics. J. Amer. Ceram. Soc. 69 [3] 265-267 (1986).

72. Yuan, Q-M., Tan, J-Q., Shen, J-Y., Zhu, X-H. and Yang, Z-F., Processing and Microstructure of Mullite-Zirconia Composites Prepared from Sol-Gel Powders. J. Amer. Ceram. Soc. 69 [3] 268-269 (1986).

73. McPherson, R., Preparation of Mullite-Zirconia Composites from Glass Powder. J. Amer. Ceram. Soc. 69 [3] 297-298 (1986).

74. Lange, F.F., Davis, B.I. and Raleigh, D.O., Transformation Strengthening of β-Al_2O_3 with Tetragonal ZrO_2. J. Amer. Ceram. Soc. 66 C50-51 (1983).

75. Viswanathan, L., Ikuma, Y. and Virkar, A.V., Transformation Toughening of β-Alumina by Incorporation of Zirconia. J. Mat. Sci. 18 109-113 (1983).

76. Binner, J.G.P. and Stevens, R., Improvement in the Mechanical Properties of Polycrystalline Beta-Alumina Via the Use of Zirconia Particles Containing Stabilizing Oxide Additions. J. Mat. Sci. 20 3119-3124 (1985).

77. Green, D.J. and Metcalf, M.G., Properties of Slipcast Transformation-Toughened β-Al_2O_3/ZrO_2 Composites. Amer. Ceram. Soc. Bull. 63 [6] 803-807 & 820 (1984).

78. Claussen, N., In Science and Technology of Zirconia II, Advances in Ceramics, vol. 12; Claussen, N., Rühle, M. and Heuer, A.H. (Eds), 325-351, American Ceramic Society (1984).

79. Ikuma, Y., Komatsu, W. and Yaegashi, S., ZrO$_2$-Toughened MgO and Critical Factors in Toughening Ceramic Materials by Incorporating Zirconia. J. Mat. Sci. Lett $\underline{4}$ 63-66 (1985).

80. Yangyun, S. and Brook, R.J., Preparation and Strength of Forsterite-Zirconia Ceramic Composites. Ceram Int'l $\underline{9}$ [2] 39-45 (1983).

81. McPherson, R., Shafer, B.V. and Wong, A.M., Zircon-Zirconia Ceramics Prepared from Plasma Dissociated Zircon. J. Amer. Ceram. Soc. $\underline{65}$ C57-58, (1982).

82. Yamaguchi, A., Densification of Cr$_2$O$_3$-ZrO$_2$ Ceramics by Sintering, J. Amer. Ceram. Soc. $\underline{64}$ C67 (1981).

83. Mussler, B.H., Shafer, M.W., Preparation and Properties of Cordierite-Based Glass-Ceramic Containing Precipitated ZrO$_2$. Amer. Ceram. Soc. Bull. $\underline{64}$ [11] 1459-1462 (1985).

84. Nogami, M. and Tomozawa, M., ZrO$_2$-Transformation-Toughened Glass-Ceramics Prepared by the Sol-Gel Process from Metal Alkoxides. J. Amer. Ceram. Soc. $\underline{69}$ [2] 99-102 (1986).

85. Nogami, W., Crystal Growth of Tetragonal ZrO$_2$ in the Glass System ZrO$_2$-SiO$_2$ Prepared by the Sol-Gel Process for Metal Alkoxides. J. Mat. Sci. $\underline{21}$ 3513-3516 (1986).

86. Morena, R., Lockwood, P.E., Evans, A.L. and Fairhurst, C.W., Toughening of Dental Porcelain by Tetragonal ZrO$_2$ Additions. J. Amer. Ceram. Soc. $\underline{69}$ [4] C75-77 (1986).

87. Lange F.F., U.S. Patent 4,640,902, May 31 (1985); assigned to Rockwell International Corporation.

88. Omori, M., Takei, H. and Ohira, K., Synthesis of SiC-ZrO$_2$ Composite Containing t-ZrO$_2$. J. Mats. Sci. Lett $\underline{4}$ 770-772 (1985).

89. Watanabe, M. and Kubuura, I., In Ceramic Science and Technology at the Present and in the Future; Somiya S. (Ed) Uchida Pokakuho Publishing Co. Ltd., Japan (1981).

90. Hojo, J, Yokoyama, H., Oono, R. and Kato, A., Sintering Behavior of Alumina-Titanium Carbide Composite Using Ultrafine Powders. Trans. JSCM $\underline{9}$ [2] 37-44 (1983).

91. Hoch, M., Preparation of Al$_2$O$_3$-TiC Ceramic Bodies by the Alkoxide Process. J. de Physique $\underline{47}$ C1-37 (1986).

92. Borom, M.P. and Lee, M., Effect of Heating Rate on Densificaiton of Alumina-Titanium Carbide Composites. Adv. Ceram. Mat. $\underline{1}$ [4] 335-40 (1986).

93. Berneburg, P.L., private communication (1986).

94. Radford, K.C., Sintering Al$_2$O$_3$-B4C Ceramics. J. Mat. Sci. $\underline{18}$ 669-678 (1983).

95. Radford, K.C., Microstructural Aspects of Moisture Adsorption of Al$_2$O$_3$-B$_4$C Compacts. J. Mat. Sci. $\underline{18}$ 679-686 (1983).

96. Cooper, C.G., Graphite Containing Refractories. Refractories Journal $\underline{6}$ 11-21 (1980).

97. Ozgen, O.S. and Rand, B., Effect of Graphite on Mechanical Properties of Alumina/Graphite Materials with Ceramic Bonds. Br. Ceram. Trans. and J. 84 138-142 (1985).

98. Cooper, C.G., Alexander, I.C. and Hampson, C.J., The Role of Graphite in the Thermal Shock Resistance of Refractories. Br. Ceram. Trans. and J. 84 57-62 (1985).

99. Prochazka, S. and Coblenz, W. S., U.S. Patent 4,081,284; August 4 (1976); assigned to General Electric Company.

100. Tanaka, H., Greil, P. and Petzow, G., Sintering and Strength of Silicon Nitride-Silicon Carbide Composites. Int. J. High Tech Ceram 1 107-118 (1985).

101. McMurtry, C.H., Boecker, W.D.G., Seshadri, S.G., Zanghi, J.S. and Garnier, J.E., Microstructure and Material Properties of SiCTiB2 Particulate Composites. Amer. Ceram. Soc. Bull. 66 [2] 325-329 (1987).

102. Becher, P.F. and Wei, G.C., Toughening Behavior in SiC Whisker-Reinforced Alumina. J. Amer. Ceram. Soc. 67 [12] C267-269 (1984).

103. Wei, G.C. and Becher, P.F., Development of SiC Whisker-Reinforced Ceramics. Amer. Ceram. Soc. Bull. 64 [2] 298-304 (1985).

104. Wei, G.C., Silicon Carbide Whisker Reinforced Ceramic Composites and Method for Making Same. U.S. Patent 4,543,345, September 24 (1985).

105. Samanta, S.C. and Musikant, S., SiC Whiskers-Reinforced Ceramic Matrix Composites. Ceram. Eng. and Sci. Proceedings 6 [7-8] 663-671 (1985).

106. Claussen, N., Weisskopf, K.L. and Ruhle, M., Tetragonal Zirconia Polycrystals Reinforced with SiC Whiskers. J. Amer. Ceram. Soc. 69 [3] 288-292 (1986).

107. Shalek, P.D., Petrovic, J.J., Hurley, G.F. and Gac, F.D., Hot-Pressed SiC Whisker/Si_3N_4 Matrix Composites. Amer. Ceram. Soc. Bull. 65 [2] 351-356 (1986).

108. Tamari, N., Kondo, I., Sodeoka, S., Ueno, K. and Toibana, Y., Sintering of Si_3N_4-SiC Whisker Composite. Yogyo-Kyokai-Shi 94 [11] 1177-1179 (1986).

109. Tiegs, T.N. and Becher, P.F., Sintered Al_2O_3-SiC whisker composites. Amer. Ceram. Soc. Bull. 66 [2] 339-42 (1987).

110. Coyle, T.W., Guyot, M.H. and Jamet, J.F., Mechanical behavior of a microcracked ceramic composite. Ceram. Eng. Sci. Proc. 7 [7-8] 947-957 (1986).

111. Vincenzini, P., Bellosi, A. and Babini, G.N., Thermal Instability of Si_3N_4/ZrO_2 Composites. Ceram. Intl. 12 [3] 133-146 (1986).

112. Watanabe, T. and Shoubu, K., Mechanical Properties of Hot-Pressed TiB2-ZrO_2 Composites. J. Amer. Ceram. Soc. 68 [2] C34-36 (1985).

113. Ogawa, K. and Miyahara, M., U.S. Patent 3,580,708, May 25 (1971); assigned to Nion Tungsten Company, Ltd.

114. Rice, R.W., Becher, P.F., Freiman, S.W. and McDonough, W.J., Thermal Structural Ceramic Composites. Ceram. Eng. Sci. Proc. 1 [7-8] 424-443 (1980).

115. Becher, P.F., Lewis III, D., Youngblood, G.E. and Bentsen, L., In Thermal Stresses in Severe Environments; Hasselman, D.P.H. and Heller, R.A. (Eds), 397-411, Plenum Press (1980).

116. Crayton, P.H. and Greene, E.E., Synthesis-Fabrication of an Al₂O₃-Tungsten Carbide Composite. J. Amer. Ceram. Soc. 56 423-426 (1973).

117. Noma, T. and Sawaoka, A., Toughening in Very High Pressure Sintered Diamond-Alumina Composite, J. of Mat. Sci. 19 2319-22 (1984).

118. Rossi, R.C., In Ceramics in Severe Environments, vol. 5; Kriegel W.W. and Palmour III, H. (Eds) 123-125, Plenum Press (1970).

119. Rossi, et al, U.S. Patent 4,007,049, February 8 (1977).

120. Murata, Y and Miccioli, B.R., Thermal and Dielectric Properties of a Boron Nitride-Silica Composite. Amer. Ceram. Soc. Bull. 49 [8] 718-723 (1970).

121. Rice, R.W., McDonough, J., Freiman, S.W. and Mecholsky, J.J., U.S. Patent 4,304,870, December 8 (1981); assigned to United States of America as represented by the Secretary of the Navy.

122. Lewis, D., Ingel, R.P., McDonough, W.J. and Rice, R.W., Microstructural and Thermomechanical Properties in Alumina- and Mullite-Boron Nitride Particulate Ceramic-Ceramic Composites. Ceram. Eng. Sci. Proc. 2 [7-8] (1981).

123. Wahi, R.P., Fracture Behaviour of Two-phase Ceramic Alloys Based on Aluminium Oxide. Trans. Indian Inst. of Metals 34 [2] 89-102 (1981).

124. Wahi, R.P. and Ilschner, B., Fracture Behaviour of Composites Based on Al₂O₃-TiC. J. Mat. Sci. 15 875-885 (1980).

125. Nakahira, A., Niihara, K. and Hirai, T., Microstructure and Mechanical Properties of Al₂O₃-SiC Composites. Yogyo-Kyokai-Shi 94 [8] 767-72 (1986).

126. Lange, F.F., Effect of Microstructure on Strength of Si₃N₄-SiC Composite System. J. Amer. Ceram. Soc. 56 [9] 445-50 (1973).

127. Mah, T-I., Mendiratta, M.G. and Lipsitt, H.A., Fracture Toughness and Strength of Si₃N₄-TiC Composites. Amer. Ceram. Soc. Bull. 60 [11] 1229-1231 (1981).

128. Noma, T. and Sawaoka, A., Fracture Toughness of High- Pressure-Sintered Diamond/Silicon Nitride Composites. J. Amer. Ceram. Soc. 68 [10] C271-273 (1985).

129. Wei, G.C. and Becher, P.F., Improvements in Mechanical Properties in SiC by the Addition of TiC Particles. J. Amer. Ceram. Soc. 67 [8] 571-574 (1984).

130. Janney, M.A., Mechanical Properties and Oxidation Behavior of a Hot-Pressed SiC 15Vol%-TiB₂ Composite. Amer. Ceram. Soc. Bull. 66 [2] 322-324 (1987).

131. Mazdiyasni, K.S. and Ruh, R., High/Low Modulus Si_3N_4-BN Composite for Improved Electrical and Thermal Shock Behavior. J. Amer. Ceram. Soc. 64 [7] 415-19 (1981).

132. Mazdiyasni, K.S., Ruh, R. and Hermes, E.E., Phase Characterization and Properties of AlN-BN Composites. Amer. Ceram. Soc. Bull. 64 [8] 1149-1154 (1985).

133. Valentine, P.G., Palazotto, A.N., Ruh, R. and Larsen, D.C., Thermal Shock Resistance of SiC-BN Composites. Adv. Ceram. Mat. 1 [1] 81-87 (1986).

134. Janney, M.A., Microstructural Development and Mechanical Properties of SiC and of SiC-TiC Composites. Amer. Ceram. Soc. Bull. 65 [2] 357-62 (1986).

135. Kalish, D., Clougherty, E.V., Kreder, K., Strength, Fracture Mode and Thermal Stress Resistance of HfB_2 and ZrB_2. J. Amer. Ceram. Soc. 52 [1] 30-36 (1969).

136. Murata, Y. and Batha, H.D., Densification and Wear Resistance of TaN-ZrB_2-ZrN and TaN-ZrB_2-WC Compositions. Amer. Ceram. Soc. Bull. 60 [8] 818-824 (1981).

137. Rice, R.W., Ingel, R.P., Bender, B.A., Spann, J.R. and McDonough, W.R., Development and Extension of Partially-Stabilized Zirconia Single Crystal Technology. Ceram. Eng. and Science Proc. 5 [7-8] 530-545 (1984).

138. Rice, R.W., In Ultrafine-Grain Ceramics; Burke, J.J., Reed N.L. and Weis, V. (Eds) 203-250, Syracuse University Press, (1970).

139. Kellett, B.J. and Lange, F.F., Hot Forging Characteristics of Fine-Grained ZrO_2 and Al_2O_3/ZrO_2 Ceramics. J. Amer. Ceram. Soc. 69 [8] C172-173 (1986).

140. Sweeting, R.B., High Temperature Deformation of Stabilized Zirconia.

141. Wakai, F. and Kata, H., Superplasticity of TZP/Al_2O_3 Composites. Adv. Ceram. Mat. 3 [1] 71-76 (1988).

142. Becher, P.F., Tiegs, T.N., Ogle, J.C. and Warwick, W.H., In Fracture Mechanis of Ceramics, vol. 7; Bradt, R.C., Evans, A.G., Hasselman, D.P.H. and Lange, F.F. (Eds), 61-73, Plenum Press, (1986).

143. Homeny, J., Vaughn, W.L. and Ferber, M.K., Processing and Mechanical Properties of SiC-whisker/Al_2O_3-matrix Composites. Amer. Ceram. Soc. Bull. 67 [2] 333-38 (1987).

144. Panda, P.C. and Seydel, E.R., Near-Net-Shape Forming of Magnesia-Alumina Spinel/Silicon Carbide Fiber Composites. Amer. Ceram. Soc. Bull. 65 [2] 338-41 (1986).

145. Gadkaree, K.P. and Chyung, K., Silicon Carbide Whisker-Reinforced Glass and Glass-Ceramic Composites, Amer. Ceram. Soc. Bull. 65 [2] 370-376 (1986).

146. Buljan, S.T., Baldoni, J.G. and Huckabee, M.L., Si_3N_4-SiC Composites, Amer. Ceram. Soc. Bull. 66 [2] 347-52 (1987).

147. Becher, P.F. and Tiegs, T.N., Toughening Behavior Involving Multiple Mechanisms: Whisker Reinforcement and Zirconia Toughening. To be published in J. Amer. Ceram. Soc.

148. Krochmal, J.J., Fiber Reinforced Ceramics: A Review and Assessment of Their Potential. Tech. Rept. AFML-TR-67-207, (October 1967).

149. Donald, I.W. and McMillan, P.W., Review Ceramic-Matrix Composites. J. Mat. Sci. 11 949-72 (1976).

150. Hove, J.E. and Davis, H.M., Assessment of Ceramic-Matrix Composite Technology and Potential DOD Alication. Institute for Defense Analyses, Paper P-1307 for Defense Advanced Research Projects Agency, DAHC15 73 C 0200, (1977).

151. Sambell, R.A.J., Bowen, D.H. and Phillips, D.C., Carbon Fibre Composites with Ceramic and Glass Matrices: Part I Discontinuous Fiber. J. Mat. Sci. 7 663-75 (1972).

152. Sambell, R.A.J., Briggs, A., Phillips, D.C. and Bowen, D.H., Carbon Fibre Composites with Ceramic and Glass Matrices. J. Mat. Sci. 7 676-681 (1972).

153. Sambell, R.A.J., Phillips, D.C. and Bowen, D.H., In Carbon Fibers - Their Place in Modern Technology; Proceedings 2nd International Congress. 105-113, Plastics Institute, London (1974).

154. Phillips, D.C., The Fracture Energy of Carbon-Fibre Reinforced Glass. J. Mat. Sci. 7 1175-1191 (1972).

155. Phillips, D.C., Interfacial Bonding and the Toughness of Carbon Fibre Reinforced Glass and Glass-Ceramics. J. Mat. Sci. 9 1847-54 (1974).

156. Phillps, D.C., In Handbook of Composites; Kelly A. and Mlenk, S.T. (Eds) 4 [7] 373-428, North Holland Publishing Company, (1963).

157. Levitt, S.R., High-Strength Graphite Fibre/Lithium Aluminosilicate Composites. J. Mat. Sci. 8 793-806 (1973).

158. Yoshikawa, M. and Asaeda, T., In Proceedings of the 19th Japan Congress on Materials Research 19 224-229, Society of Materials Science, Japan (1976).

159. Yasuda, E. and Schlichting, J., Carbon Fibre Reinforced Al_2O_3 and Mullite. Z. Werkstofftech 9 310-315 (1978).

160. Fitzer, E. and Schlichting, J., Fiber-Reinforced Refractory Oxides. High Temp. Sci. 13 149-172 (1980).

161. Brennan, J.J., Development of Fiber Reinforced Ceramic Matrix Composites. United Technologies Research Center Report No. R911848-4 for U.S. Naval Air Systems Command, Contract N62269-74-0359, March 1975.

162. Shetty, D.K., Pascucci, M.R., Mutsuddy, B.C. and Willis, R.R., SiC Monofilament-Reinforced Si_3N_4 Matrix Composites. Ceram. Eng. and Sci. Proc. 6 [7-8] 632-645 (1985).

163. Prewo, K.M. and Brennan, J.J., High-Strength Silicon Carbide Fibre-Reinforced Glass-Matrix Composites. J. Mat. Sci. 15 463-468 (1980).

164. Yajima, S., Hasegawa, Y., Okamura, K. and Matsuzawa, T., Development of High Tensile Strength Silicon Carbide Fibre Using an Organosilicon Polymer Precursor. Nature 273 525-527 (June 1978).

165. Wynne, K.J. and Rice, R.W., Ceramics Via Polymer Pyrolysis. Ann. Rev. Mat. Sci. 14 297-334 (1984).

166. Prewo, K.M., Brennan, J.J. and Layden, G.K., Fiber Reinfroced Glasses and Glass-Ceramics for High-Performance Applications. Amer. Ceram. Soc. Bull. 65 [2] 305-313 (1986).

167. Prewo, K.M. and Brennan, J.J., Silicon Carbide Yarn Reinforced Glass Matrix Composites. J. Mat. Sci. 17 1201-1206 (1982).

168. Prewo, K.M. and Brennan, J.J., In Reference Book for Composites Technology; Lee, S.M. (ed), Technomic Publishing Company, Inc., (in press).

169. Stewart, R.L., Chyung, K., Taylor, M.P. and Cooper, R.F., In Fracture Mechanics of Ceramics; Bradt, R.C, Evans, A.G., Hasselman D.P.H. and Lange, F.F. (Eds) 7 33-52, Plenum Press, (1986).

170. Mah, T-I., Mendiratta, M.G., Katz, A.P. and Mazdiyasni, K.S., Recent Developments in Fiber-Reinforced High Temperature Ceramic Composites. Amer. Ceram. Soc. Bull. 66 [2] 304-307 (1987).

171. Rice, R.W., Matt, C.V., McDonough, W.J., McKinney, K.R. and Wu, C.C., Refractory-Ceramic-Fiber Composites: Progress, Needs and Oortunities. Ceram. Eng. and Sci. Proc. 3 [9-10] 698-713 (1982).

172. Rice, R.W., Spann, J.R., Lewis, D. and Coblenz, W., The Effect of Ceramic Fiber Coatings on the Room Temperature Mechanical Behavior of Ceramic-Fiber Composites. Ceram. Eng. and Sci. Proc. 5 [7-8] 443-474 (1984).

173. Bender, B., Shadwell, D., Bulik, C., Incorvati, L. and Lewis III, D., Effect of Fiber Coatings and Composite Processing on Properties of Zirconia-Based Matrix SiC Fiber Composites. Amer. Ceram. Soc. Bull. 65 [2] 363-369 (1986).

174. Lewis, D. and Rice, R.W., Further Assessment of Ceramic Fiber Coating Effects on Ceramic Fiber Composites. NASA Conf. Pub. No. 2406, Proc. of Joint NASA/DOD Conf., 13-26 (January 1985).

175. Bender, B.A., Lewis, D., Coblenz, W.S. and Rice, R.W., Electron Microscopy of Ceramic Fiber-Ceramic Matrix Composites - Comparison with Processing and Behavior. Ceram. Eng. and Sci. Proc. 5 [7-8] 443-474 (1984).

176. Rice, R.W. and Lewis, D., In Reference Book for Composite Technology; Lee, S.M. (Ed), Technomic Publishing Company, Inc. pp 117-142 (1989).

177. Rice, R.W., U.S. Patent 4,642,271, February 10 (1987); assigned to The United States of America as represented by the Secretary of the Navy.

178. Lundberg, R., Kahlman, L., Pompe, R., Carlsson, R. and Warren, R., SiC Whisker-Reinforced Si_3N_4 Composites. Amer. Ceram. Soc. Bull. 66 [2] 330-333 (1987).

179. Yangyun, S. and Brook, R.J., Preparation and Strength of Forsterite-Zirconia Ceramic Composites. Ceram. Int'l. 9 [2] 39-45 (1983).

180. Pena, P., Miranzo, P., Moya, J.S. and DeAza, S., Multicomponent Toughened Ceramic Materials Obtained by Reaction Sintering Part 1 ZrO_2-Al_2O_3-SiO_2-CaO System. J. Mat. Sci. 20 2011-2022 (1985).

181. Melo, M.F., Moya, J.S., Pena, P. and DeAza, S., Multicomponent Toughened Ceramic Materials Obtained by Reaction Sintering Part 3 System ZrO_2-Al_2O_3-SiO_2-TiO2. J. Mat. Sci. 20 2711-2718 (1985).

182. Rice, R.W., Processing of Advanced Ceramic Composites. Materials Research Society Symposium Proceedings, 32 337-345, Elsevier Science Publishing Co. (1984).

183. Lambe, K.A.D., Mattingley, N.J. and Bowen, D.H., Precipitation of Whiskers and Their Effect on Cleavage and Fracture in Magnesium Oxide. Fibre Sci. and Techn. 2 59-74 (1969).

184. Hori, S., Kaji, H., Yoshimura, M., Somiya, S., Deflection-Toughened Corundum-Rutile Composites. Mater. Res. Soc. Symp. Proc. 78 283-288 (1987).

185. Abramovici, R., Composite Ceramics in Powder or Sintered Form Obtained by Aluminothermal Reactions. Mat. Sci. and Eng. 71 313-320 (1985).

186. Rice, R.W. and McDonough, W.J., Intrinsic Volume Changes of Self-Propagating Synthesis. J. of Amer. Ceram. Soc. 68 [5] (1985).

187. Richardson, G.Y., Rice, R.W., McDonough, W.J., Kunetz, J.M. and Schroeter, T., Hot Pressing Ceramics Using Self- Propagating Synthesis. Ceram. Eng. and Sci. Proc. 7 [7-8] 761-770 (1986).

188. Rice, R.W., McDonough, W.J., Richardson, G.Y., Kunetz, J.M. and Schroeter, T., Hot Rolling of Ceramics Using Self-Propagating High-Temperature Synthesis. Ceram. Eng. and Sci. Proc. 7 [7-8] 751-760 (1986).

189. Rice, R.W., Richardson, G.Y., Kunetz, J.M., Schroeter, T. and McDonough, W.J., Effects of Self-propagating Synthesis Reactant Compact Character on Ignition Propagation and Resultant Microstructure. Ceram. Eng. and Sci. Proc. 7 [7-8] 737-750 (1986).

190. Newkirk, M.S., Lesher, H.D., White, D.R., Kennedy, C.R., Urquhart, W.W., Claar, T.D., Preparation of Lanxide Ceramic Matrix Composites: Matrix Formation by the Directed Oxidation of Molten Metals. To be published in Ceram. Eng. and Sci. Proc.

191. Walker, Jr., B.E., Rice, R.W., Becher, P.F., Bender, B.A. and Coblenz, W.S., Preparation and Properties of Monolithic and Composite Ceramics Produced by Polymer Pyrolysis. Amer. Ceram. Soc. Bull. 62 [8] 916-923 (1983).

192. Jamet, J., Spann, J.R., Rice, R.W., Ldwis, D. and Coblenz, W.S., Ceramic-Fiber Composite Processing Via Polymer-Filler Matrices. Ceram. Eng. and Sci. Proc. 5 [7-8] 443-474 (1984).

193. Dr. F. Chi, then at Dow Corning - private communication 1984.

194. Pierson, H.O., (Ed.), Chemically Vapor Deposited Coatings. The American Ceramic Society, Inc. (1981).

195. Kaae, J. L. and Gulden, T. D., Structure and Mechanical Properties of Codeposited Pyrolytic C-SiC Alloys. J. Amer. Ceram. Soc. 54 [12] 605-609 (1971).

196. Ogawa, T., Ikawa, K. and Iwamoto, K., Effect of Gas Composition on the Deposition of ZrC-C Mixtures: The Bromide Process. J. Mat. Sci. 14 125-132 (1979).

197. Stinton, D.P., Lackey, W.J., Lauf, R.J. and Besmann, T.M., Fabrication of Ceramic-Ceramic Composites by Chemical Vapor Deposition. Ceram. Eng. and Sci. Proc. 5 [7-8] 443-474 (1984).

198. Hirai, T. and Hayashi, S., Preparation and Some Properties of Chemically Vapour-Deposited Si_3N_4-TiN Composite. J. Mat. Sci. 17 1320-1328 (1982).

199. Hirai, T., Goto, T. and Sakai, T., In Emergent Process Methods for High-Technology Ceramics; Davis, R.F., Palmour III, H. and Porter, R.L. (Eds) 17 347-369 Plenum Press, (1984).

200. Pierson, H.O., Boron Nitride Composites by Chemical Vapor Deposition. J. Composite Materials 9 228-239 (1975).

201. Dauchier, M., Lamicq, P. and Mace J., Comportement Thermomecanique des Composites Ceramiques-Ceramiques. Rev. Int. Hautes Temp. Refract. 19 [4] 285-299 (1982).

202. Naslain, R., Quenisset, J.M., Rossignol, J.Y. and Hannache, H., In International Conference of Composite Materials, ICCM-V Conference Proceedings, 5th Edition; Harrison Jr., W.C., Strife, J. and Ghingra, A.K. (Eds) 499-514, Metal. Soc. AIME, (1985).

203. Dauchier, M., Bernhart, G. and Bonnet, C., Properties of Silicon Carbide Based Ceramic-Ceramic Composites. Proc. of the 30th Natl. SAMPE Symp., 1519-1525 (1985).

204. Lamicq, P.J., Bernhart, G.A., Dauchier, M.M. and Mace, J.G., SiC/SiC Composite Ceramics. Amer. Ceram. Soc. Bull. 65 [2] 336-338 (1986).

205. Fisher, R.E., Burkland, C.V. and Bustamante, W.E., Fiber Reinforced Silicon Carbide for Heat Exchanger Alications. Amer. Ceram. Soc. 87th Ann. Mtg. (1985).

206. Fisher, R.E., Burkland, C.V. and Bustamate, W.E., Ceramic Composite Thermal Protection Systems. Ceram. Eng. and Sci. Proc. 6 [7-8] 806-819 (1985).

207. Warren, J.W., Fiber and Grain-Reinforced Chemical Vapor Infiltration [CVI] Silicon Carbide Matrix Composites. Ceram. Eng. and Sci. Proc. 6 [7-8] 684-693 (1985).

208. Caputo, A.J. and Lackey, W.J., Fabrication of Fiber-Reinforced Ceramic Composites by Chemical Vapor Infiltration. Ceram. Eng. and Sci. Proc. 5 [7-8] 443-474 (1984).

209. Caputo, A.J., Lackey, W.J. and Stinton, D.P., Development of a New, Faster Process for the Fabrication of Ceramic Fiber-Reinforced Ceramic Composites by Chemical Vapor Infiltration. Ceram. Eng. and Sci. Proc. 6 [7-8] 694-706 (1985).

210. Stinton, D.P., Caputo, A.J. and Lowden, R.A., Synthesis of Fiber-Reinforced SiC Composites by Chemical Vapor Infiltration. Amer. Ceram. Soc. Bull. 65 [2] 347-350 (1986).

211. Caputo, A.J., Stinton, D.P., Lowden, R.A. and Besmann, T.M., Fiber-Reinforced SiC Composites with Improved Mechanical Properties. Amer. Ceram. Soc. Bull. 66 [2] 368-372 (1987).

212. Fitzer, E. and Gadow, R., Fiber-Reinforced Silicon Carbide. Amer. Ceram. Soc. Bull. 65 [2] 326-335 (1986).

213. Colmet, R., Lhermitte-Sebire, I. and Naslain, R., Alumina Fiber/Alumina Matrix Composites Prepared by a Chemical Vapor Infiltration Technique. Adv. Ceram. Matls. 1 [2] 185-191 (1986).

214. Rasmussen, J.J., Surface Tension, Density and Volume Change on Melting of Al_2O_3 System, Cr_2O_3 and Sm_2O_3, J. Amer. Ceram. Soc., 55 [6] 326 (1972).

215. Spann, J.R., Rice, R.W., Coblenz, W.S. and McDonough, W.J., In Emergent Process Methods for High-Technology Ceramics; Davis, R.F., Palmour III, H. and Porter, R.L. (Eds) 17 473-503, Plenum Publishing Corporation, (1984).

216. Rice, R.W., Bender, B.A., Ingel, R.P., Coyle, T.W. and Spann, J.R., In Ultrastructure Processing of Ceramics, Glasses and Composites; Hench. L.L. and Ulrich, D.R. (Eds) 507-523, Wiley (1984).

217. Kuriakose, A.K. and Beaudin, L.J., Tetragonal Zirconia in Chilled-Cast Alumina-Zirconia. J. Canad. Ceram. Soc. 46 45-50 (1977).

218. Claussen, N., Lindemann, G. and Petzow, G., Rapid Solidification of Al_2O_3-ZrO_2 Ceramics. 5th CIMTEC, Lignano-Sabbiadoro, Italy (1982).

219. Claussen, N., Lindemann, G. and Petzow, G., Rapid Solidification in the Al_2O_3-ZrO_2 System. Ceram. Int'l. 9 [3] 83-86 (1983).

220. Krohn, U., Olapinski, H and Dworak, U., U.S. Patent 4,595,663, June 17 (1986); assigned to Feldmühle Aktiengesellschaft.

221. Rice, R.W., McKinney, K.R. and Ingel, R.P., Grain Boundaries, Fracture and Heat Treatment of Commercial Partially Stabilized Zirconia. J. Amer. Ceram. Soc. 64 [12] C175-177 (1981).

222. Hillig, W.G., Mehan, R.L., Morelock, C.R., DeCarlo, V.J. and Laskow, W., Silicon/Silicon Carbide Composites, Amer. Ceram. Soc. Bull. 54 [12] 1054-1056 (1975).

223. Hillig, W.B., Ceramic Composites by Infiltration. Ceram. Eng. and Sci. Proc. 6 [7-8] 674-683 (1985).

224. Apler, A.M., Doman, R.C. and McNally, R.N., In Science of Ceramics; Stewart, G.H. (Ed) 4 389-419, The British Ceramic Society, (1968).

225. Rowcliffe, D.J., Warren, W.J., Elliot, A.G. and Rothwell, W.S., The Growth of Oriented Ceramic Eutectics. J. Mat. Sci. 4 902-907 (1969).

226. Viechnicki, D.J., In Materials Science Monographs, Frontiers in Materials Technologies; Meyers, M.A. and Inal, O.T. (Eds) 26 280-303, Elsevier Press, (1985).

227. LaBell, Jr., H.E., Growth of Controlled Profile Crystals from the Melt: Part II - Edge-Defined, Film-Fed Growth [EFG]. Mat. Res. Bull. 6 581-590 (1971).

228. Harada, Y., Bortz, S.A. and Blum, S.L., In High Temperature Materials, 6th Plansee Seminar; Benesovsky, F., (ed) 608-629, Springer-Verlay (1969).

229. Hulse, C.O. and Batt, J.A., The Effect of Eutectic Microstructure on the Mechanical Properties of Ceramic Oxides. U.S. Navy Report No. N910803-10, Contrct No. N00014-69-0073, May 1974.

230. Michel, D., Rouaux, Y. and Perez Y Jorba, M., Ceramic Eutectics in the Systems $ZrO_2-Ln_2O_3$ [Ln=Lanthanide]: Unidirectional Solidification, Microstructural and Crystallographic Characterization. J. Mat. Sci. 15 61-66 (1980).

231. Fragneau, M. and Revcolevschi, A., Crystallography of the Directionally Solidified NiO-CaO Eutectic. J. Amer. Ceram. Soc. 66 C162-163 (1983).

232. Mazerolles, L., Michel, D. and Portier, R., Interfaces in Oriented Al_2O_3-ZrO_2 (Y_2O_3) Eutectics. J. Amer. Ceram. Soc. 69 [3] 252-255 (1986).

233. Echigoya, J., Takabayashi, Y., Suto, H. and Ishigame, M., Structure and Crystallography of Directionally Solidified $Al_2O_3-ZrO_2-Y_2O_3$ Eutectic by the Floating Zone Melting Method. J. Mat. Sci. Lett 5 150-152 (1986).

234. Echigoya, J., Suto, H., Microstructure of ZrO_2-2.5 mol% Y_2O_3-8.5 mol% Al_2O_3 in Directionally Solidified $Al_2O_3-ZrO_2$ (Y_2O_3) Eutectic. J. Mat. Sci. Lett 5 949-950 (1986).

235. Echigoya, J., Takabayashi, Y. and Suto, H., Hardness and Fracture Toughness of Directionally Solidified $Al_2O_3-ZrO_2(Y_2O_3)$ Eutectics. J. Mat. Sci. Lett 5 153-154 (1986).

236. Ito, S., Kokubo, T. and Tashiro, M., Unidirectionally Solidified Transparent Ceramics in the System $NaNbO_3-BaTiO_3$. Amer. Ceram. Soc. Bull. 58 [6] 591-594 (1979).

237. Hong, J.D., Spear, K.E. and Stubican, V.S., Directional Solidification of $SiC-B_4C$ Eutectic: Growth and Some Properties. Mat. Res. Bull. 14 775-783 (1979).

238. Sorrell, C.C., Beratan, H.R., Bradt, R.C. and Stubican, V.S., Directional Solidification of [Ti,Zr] Carbide-[Ti,Zr] Diboride Eutectics. J. Amer. Ceram. Soc. 67 [3] 190-194 (1984).

239. Sorrell, C.C., Stubican, V.S. and Bradt, R.C., Mechanical Properties of $ZrC-ZrB_2$ and $ZrC-TiB_2$ Directionally Solidified Eutectics. J. Amer. Ceram. Soc. 69 [4] 317-321 (1986).

240. Yoshimura, M., Kaneko, M. and Somiya, S., Preparation of Amorphous Materials by Rapid Quenching of Melts in the System $ZrO_2-SiO_2-Al_2O_3$. J. Mat. Sci. Lett 4 1082-1084 (1985).

241. McCoy, M., Lee, W.E. and Heuer, A.H., Crystallization of $MgO-Al_2O_3-SiO_2-ZrO_2$ Glasses. J. Amer. Ceram. Soc. 69 [3] 292-296 (1986).

242. Mussler, B.H. and Shafer, M.W., Preparation and Properties of Cordierite-Based Glass-Ceramic Containing Precipitated ZrO_2. Amer. Ceram. Soc. Bull. 64 [11] 1459-1462 (1985).

243. Leatherman, G.L., Transformation Toughening in Glass-Ceramics. Thesis submitted to Graduate Faculty of Rensselaer Polytechnic Institute for Ph.D., Troy, New York (December 1986).

244. Abe, Y., Kasuga, T., Hosono, H. and de Groot, K., Preparation of High-Strength Calcium Phosphate Glass-Ceramics by Unidirectional Crystallization. J. Amer. Ceram. Soc. 67 C142-144 (1984).

245. Yoshimura, M., Kaneko, M. and Somiya, S., Preparation of Amorphous Materials by Rapid Quenching of Melts in the System ZrO_2-SiO_2-Al_2O_3. J. Mat. Sci. Lett. 4 1082-1084 (1985).

246. Yoshimura, M., Kaneko, M. and Somiya, S., Crystallization of Mullite-Zirconia Amorphous Materials Prepared by Rapid-Quenching. Yogyo-Kyokai-Shi 95 [2] 64-70 (1987).

247. Weston, R.M. and Rogers, P.S., In Mineralogy of Ceramics; Taylor, D. and Rogers, P.S. (Eds) 28 37-52, British Ceramic Society, (1979).

6

Injection Moulding of Fine Ceramics

J.R.G. Evans

Department of Materials Technology, Brunel University, Uxbridge, Middlesex, UB8 3PH, UK.

1. INTRODUCTION

The use of polycrystalline high temperature ceramics in reciprocating[1,2] and turbine engine[3,4] applications has been made possible by considerable developments in the fabrication of fine powders[5,6] and in understanding the sintering process[7,8]. Covalent ceramics such as SiC[9] and Si_3N_4[10] as well as oxides such as ZrO_2[11] and ZrO_2- or SiC-toughened Al_2O_3[12] have been used in experimental engines. There are many less exotic but important, applications for fine ceramics in wire drawing dies, cutting tools, molten metal handling and in agriculture.

Fine technical ceramics can be fabricated in three stages:
1) preparation of a suitable powder,
2) creation of shape and form in an assembly of particles, and
3) sintering of particles.

Historically, it was the development of the last stage in the sequence which was investigated first and in greatest detail. Important developments were the ability to sinter fine powders of SiC and Si_3N_4 to near full density in the 1970s[13,14]. Subsequently, attention focussed on the characteristics of starting powders; largely due to the influence of H.K. Bowen. This has led to fine TiO_2[15], Al_2O_3[16], ZrO_2[17], SiC[18] and Si_3N_4[19] powders which have in common, submicron, uniformtly sized particles of near spherical shape. Such powders have a lower space filling efficiency than wide size distribution powders but are capable of forming uniform assemblies free from large pores. The absence of differences in particle size minimizes the

215

driving force for grain growth during sintering (see also Chapters 1 & 2).

The second stage in the sequence has been left until last but now, as ceramic mechanical devices move closer to mass production, it becomes unavoidable. Of the methods available for creating shape and form in ceramic particle assemblies two in particular offer versatility to the engineering designer; slip casting and injection moulding.

Slip casting has a strong tradition in the ceramics industry and has the advantage over injection moulding that it needs no binder removal stage. Neither does it require the capital associated with injection moulding equipment. Slip casting tends to require greater operator skill and to be less conveniently automated (see also Chapter 3).

Figure 1 shows a silicon nitride precombustion chamber manufactured by Kyocera. The shape appears simple but contains an oblique oval hole surrounded by sharp knife edges. It is an example of a shape which would be very difficult to mass produce by slip casting.

Turbine blades have been made by injection moulding silicon carbide powders and are accurate to 50µm over the blade profile after 18% linear

Figure 1: Silicon nitride precombustion chamber.

shrinkage[20]. In many applications e.g. exhaust valves, turbine blades and precombustion chambers, a minimal amount of final grinding is necessary after sintering. The aim of the moulding process is therefore to produce an unsintered particle assembly which will shrink isotropically to yield a shape slightly oversize for final machining. Distortion of the body during moulding, binder removal or sintering may render the component useless.

The selection of materials for injection moulding has recently been reviewed[21] as have injection moulding techniques[22]. The purpose of this chapter is to review the general principles that govern the injection moulding of ceramic suspensions in order that the reader may readily access the extensive literature which offers insights into a very complex process. Probably the most serious unsolved problems are i) the control over residual stresses which are established during solidification in the cavity in large sclections, ii) the removal of organic vehicle from large sections, particularly where fine powders are employed and iii) the control of machine wear. The latter problem is beyond the scope of the present work.

In considering ceramics fabrication by plastic processing methods considerable information is drawn from polymer processing technology and the interdisciplinary nature of the work provokes the ceramics community to embrace meritorious developments in this area.

The small contribution that the Brunel University Ceramics Fabrication Group has made to the understanding of this process has been made possible by the wide explorative remit afforded by the UK Science and Engineering Research Council which has permitted stepwise investigation without a demand for perfect finished products.

2. COMPOUNDING

The initial stage in the moulding process, which is also a stage that has received scant attention until recently, involves dispersing fine ceramic powder in a polymer blend. The matrix phase is referred to either as a vehicle, a term drawn from the paint and printing ink industry, because its purpose is to render the ceramic fluid, or as a binder because it also serves as an adhesive between particles. The objective of the compounding stage is to disperse agglomerates and to distribute particles uniformly throughout the matrix. The measure of uniformity is the <u>characteristic volume</u> which is the amount of material at every location throughout the mixture within which the position of individual particles is unimportant. The dimensions of the characteristic volume can be found from the critical flaw size needed to produce a desired strength in the sintered product.

The compounding stage may provide a material that is suitable for ceramics fabrication by various plastic forming processes. These could include injection moulding[21,22], blow moulding[23], extrusion[24], calendering, melt spinning or even vacuum forming.

Considerable work on the dispersive mixing of polymer-ceramic blends has been carried out by the polymer composite industry. Carbon blacks are used as fillers in rubber[25], and calcium carbonate, mica or sand are added to thermoplastics for reinforcement or bulking[26]. Minerals such as aluminium trihydrate[27] or magnesium hydroxide[28] are added to polymers as flame retardants[29] and magnetic[30] or dielectric[31] powders have been blended in high volume loadings to produce cheap magnetic or dielectric composites, often with special mechanical properties.

In low viscosity fluids, considerable effort has been devoted to the dispersion of concentrated suspensions for slip casting or tape casting. In aqueous media it is the electrical behaviour of the solid-liquid interface which is used to preserve dispersion[32]. In non-aqueous media dispersion and deflocculation have been achieved by employing strongly polar solvents[33] or by adding surfactants[34].

In polymer systems, however, the high non-Newtonian viscosity of the fluid would make dispersion based on these forces a slow process. Dispersive mixing is achieved by the application of shear stress to the mixture and the efficiency of dispersion therefore depends on machine design[35].

3. MIXING DEVICES

Blending equipment is available with zero, one or two principal moving parts. The former are motionless mixers in which fluids are pumped through contoured passages which apply strain to the mixture.   These tend to be unsuitable for high viscosity systems.   Single blade type mixers[36] are effective in imposing strain and hence producing distributive mixing but cannot apply a shear stress uniformly to the mixture.   Single screw extruders are not efficient mixing devices[37].

Double blade mixers such as the 'Z' blade mixer shown in Figure 2, have been used for the preparation of wax based injection moulding mixtures[38,39] but have the disadvantage that the material can find 'dead' spaces in the vessel.   The greatest success has therefore been achieved with double cam type mixers[40], twin roll mills[41] and twin screw extruders[42].

The latter have been shown to be particularly suitable for the preparation of filled polymers[43] and engineering ceramics[44].   Twin screw extrusion was developed in Italy in 1949[45] and is now well established as a polymer compounding technique[46].   Four types of extruder are available depending on whether the screws intermesh or not and on whether they counter- or co-rotate.   Intermeshing co-rotating extruders are found to provide intensive mixing and are available in designs which allow rapid modification of modular screw configuration to suit different materials[47], an example

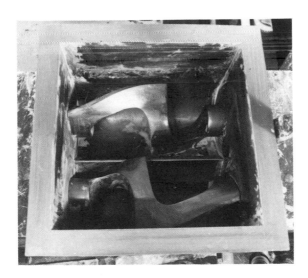

Figure 2: An oil-jacketed 'Z' blade mixer.

Figure 3: Co-rotating twin screw extruder used for compounding high filled polymers(44).

of which is shown in Figure 3.   They have been shown to be effective in dispersing soft agglomerates in conventional ceramic powders[44]. The screw configuration shown in Figure 4 incorporates a cascade of mixing discs to aid dispersion. Prefired densities of dispersed powders were found to be higher than the cold die pressed powder and this is another indication that agglomerates have been dispersed[44].

Techniques for assessment of dispersion in real polymer systems are sparse but Hornsby and coworkers[48] have investigated contact microradiography, acoustic microscopy, transmitted and reflected light microscopy and scanning electron microscopy and automatic image analysis to quantify the state of dispersion. In ceramic systems, removal of the polymer followed by partial sintering and scanning electron microscopy of polished sections or fracture surfaces reveals the existence of undispersed agglomerates[44]. The incorporation of small amounts of carbon black aids in optical microscopical identification of some undispersed oxide ceramic or polymer components[44].

The importance of dispersion can be deduced from Lange's work on the

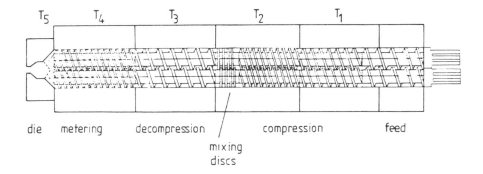

T_5 T_4 T_3 T_2 T_1

die   metering   decompression   compression   feed

mixing
discs

Figure 4: Diagram of the twin screw extruder configuration used for processing ceramics at Brunel University.

role of agglomerates as a source of strength-limiting defects. In general, different sintering rates occur when agglomerate prefired densities differ from the matrix prefired density and for normal sintering the densification rate is inversely proportional to the prefired density of the powder[49] (see also Chapter 1, this volume). For subnormal sintering, i.e exceptionally low prefired densities, lower sintering rates are noted for low prefired density[50]. Direct experiments on bulk samples have shown that differential sintering due to different prefired density sets up stresses in the composite[51]. In the case of agglomerates in ceramic assemblies these stresses can cause circumferential cracks (Fig. 5a) if the agglomerate sinters faster than the matrix or radial cracks if the matrix sinters faster (Fig. 5b)[52]. A further disadvantage of agglomerates is that their presence may lead to pores with large grain co-ordination numbers which may become stable during sintering and this tends to limit the final density[53].

Fine particles frequently show agglomeration under the influence of London dispersion forces[54], hydrogen bonding between adsorbed water layers, hydrostatic tension within bulk liquid lenses[55], or solid bridges arising from the manufacturing process or from added sintering aids.

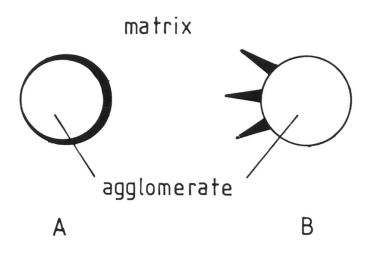

Figure 5: Defects originating from undispersed agglomerates caused by a) agglomerate sintering faster than the matrix ($\rho_a < \rho_m$) and b) matrix sintering faster than the agglomerate ($\rho_m < \rho_a$).

Agglomerates of the latter type may be very difficult to disperse and such powders may require milling. Soft agglomerates are generally dispersable for the purpose of slip casting and preliminary work has shown that they can be dispersed in intensive mixers prior to injection moulding[44].

The force needed to separate two spherical particles of radii $r_1 r_2$ initially in contact, by dispersive mixing in simple shear flow in a fluid of viscosity, η, is given by[56];

$$F_{max} = 3\,\eta\dot{\gamma}\,r_1 r_2 \qquad\qquad 1$$

where $\dot{\gamma}$ is the shear rate. Thus, the force on the particles can be increased by increasing the shear stress $\eta\dot{\gamma}$. This can be achieved by increasing the viscosity perhaps by lowering the processing temperature or by increasing the shear rate by increasing the machine speed or decreasing the flow gap. It is unfortunately the case that the product of particle radii, $r_1 r_2$, determines the force to separate them. Thus small particles are less easily separated in shear flow than large particles. The dispersion of very fine zirconia powders in shear flow has been shown to be limited for these reasons[57].

4. PARTICLE SURFACE MODIFICATION

There has been extensive interest in the surface modification of mineral fillers for incorporation into polymer systems. Stearic acid has been used as a component of injection moulding blends[58,59] and the acid groups are thought to chemisorb on basic mineral surfaces. It is possible that the stearic acid also acts as a plasticizer or lubricant for the polymer. The importance of acid-base interactions in the formation and stabilization of deflocculated polymer-mineral suspension is highlighted by Fowkes and co-workers[60] suggesting that dispersion and adhesion are enhanced by matching a basic filler with a polymer containing electron acceptor groups. The treatment of particle and fibre reinforcements for polymers with so-called coupling agents is commonplace. In particular silanes[61], titanates[62], zirconates[63] and zircoaluminates[64] are used to improve dispersion, reduce viscosity or increase adhesion.

Silane coupling agents have been used for many years to pretreat fibres for incorporation in composite materials to improve mechanical properties and resistance to aqueous environments and they are also effective in enhancing adhesive joint strength[65]. Subsequently, organic titanates were used as adhesion promoters for coatings[66]. The mechanical properties of as-moulded ceramic bodies may be important in determining the incidence of defects during solidification, ejection and handling as described below, but much interest in the use of coupling agents focuses on improvements in melt fluidity and dispersion of powders.

Silane coupling agents are generally of the form:

$$XSi(OR)_3$$

where (OR) is an alkoxy group, readily hydrolysed in the presence of water to yield a silanol. X is a group which is selected to react with, or to be compatible with, the polymer. After hydrolysis:

$$XSi(OR)_3 + 3H_2O \rightarrow XSi(OH)_3 + 3ROH$$

the silanol may condense with hydrated metal (M) oxides to give an -Si-O-M bond or it may form a hydrogen bond with adsorbed OH groups:

$$-\overset{|}{\underset{|}{Si}} - O - H - - - O - \overset{|}{\underset{|}{M}} -$$
$$\underset{H}{|}$$

Organic titanates are thought to adsorb in a similar way. The coupling agents, may be added directly to the polymer-ceramic blend, blended directly with the powder or used to treat the powder from solution[67]. Additions are generally made on an **ad hoc** basis at the level of 0.2 - 2 wt.% based on the filler. In general this produces more than monolayer coverage of the filler surface and the structure of the initial chemisorbed layer and subsequent physically adsorbed layers is a matter of considerable interest[68].

There are a number of examples of the addition of organic titanates reducing the viscosity of filled systems and thus allowing higher volume fractions of filler[69]. Han and co-workers[70] report that the viscosity of polypropylene and high density polyethylene composites with $CaCO_3$, talc or glass fibre fillers was reduced by the addition of 0.5 or 1 wt.% of an appropriate titanate based on the filler. However, the mechanical properties of the solid composite were adversely affected. In a similar study[71] they showed that a silane coupling agent with a paraffinic X-group decreased the viscosity of polypropylene filled with glass or $CaCO_3$ whereas an amino-functional silane decreased the viscosity of the $CaCO_3$ system but increased the viscosity of glass systems. In a study of nylon-filled systems, Han and co-workers[72] found that silane and titanate coupling agents increased the viscosity of nylon-$CaCO_3$ but that an amino-functional silane and a titanate reduced the viscosity of nylon-wollastonite systems.

A basic study of surface active species on the properties of filled systems was performed by Bigg[73] using coarse (5-44μm) steel spheres at volume fractions up to 0.7. This showed that addition of both a silane and a titanate coupling agent increased the viscosity of polyethylene-steel composites.

Coupling agents with unreactive paraffinic X groups have been shown to reduce the viscosity of ceramic injection moulding suspensions incorporating coarse silicon powder and fine silicon nitride[74]. However, it should be remembered that coupling agents may have several effects; they may enhance wetting of the filler or reduce particle attraction thus improving dispersion. They may chemically combine a polymer layer to the particle surface thus increasing the effective volume fraction of filler and this would tend to increase viscosity. They may act as a low molecular weight plasticizer for the polymer phase; an effect which may be substantially independent of the filler altogether. Thus titanate and zirconate coupling agents were shown to reduce the viscosity of zirconia-polypropylene suspensions but also to reduce the viscosity of the organic vehicle in the absence of powder[75]. Furthermore, a triglyceride, similar in molecular structure to the coupling agents selected but without the coupling group, performed in a similar way[75]. Contact angle experiments showed that the titanate did not enhance the wettability of the ceramic surface[75].

5. FILLING THE CAVITY

The injection moulding process originated with metal die casting and was first used for polymers in 1878[76]. The technique is recorded as being used for ceramics fabrication in 1937[38] but development of the ceramics process over the last fifty years has not been systematic and reporting of experimental detail has been sparse. In the last ten years there has been renewed interest in the process as ceramics are used increasingly in mass produced devices.

Two principle designs for injection moulding machines are available and are characterized according to the method of plasticization. A simple plunger machine is shown in Figure 6 and it is held that such designs suffer less from abrasive wear when used with ceramics[77]. In the plunger technique, the process of bringing the material uniformly to the desired temperature in the shortest possible time is hindered by the limited forced convective flow in the chamber. Thus the pressure in the nozzle of the machine can vary widely as materials of widely different viscosity flow into the mould. In order to improve heat transfer a torpedo or spreader is incorporated into the barrel.

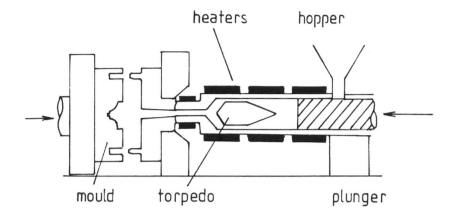

Figure 6: A plunger injection moulding machine.

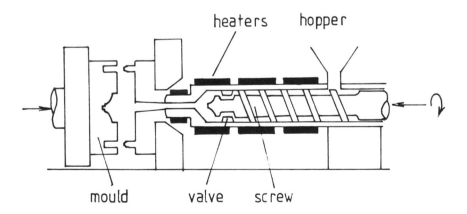

Figure 7: A reciprocating screw injection moulding machine.

A reciprocating screw machine is shown in Figure 7 in which material flows in the channel of an helical screw the entire length prior to injection. The rotational flow and close contact with the heated barrel bring the material to a uniform temperature and metering, homogeneity, injection pressure control, reproducibility and cycle time are thereby improved[78].

Alternative designs of low pressure equipment are available[79] wherein a ceramic suspension in a wax based binder is injected into a mould by pneumatic pressure and this is claimed to avoid some of the wear associated with high pressure devices.

The advantages of the plunger machine and the excellent plasticizing and compounding effects of the twin screw extruder have been combined into one device by U.K. inventors[80]. The in-line compounding/moulding machine shown in Figures 8 and 9 uses a twin screw extruder to compound highly filled materials for ceramic injection moulding and for artificial bone. The feedstock is premixed or metered ceramic and polymer powder. The compounded and plasticized material is fed through a non-return valve into a plunger barrel from where it is injected into the cavity. The oscillating packing device allows control over solidification in the cavity and this aspect is discussed in further detail below.

The plunger and screw machines behave in a quite different way in the plasticizing stage but essentially the mould filling stage is the same. For thermoplastic systems the fluid suspension is forced under pressure through passages of different cross section into a colder mould. The fluid is frequently non-Newtonian and the temperature is non-uniform so that mass flow behaviour is extremely complex.

The incorporation of a ceramic powder into a polymer tends to reduce the specific heat and increases thermal conductivity. The resulting increase in thermal diffusivity results in rapid chilling in the sprue, runner and cavity. Viscosity of filled polymers is generally considerably higher[81] and filled system often show yield point behaviour.

Rather than understand the system, the approach to moulding technology has tended to be to adjust moulding conditions according to general rules until satisfactory mouldings are obtained. In this context there are a set of 'machine variables', listed in Table 1 and a set of 'material variables'

Figure 8: The in-line direct blending injection moulder developed at Brunel University for fabrication of engineering ceramics and artificial bone.

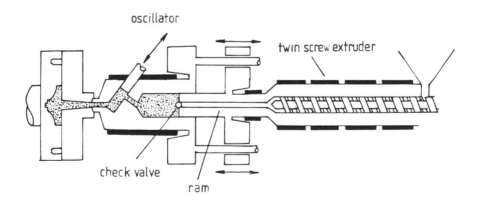

Figure 9: Schematic diagram of the in-line compounding-injection moulding machine.

Main Machine Variables

    Barrel temperatures: feed zone to nozzle

    Plasticization variables: screw rotation speed

                               plasticization counter pressure

    Mould temperature

    Mould design (Sprue runner and gate geometries)

    Injection speed

    Injection pressure-time profile

    Post injection hold pressure

    Post injection hold time

Material Variables - relevant to mould filling and solidification.

    Viscosity at shear rates in the sprue runner and mould ($100s^{-1}$)

    Flow behaviour index

    Temperature dependence of viscosity in the region just below the injection temperature

    Solidification temperature

    Volumetric shrinkage associated with crystallinity

    Equation of state

    Thermal diffusivity

    Coefficient of thermal expansion in the solid state

    Mechanical strength

    Elastic modulus

Table 1: Machine and material variables relevant to ceramic injection moulding.

discussed below. In general, the guidelines for polymer moulding[82] can be applied, **mutatis mutandis** to the ceramic moulding situation.

    Thus the failure to fill the cavity (short shots) can be attributed to low barrel temperture, mould temperature, injection pressure or injection speed. The incidence of weld lines in mouldings, which diminishes the toughness of polymers[83] and is more severe for ceramics, can be avoided by proper mould design. The suspension should not have to split to overcome an obstacle and the cavity should be side gated so that a 'plug' of material flows into the main cavity[84]. An example of such an arrangement is shown in Figure 10 where a step wedge is used to investigate defects in mouldings of different cross section[85]. The gate is situated on the side of the

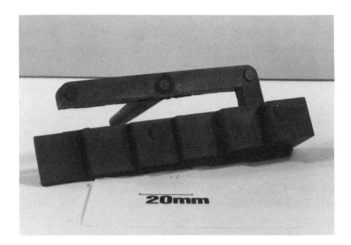

Figure 10: Side gated step wedge moulding used to study defects in various mould sections.

largest section of the cavity. The nature of flow into the cavity can easily be discerned by producing a series of short shots.   In this way the appearance of 'jetting'; producing a coil of material and hence a body replete with internal weld lines can be noted. Die swell, observed in high polymers as the melt front emerges from a narrow channel helps to prevent the phenomenon of jetting into a cavity for unreinforced polymers, but is largely absent when the polymer contains large amounts of filler[86].

Mould filling is also facilitated by the incorporation of small vents machined on the face, to allow air to escape as the material enters. Frequently, the ejector pin clearances are used for this purpose but they can become blocked especially if waxes are used in the formulation or release agents employed.

Thus by a combination of the viscosity characteristics discussed below and the machine parameters, a moulded body can be produced which is slightly smaller in dimension (by < 1% linear) than the cavity.   The slight contraction, caused by the higher thermal expansion of polymers compared with most metals and ceramics, assists in ejection of the component.

6. FORMULATION OF MOULDING COMPOSITIONS

The approach to formulation of the injection moulding vehicle has recently been reviewed[21]. The polymer blend is often composed of three components: **major binder,** which determines the overall range of properties; **minor components,** which are often low molecular weight species such as plasticizers for the main binder; and **processing aids** which modify the particle-binder interface or act as lubricants.

The blend should have the following characteristics:

- A capacity for dispersing the ceramic powder in the chosen mixer by breaking up agglomerates, removing entrapped gas and distributing the particles.

In this respect it should have a sufficiently high viscosity at the mixing temperature for the imposition of shear stress on particle clusters under the influence of machine-imposed shear rates. Conditions of polymer-ceramic wetting may be desirable for stability of the dispersion but the long time required for establishment of the equilibrium contact angle of a high polymer on a mineral surface[87] compared with processing times probably means that equilibrium surface energetics have little importance during the actual mixing process.

- Stability under mixing and moulding conditions.

It is often the case that highly filled polymers, $V > 0.5$, must be processed at 10 - 20°C above the normal temperatures for the polymer. Some degradation is therefore to be expected. The use of antioxidants may be undesirable because it may interfere with the subsequent burnout stage. Under no circumstances should degradation proceed to the stage when volatile degradation products are liberated during processing. Mixing procedure has been shown to influence degradation of the organic vehicle and hence viscosity of the ceramic suspension and polarity of the organic vehicle[88].

- Confer fluidity on the powder sufficient for complete filling of the cavity.

This aspect is discussed below.

-   Confer adequate strength to the body during ejection from the mould, handling and in the initial stages of burnout.

Distortion during ejection may be caused by low yield stress in the solidified body[89] and is avoided by employing semicrystalline polymers or waxes or species with a glass transition temperature above ambient.

-   Leave a low residue after polymer pyrolysis.

It is important for high performance ceramic applications that the ash content of the polymer blend should be low in silicates and alkali metals. For moulding of non-carbide ceramics it is desirable to have a low residual carbon content.

-   Be readily available at acceptable cost.

Where special polymers are incorporated into the blend their availability and reproducibility of composition may cause problems.   The pricing of the polymer blend is clearly a function of the added value it confers on the ceramic.

The viscosity of polymer-ceramic blends is dependent on the volume fraction, size distribution and shape of the particles[21] and partly on the nature of the organic component. Volume fraction of powder of the order of 70 vol.% can be obtained with wide particle size distribution[90] but the increasing emphasis on fine monosized powders means that volume loading and hence prefired densities need be much lower; typically in the 55-60 vol.% range. Although the critical powder volume concentration (CPVC) is a useful indicator of the approximate volume fraction of an unknown powder which can be incorporated in an organic vehicle[91] it is preferable to have a knowledge of the relative viscosity - volume loading curve for a given powder.   Of the many semi-empirical relationships which are approximately valid at high powder loadings, a modification of that due to Chong, Christiansen and Baer[92] has been found to be reasonably general for ceramic suspensions[93]:

$$\eta_r = \left[\frac{V_m - CV}{V_m - V} \right]^2$$

2

Here η_r is the suspension viscosity divided by the viscosity of the unfilled organic vehicle measured under identical conditions, C is a constant which in Chong's original equation was 0.25 and V_m is the maximum packing fraction of ceramic powder at which viscosity approaches infinity. If equation 2 is rewritten as:

$$\frac{1}{\sqrt{\eta_r} - 1} = \frac{V_m}{1 - C} \cdot \frac{1}{V} - \frac{1}{1 - C} \qquad 3$$

it can be seen that a minimum of two viscosity measurements at different volume loadings can be plotted as a straight line to give the constant C and the maximum packing fraction V_m.

A number of methods are available for production of maximum packing efficiency from particle characteristics[94-96] but the accuracy with which such procedures could predict V_m for the purpose of predicting viscosity of suspensions is questionable. Note that the denominator in equation 2, as in other relative viscosity-volume fraction relationships for use at high volume fractions, is a function of the 'free volume' fraction of organic vehicle over and above that needed to fill the interstices between contacting particles.

Mutsuddy has attempted to relate capillary rheometer measurements of viscosity to moulding trials and others have done the same for unfilled polymers[98]. Weir[99] developed a criterion for acceptable rheology for injection moulding. In the first place he drew attention to the absolute viscosity at low shear rates, typical of flow in the nozzle, sprue and cavity. Mutsuddy selected a shear rate of 100 s^{-1} [97] at which to measure viscosity. In a reciprocating screw machine, shear rates vary between 100-1000 s^{-1}. Mutsuddy[97] suggested that viscosity at 100 s^{-1} should be less than 1000 Pas at a suitable nozzle temperature. Dilatancy should be absent over the entire shear rate range encountered in the injection moulding machine. Furthermore, the temperature dependence of viscosity expressed either as an activation energy or as d log/dT should be low in the region just below the moulding temperature[59,99,100]. The reason for this is that the final stage of mould filling, immediately before complete solidification of the sprue, involves a competition between shrinkage of the molten core and flow along a capillary of diminishing radius. Pressure drop in the cavity occurs when:

$$\left[\frac{dV}{dt} \right]_{shrinkage} \geq \left[\frac{dV}{dt} \right]_{flow} \qquad\qquad 4$$

The left hand side of the inequality is governed by thermal contraction and shrinkage associated with phase changes whereas the right hand side is increased by a low temperature dependence of viscosity, a large sprue radius and a high hold pressure.

Table 2 shows the flow characteristics of a range of suspensions each conveying 65 vol.% silicon powder[59,100]. Compositions based on polypropylene have found favour with a number of practitioners[101-103].

7. SOLIDIFICATION IN THE CAVITY

It is interesting to note that successful injection moulding of engineering ceramics has been accomplished in the past with parts of small cross section such as individual small ceramic turbine blades or thread guides. When moulded parts contain large sections such as the hub of a rotor, problems are likely to be encountered, not just at the polymer removal

Major Binder	Minor Binder	Temp °C	$1/\eta$ Pa^{-1}s^{-1}	n	$d\log\eta/dt$ Pa s K^{-1} ×10^{-3}
Wax blend	–	90	7.9	0.57	−31.6
Polypropylene	–	225	0.7	0.31	− 1.7
Polypropylene	wax	225	1.1	0.25	− 0.3
Polypropylene	dioctylphthalate	225	1.2	0.36	− 2.3
Polypropylene	wax	195	0.9	0.51	−10.5
Ethylene-vinyl-acetate	wax	195	0.9	0.57	−10.2

Table 2: Rheological characteristics of ceramic moulding compositions based on 65 vol% silicon powder and 2 wt% stearic acid.

stage but during the moulding of thick sections. For these reasons some organizations have adopted a two-stage manufacturing route for turbo rotors[104].

The sequence of events as the thermoplastic moulded part cools in the cavity exerts a subtle but important influence on the production of sound mouldings. Unfortunately, the shrinkage of the moulded component is non-uniform; the material at the cavity wall solidifies first and residual stresses are generated during solidification. In the early stages of mould filling the sprue remains molten and the centre of the moulding continues to fill. After solidification of the sprue the centre of the moulding solidifies and shrinks in isolation.

To prolong pressure transmission to the material in the cavity, the sprue, runners and gate should be as large as possible and temperature dependence of viscosity should be low. When the largest moulded section is much larger than the sprue, other methods are needed to keep the sprue molten and these are discussed below.

Much of the shrinkage of a moulded body may be taken up by mould packing under the influence of post injection pressure, but residual shrinkage may give rise to tensile stresses in the centre of the component[105]. A number of techniques have been evolved to measure these stresses experimentally[106–108] and for simple shapes they can be calculated[109]. The application of very high pressures during moulding can reverse the sign of these residual stresses[110] but such pressures are beyond the scope of most moulding equipment. Residual tensile stresses in the interior of unfilled thermoplastic polymer mouldings are frequently relieved by deformation and the appearance of characteristic sink marks at the surface. Unfortunately, in highly filled polymers the material rigidity and faster cooling rates sometimes prevent the relaxation of stresses in this way and defects may appear in the centre of mouldings. These may take the form of voids or clusters of micro-voids or cracks (Figure 11). Thomas[111] has shown how voids can be prevented by the manipulation of hold pressure, but even when voids were absent, residual stress-induced cracks appeared in large sections and were most pronounced after removal of the organic vehicle. Hunt and co-workers[112] have modelled the conditions for void formation in an injection moulded ceramic body from knowledge of the thermal properties of the suspension.

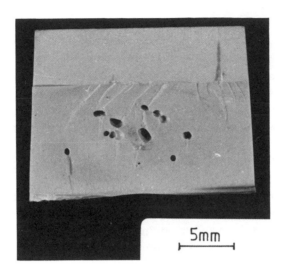

Figure 11: Optical micrograph of shrinkage voids in an alumina moulded part.

The resistance to cracking is expected to be influenced by the bulk mechanical strength of the composite and hence dependent on particle-polymer adhesion[113]. Such adhesion is known to be influenced by the polarity of the polymer[114] and therefore polymers such as vinyl acetate copolymers or acrylics are advantageous. On the other hand the adhesion of polyolefins, used in the form of waxes or high polymers for ceramic injection moulding is likely to be influenced by the extent of oxidation; not least the oxidation brought about during processing.

Other factors that affect the incidence of defects during solidification are thermal shrinkage from the solidification point to room temperature and shrinkage due to crystallization of semicrystalline polymers[85]. In this respect it was found that a low solidification temperature was desirable and that the minimum of crystallinity compatible with rigidity of the moulded component was advantageous[85]. Rigidity is especially important in moulding of high aspect ratio components where irreversible distortion could result in scrap. It can be achieved either by the use of an amorphous polymer with Tg greater than ambient temperature or by semicrystalline polymers or waxes.

Avoidance of shrinkage-related defects in mouldings can be achieved by attention to the material properties described above or by machine techniques. In this latter respect there are several ways of controlling the solidification and shrinkage in the cavity. Perhaps the simplest, used in the polymer processing industry, is the use of hot runner moulds. Rather than allow the sprue and runner to solidify before the larger sections of the moulding, they are kept molten by electrically heated inserts. Meanwhile, the screw or plunger of the moulding machine continues to pack the centre of the cavity and compensates for shrinkage[115]. Hunt[116] shows how a sprue which is alternately heated and cooled can improve mould packing while avoiding the problem of leaving part of the next charge in the heated sprue.

Another method for overcoming this type of defect is the use of combined injection/compression moulding device such as that described by Thomas[117]. The simplest type of mould consists of three parts; a mould cavity, an upper and a lower platten. The injection barrel and nozzle register with a sprue bush and discharge a metered quantity of molten material into the cavity. A press ram then advances, shearing and blocking the gate and compressing the material into the cavity. It is claimed that such mouldings are free from bubbles and sink marks because pressure is applied to the entire moulded part during solidification. There are, however, limitations in component shape which are not inherent in injection moulding.

An alternative method of controlling solidification is to apply an oscillating pressure to the material in the cavity. Techniques to achieve this have been devised[118,119] and employed with great success in the moulding of thermoplastics[120] and particle or fibre reinforced polymers[121]. After the cavity has been filled, an hydraulically driven piston applies oscillating pressure to the material in the sprue and runner with a cushion pressure held on the screw or plunger (Figure 12). The energy transmitted through the fluid is partially expended as heat in the centre of the sprue and runner and in this way the entire solidification and mould packing cycle is controlled. This technique has been successfully applied to ceramic injection moulding[122]. Cracks which appeared in mouldings of silicon powder using a wax binder were overcome by applying a post injection oscillating pressure to the cavity. The in-line compounding injection moulding machine shown in Figures 9 and 10 has this facility built into the plunger barrel. Early work on the technique identified severe localized wear in the oscillator barrel and piston and a device was constructed to receive interchangeable parts to assess wear resistant inserts[123,124].

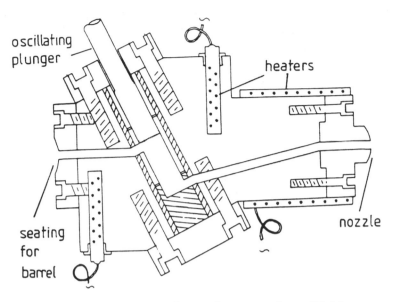

Figure 12: Diagram of equipment for application of oscillating pressure during solidification.

The influence of modulated hold pressure on the sprue solidification time and quality of mouldings has been demonstrated with silicon nitride and silicon suspensions and a rotor mould[125]. An increase in pressure amplitude was found to eliminate cracks in the moulded state and rotors could be heated near to the softening point of the suspension without the appearance of cracks. At very high pressure amplitudes cracks again appeared. Clearly the residual stress distribution is a complex function of pressure amplitude and at present can only be adjusted empirically. The combined effect of mould temperature adjustment and modulated hold pressure has shown how sprue solidification can be adjusted and it also allowed an estimate of sprue power input[126]. The power per unit volume disipated in viscous flow is proportional to the product of viscosity and the square of shear rate. The dependence of sprue heating by modulated pressure on viscosity was demonstrated by modifying viscosity with coupling agents without affecting thermal diffusivity[125].

It is hoped that this section emphasises that it is not enough to select materials and machine conditions which will allow the mould cavity to be filled uniformly and without the incidence of weld lines, but it is also necessary, in mouldings which incorporate thick sections, to be able to control the progress of solidification in the cavity by prolonging access to the moulded part during solidification. This can be achieved in several ways:

i)   attention to material properties;

ii)   use of large sprue, runner and gating systems;

iii) use of heated sprues or hot runners;

iv)   use of appropriate static hold pressures;

v)   use of machine techniques which combine compression and injection moulding; and

vi)   use of modulated hold pressures.

8. REMOVAL OF ORGANIC VEHICLE

The removal of organic species from ceramic injection mouldings is frequently referred to as the most problematic stage in the process, not least because it has received so little systematic attention. Once again, the challenge refers specifically to mouldings with large cross sections where the non-catastrophic removal of polymer could take weeks to accomplish. Yet techniques for removal of the vehicle flourish in published work[22] and can be broadly classified into four categories:

i)   fluid flow by capillary action;

ii)   solvent washing;

iii) evaporation or sublimation; and

iv)   oxidative or thermal degradation.

The former two methods have the considerable advantage that organic vehicle is removed in the liquid state, obviating the need to convert it into large volumes of gaseous product. The former method is particularly attractive and has been widely practised for ceramic[127] and metal powders[128].

In the author's experience it was possible to remove 49 wt.% of a wax blend from a coarse silicon powder-wax moulding measuring 6mm x 12.5mm in cross section by heating in air for 180 minutes surrounded by a fine alumina

(Alcoa A16) powder bed. This procedure overcomes the dangerous early stage when the pores in the ceramic body are completely blocked. In Wiech's work the porous body need only touch the component over one small region to withdraw the wax[129]. There seem to be two disadvantages to the process. In the first place it appears to be most successful with very low viscosity fluids such as waxes and may be less efficient with higher polymers. The second disadvantage centres around the type of powder needed for the surrounding powder bed. The partitioning of vehicle between moulded body and powder bed involves a balance of capillary forces well known to ceramists in the partitioning of glass between alumina and its metallized coating[130] and controlled principally by the respective pore sizes. Thus assuming similar contact angles for the fluid on the material of the moulding and that of the powder bed, smaller pores are required for the powder bed. Clearly a problem arises when fine ceramic powders are used for the body. Capillary pressure, p, is inversely proportional to pore radius, r, so that the pressure difference is:

$$\Delta p = 2\gamma_{LV}\left[\frac{\cos \theta_1}{r_1} - \frac{\cos \theta_2}{r_2}\right] \qquad 5$$

where γ_{LV} is the surface energy of the organic liquid, θ is the contact angle on the ceramic surface and subscripts 1 and 2 refer to powder bed and ceramic body respectively. However, the flow rate of liquid in a capillary, Q, is proportional to the fourth power of radius;

$$Q = \frac{\pi \Delta p r^4}{8\eta L} \qquad 6$$

where η is the fluid viscosity. This disparity is well known in the filling of porous substrates by adhesives[131]. A further difficulty occurs with fine powders as a result of adsorption of organic vehicle on the ceramic surface. A static hydrodynamic layer approximately equal in thickness to the dimensions of a random coiled molecule impedes the flow in fine capillaries[132].

The use of solvent extraction is described in a patent by Strivens[133]. The binder is a mixture of two components, one soluble and one insoluble in the chosen solvent. In the example given, a blend of a coumarone-indene resin and an oxidised wax was used. The wax component could be substantially

removed by solvent extraction in methylated spirits for 12 hours, leaving a
partially porous body from which the remaining binder could be removed by
pyrolysis. Once again, the technique addresses the problem of the initial
stage of binder removal when all pores are filled with resin; thermal methods
of extraction being suitable for the latter stage when mobility of gaseous
decomposition products in the body is enhanced by continuous porosity.
Wiech[134] finds it preferable to perform solvent washing when the wax-
ceramic body is heated above its softening point. One disadvantage of this
process in mass production is the problem of solvent handling under modern
safety standards.

The majority of work has used pyrolysis to remove organic binders.
Three mechanisms of weight loss can be identified[135] and their activation
energies can be measured by thermogravimetry. Low molecular weight species
are lost by evaporation which is not preceded by degradation processes. This
is the case for paraffin wax, for example. Plasticizers such as phthalates
are frequently employed in thermoplastic binder systems[136] as are oils[137]
and these are likely to be removed by evaporation. Wiech[138] prefers to
remove volatile species under an applied isostatic pressure from inert gas
which inhibits boiling. The sublimation of such materials as napthalene,
paradichlorobenzene and camphor has been used to bring the binder removal
time down to 12 hours in the moulding of precision cores for metal
casting[139].

The ceramics industry is accustomed to handling water based systems and
therefore there is considerable interest in water soluble polymer
compositions for oxide ceramics. The thermo-gelling properties of methyl
cellulose solutions have produced a successful moulding process for metal
powders[140] which could be adapted to ceramic suspensions. There is some
interest in moulding processes which use chilled mould conditions below
ambient temperatures to produce solidification in a range of liquids. The
moulded article is ejected and the liquid component removed without
remelting, by sublimation under freeze-drying conditions.

When higher molecular weight organic species are used in order to
improve flow properties and mechanical strength in the moulded body, either
oxidative or thermal degradation processes precede evaporative loss and these
processes are reflected in activation energy measurements in thermogravimetric analysis[135]. Many investigators have used thermal degradation
of thermoplastic polymers or waxes in an inert atmosphere or vacuum as the

preferred binder removal process[40,58,141]. Thermal degradation processes proceed uniformly throughout the body generating low molecular weight products of decomposition which must in turn diffuse, in solution, to the surface and evaporate. The operator aims to control temperature and pressure to prevent boiling of the degradation product solution. It follows that there is a critical balance between the rate of generation of degradation products and diffusion in the polymer-ceramic suspension. Polymers which yield high boiling point products at first appear to be desirable. Indeed polymers which decompose at temperatures below the boiling point of their products would be ideal. Unfortunately such species are rare and frequently present other disadvantages. However, high boiling point products have in general, large molecular dimensions and hence they present lower binary mutual diffusion coefficients in the parent polymer melt.

Many binder systems are composed of olefins either as waxes or high polymers[21] and in this respect it is fortunate that olefin thermal degradation is well researched[142-146]. From the work of Kiang and co-workers[144] the thermal degradation product range of isotactic polypropylene is known (Table 3) and can be used to derive the total volume of gas to be evolved from each moulding. From Table 3 it can be shown that 1g isotactic polypropene can be expected to yield $0.235 \times 10^{-3} m^3$ of gaseous product at S.T.P. For a typical polypropene of density 905 kgm^{-3}, and correcting to 388°C where this particular pyrolysis occurred, this becomes $0.569 \times 10^{-3} m^3 g^{-1}$ or 513 times its own volume. Since the polypropene is likely to make up 40 vol% of injection moulded ceramic, some 205 times the volume of moulded product must be lost as gaseous decomposition product at 388°C.

A similar experiment by Tsuchiya and Sumi[146] gives a slightly different decomposition product distribution to that given in Table 3. Similar data is available for other polymers which undergo random chain scission such as polyethylene[142], waxes and polyisobutylene[145]. Kiang's paper also gives rate constant and activation energies for the pyrolysis reaction (Table 3) and in principle this makes it possible to estimate the rate of production of gases during decomposition.

When furnace atmospheres contain oxygen during the binder removal stage the thermal degradation process is accompanied by, and frequently overwhelmed by, a complex series of chemical reactions involving oxygen and oxygen-containing free radicals[147]. However, oxidative degradation tends to be confined initially to the surface region by the rate limiting step of oxygen

Degradation Product	m.wt	mole % (388°C for 60 min)	
		Isostatic P.P.	Atactic P.P.
Methane	16	0.5	0.5
Ethane	30	3.3	2.7
Propene	42	15.7	19.3
Isobutene	56	3.0	4.4
2-pentene	70	18.9	19.4
2-methyl-1-pentene	84	12.3	12.9
3-methyl-3,5-hexadiene	96	1.0	1.0
2,4-dimethyl-1-heptene	126	33.6	30.8
2,4,6-trimethyl-1-heptene	140	1.0	1.1
4,6-dimethyl-2-nonene	154	1.9	1.4
2,4,6-trimethyl-1-nonene	168	7.8	5.9
$C_{13}H_{22}$	178	0.8	0.7
First order rate constant sec^{-1}		3.7×10^4	4.0×10^{-4}
Activation energy $kJmol^{-1}$		234 ± 25	213 ± 21

Table 3: Thermal degradation product range for isotactic and atactic polypropene.

diffusion in the melt[140]. Thermal degradation processes, needless to say, continue in the interior and the overall weight loss is a complex function of component shape. This has been pointed out by Mutsuddy[149] who has preferred binder removal in oxidising atmospheres. Others[150] have found pure oxygen to be advantageous.

Process control is facilitated by a weight loss control loop in a thermogravimetric furnace designed by Johnsson and co-workers[141]. Using silicon nitride at 57wt% loading in a polyethylene binder system the weight loss was controlled at an overall linear ramp of 5×10^{-3} g min^{-1} which represents approximately 0.06% wt. loss min^{-1} in order to produce 20mm thick cylinders. This reduced the total removal time, included heating up to the initiation of weight loss, and removal of residual carbon to 33 hours. This type of process control makes the manufacturing route much more attractive. It is not restricted to weight loss sensors; gas detectors, gas pressure sensors and d.t.a. devices are used in polymer degradation studies and could conceivably be applied to binder removal process control.

The idea that steady weight loss as a function of temperature in thermogravimetric measurements is desirable for injection moulding suspensions has led to the design of molecular weight distributions for this purpose[152]. Stedman[153] shows how organic species can be blended with this objective by computer selection. While there is no doubt that the thermogravimetric behaviour of organic vehicle can be tailored to assist process control, it was found that no clear correlation existed between thermogravimetric traces for a range of similar organic vehicles and the defects in moulded bodies after pyrolysis[148].

Although in principal, injection moulding should be capable of servicing a wide range of ceramic powders provided attention is paid to optimum volume loading, there are several ways in which the powder chacteristics can interfere with the binder removal. In the first place the powder can act as a catalyst for thermal and oxidative degradation of some polymers[154]. In the second place, adsorbed layers may increase the effective volume fraction of ceramic and impede diffusion of degradation products through the suspension or fluid flow of the liquid binder in capillary extraction.

Many types of defect are produced during removal of the organic vehicle and these have been catalogued[89]. The diagnosis of these defects is not simple. As it is reheated, the moulded body passes through a region where delayed failure can cause cracking under the influence of residual stresses[125]. Deformations as a result of residual stresses may also occur. Prolonged heating at low temperatures may cause cracks because organic degradation may take place at low temperatures[155]. Slumping may occur in unsupported bodies once the softening point of the binder is reached. This is particularly the case where the composition falls short of the maximum volume fraction of ceramic powder[136]. Many investigators prefer to pack the injection moulded part into a powder to support it against this kind of deformation[40,58]. Once the softening point is reached rapid heating may cause cavities to form in the melt. However, fluid properties are subsequently lost as the effective volume fraction of ceramic increases beyond that needed for fluid flow and cracking is thereafter possible. If the shrinkage of the particle assembly as organic vehicle departs from particle junctions, occurs at different times throughout the body then cracks may occur which cannot be attributed to the evolution of gases. This has been a particular problem in the fabrication of ceramic windings[156] and a similar problem has been described by Quackenbush[40].

Much work continues to address these problems and much debate centres around the origin of defects which appear after binder removal. Thomas[111] has shown that under certain circumstances such defects can be traced back to the solidification stage of injection moulding. Carefully designed experiments are needed to isolate the causes of the many defects that beset the ceramic injection moulding process.

9. CONCLUSIONS

Ceramic injection moulding is now recognized to be far more complicated than was at first thought. It can no longer be regarded as a straight technology transfer from the polymer industry. Neither is it likely to be developed to perfection by ceramists alone and the polymer industry still has a part to play. The design of polymer blends and perhaps novel copolymers specifically for the transient application in ceramic fabrication is a potentially rewarding area because of the high added value in fine technical ceramic components, but it has been little exploited to date.

The activity of making artefacts has always been regarded as the poor relation to scientific study. In ceramic fabrication by plastic forming methods, the two are closely interwoven by necessity. This is evidenced by the following observation: for the last fifty years the work-equivalent of many human lifetimes has been devoted to ceramic injection moulding, yet still we do not know the properties of organic binders, powders or machine techniques which are needed for reliable mass production. Most work has been, and much work continues to be, assessed by the production of a token artefact with scant regard for the many complicated phenomena that take place in its manufacture. History suggests that success emerges not merely from the vagueries of chance but from the ordering of knowledge!

REFERENCES

1. Timoney, S.G., Survey of Technological Requirements for High Temperature Materials: Diesel Engine. EUR 7660 EN, EEC Brussels (1981).

2. Woods, M.E., Mandler, W.F. and Scofield, T.L., Designing Ceramic Insulated Components for the Adiabatic Engine. Amer. Ceram. Soc. Bull., 64 [2] 287-293 (1985).

3. Richerson, D.W., Evolution in the U.S. of Ceramic Technology for Turbine Engines. Amer. Ceram. Soc. Bull. 64 [2] 282-286 (1985).

4. Helms, H.E., Johnson, R.A. and Groseclose, L.E., AGT 100 Advanced Gas Turbine Technology Development Contractors Co-ordination Meeting, NASA Contract DEN 3-168 Oct. 1985.

5. Bowen, H.K., Ceramics as Engineering Materials: Structure-Property Processing. In Defect Properties and Processing of High Technology Non-metallic Materials; Crawford, J.H., Chen, Y. and Sibley, W.A. (Eds). M.R.S. Symposium Proceedings 24, North Holland N.Y. (1984).

6. Johnson, D.W., Non-conventional Powder Preparation Techniques. Bull. Amer. Ceram. Soc. 60 [2] 221-224 cont. 243 (1981).

7. Brook, R.J., Fabrication Principles for the Production of Ceramics with Superior Mechanical Properties. Proc. Brit. Ceram. Soc. 32 7-24 (1982).

8. Brook, R.J., Processing Technology for High Performance Ceramics. Mat. Sci. Eng. 71 305-312 (1985).

9. Storm, R.S., Net Shape Fabrication of Alpha Silicon Carbide Turbine Components. ASME Publication 82-GT-216 (1982).

10. Engel, W., Lange, E. and Muller, N., Injection Moulded and Duo-density Silicon Nitride. In Ceramics for High Performance Applications II; Burke, J.J., Lenoe, E.N. and Katz, R.N. (Eds), 527-538, Brook Hill (1978).

11. Woods, M.E., Mandler, W.F. and Scofield, T.L., Designing Ceramic Insulated Components for the Adiabatic Engine. Bull. Amer. Ceram. Soc. 64 [2] 287-293 (1985).

12. Johnson, D.R., Schaffhauser, A.C., Tennery, V.J. and Long, E.L., Ceramic technology for advanced heat engines project. Bull. Amer. Ceram. Soc. 64 [2] 276-281 (1985).

13. Prochazka, S., Sintering of Silicon Carbide. In Ceramics for High Performance Applications; Burke, J.J., Gorum, A.E. and Katz, R.N. (Eds), 239-252, Brook Hill (1974).

14. Terwilliger, G.R. and Lange, F.F., Pressureless Sintering of Silicon Nitride. J. Mat. Sci. 10 1169-74 (1975).

15. Barringer, E.A. and Bowen, H.K., Formation, Packing and Sintering of Monodisperse TiO_2 Powders. Comm Amer. Ceram. Soc. 65 C199-C201 (1982).

16. Tani, E., Yoshimura, M. and Somiya, S., Formation of Ultrafine Tetragonal ZrO_2 Powder Under Hydrothermal Conditions. J. Amer. Ceram. Soc. 66 11-14 (1983).

17. Morgan, P.E.D., Synthesis of 6nm Ultrafine Monoclinic Zirconia. Comm. Amer. Ceram. Soc. 64 C204 (1984).

18. Wei, G.C., Kennedy, C.R. and Harris, L.A., Synthesis of Sinterable SiC Powders by Carbothermal Reduction of Gel-derived Precursors and Pyrolysis of Polycarbosilane. Bull. Amer. Ceram. Soc. 63 [8] 1054-1061 (1984).

19. Japan Patent 80-116604, Sept. 8, 1980; assigned to Asahi Chem. Co. Ltd. Tokyo.

20. Storm, R.S., Ohnsorg, R.W. and Frechette, F.J., Fabrication of Injection Moulded Sintered Alpha SiC Turbine Components. J. Engineering for Power 104 601-606 (1982).

21. Edirisinghe, M.J. and Evans, J.R.G., Review: Fabrication of Engineering Ceramics by Injection Moulding I: Materials Selection. Int. J. High Tech. Ceramics 1-31 (1986).

22. Edirisinghe, M.J. and Evans, J.R.G., Review: Fabrication of Engineering Ceramics by Injection Moulding II: Techniques. Int. J. High Tech. Ceramics 2 249-278 (1986).

23. Kobayashi, K., Furuta, M. and Maeno, Y., European Patent 0034056, Aug. 19, 1981; assigned to NGK Insulators Ltd.

24. Capriz, G. and Loratta, A., Extrusion of non-Newtonian bodies. Proc. Brit. Ceram. Soc. 3 117-133 (1965).

25. Bigg, D.M., Conductive Polymeric Compositions. Polym. Eng. Sci. 17 [12] 842-847 (1977).

26. Riscer, J. and Anderson, E., Composites with Planar Reinforcements. Polym. Eng. Sci. 19 [11] 1-11 (1979).

27. Atkins, K.Z., Gentry, R.R., Gandy, R.C., Berger, S.E. and Schwarz, E.G., Silane Treated Aluminium Trihydrate. Polym. Eng. Sci. 18 73-77 (1978).

28. Miyata, S., Imahashi, T. and Anabuki, H., Fire retarding polypropylene with magnesium hydroxide. J. Appl. Polym. Sci. 25 415-425 (1980).

29. Chang, E.P., Kirsten, R. and Salovery, R., Dynamic Mechanical and Thermal Properties of Fire Retardant Polypropylene. Polym. Eng. Sci. 15 [10] 697-701 (1975).

30. Kerckes, Z.E., Magnetic fillers. In Handbook of fillers and reinforcements for plastics; Katz, H.S. and Milewski, J.V. (Eds), Van Nostrand Rheinhold Co. (1978).

31. Calderwood, H. and Scaife, B.K.P., On the Estimation of Relative Permittability of a Mixture. Proc. 3rd Int. Conf. on Dielectric Materials Measurement and Application (1979).

32. Van Olphen, H., An Introduction to Clay Colloid Chemistry, 2nd Ed, Chapt 7., Wiley, New York, (1977).

33. Parish, M.V., Garcia, R.R. and Bowen, H.K., Dispersion of Oxide Powders in Organic Liquids. J. Mat. Sci. 20 996-1008 (1985).

34. Napper, D.H., Polymeric Stabilization of Colloid Dispersions. Academic Press (1983).

35. Tadmor, Z. and Gogos, C.G., Principles of Polymer Processing, Wiley, New York (1979).

36. Middleman, S., Fundamentals of Polymer Processing, Wiley, New York (1977).

37. Irving, H.F. and Saxton, R.L., Mixing of High Viscosity Materials. In Mixing Theory and Practice II; Uhl, V.W. and Gray, J.B. (Eds), p 189, Academic Press, New York (1967).

38. Schwartzwalder, K., Injection Moulding of Ceramic Materials. Amer. Ceram. Soc. Bull. 28 459-461 (1949).

39. Taylor, H.D., Injection Moulding Intricate Ceramic Shapes. Amer. Ceram. Soc. Bull. 45 768-770 (1966).

40. Quackenbush, C.L., French, K. and Neil, J.T., Fabrication of Sinterable Silicon Nitride by Injection Moulding. Ceram. Eng. Sci. Proc. 3 20-34 (1982).

41. Birchall, J.D., Howard, A.J. and Kendall, K., European Patent 0021682, Jan. 7, 1981; assigned to Imperial Chemical Industries Ltd.

42. Janssen, L.P.B.M., Twin Screw Extrusion, Elsevier (1978).

43. Abram, J., Bowman, J., Behiri, J.C. and Bonfield, W., The Influence of Compounding Route on the Mechanical Properties of Highly Loaded Particulate Filled Polyethylene Composites. Plast. and Rubb. Proc. and Appl. 4 261-269 (1984).

44. Edirisinghe, M.J. and Evans, J.R.G., Compounding Ceramic Powders Prior to Injection Moulding. Proc. Brit. Ceram. Soc. 38 67-80 (1986).

45. U.K. Patent 629, 109, Sept. 13, 1949; assigned to SpA. Plastics Materials Laboratory.

46. Martelli, F., Twin Screw Extruders, Van Nostrand Rheinhold Co. (1983).

47. U.K. Laboratory Compounding Extruder Sets High Standards. European Plastic News p9 June (1982).

48. Ess, J.W., Hornsby, P.R., Lin, S.Y. and Bevis, M.J., Characterization of Dispersion in Mineral Filled Thermoplastics Compounds. Plast. and Rubb. Proc. and Appl. 4 7-14 (1984).

49. Lange, F.F. and Metcalfe, F., Processing Related Fracture Origins II Agglomerate Motion and Cracklike Internal Surfaces. J. Amer. Ceram. Soc. 66 398-406 (1983).

50. Lange, F.F., Processing Related Origins I Observations in Sintered and Isostatically Hot Pressed Al_2O_3-ZrO_2 Composites. J. Amer. Ceram. Soc. 66 393-397 (1983).

51. Kellet, B. and Lange, F.F., Stresses Induced by Differential Sintering in Powder Compacts. J. Amer. Ceram. Soc. 67 369-371 (1984).

52. Lange, F.F., Davis, B.I. and Aksay, I.A., Processing Related Fracture Origins III Differential Sintering of ZrO_2 Agglomerates. J. Amer. Ceram. Soc. 66 407-408 (1983).

53. Lange, F.F., Sinterability of Agglomerated Powders. J. Amer. Ceram. Soc. 67 83-89 (1984).

54. Langbein, D., Van der Waals Attraction between Macroscopic bodies. J. Adhesion 1 237-245 (1969).

55. Temperley, H.N.V., Properties of matter, University Tutorial Press p.185 (1965).

56. Loc. cit. 35 p.436.

57. Zhang, J.G., Edirisinghe, M.J. and Evans, J.R.G., On the Dispersion of Unary and Binary Ceramic Powders in Polymer Blends. Proc. Brit. Ceram. Soc. 42 91-99 (1989).

58. Mann, D.L., Injection Moulding of Sinterable Silicon-base Non-oxide Ceramics. Technical Report AFML - TR-78-200 Dec. (1978).

59. Edirisinghe, M.J. and Evans, J.R.G., Properties of Ceramic Injection Moulding Formulations I Melt Rheology. J. Mat. Sci. 22 269-277 (1987).

60. Mormo, M.J., Mostafa, M.A., Ilideo, J., Fowkes, F.M. and Manson, J.A., Acid-base Interaction in Filler-Matrix Systems. Ind. Eng. Chem. Prod. Res. Dev. 15 [3] 206-211 (1976).

61. Plueddemann, E.P., Coupling Agents Plenum Press, New York (1982).

62. Monte, S.J. and Sugarman, G., Processing Composites with Titanate Coupling Agents - A Review. Polym. Eng. and Sci. 24 1369-1382 (1984).

63. Technical Data Kenrich Petrochemicals Inc. Bayonne N.J. (1985).

64. Cohen, L.B., U.S. Patent 4,539,049, Sept. 3, 1985.

65. Plueddemann, E.P., Adhesion Through Silane Coupling Agents, J. Adhesion 2 184-210 (1970).

66. Anon., Study of Organic Titanates as Adhesion Promoters. J. Coatings Technology 51 [655] 38-43 (1979).

67. Rosen, M.R., From Treating Solution to Filler Surface and Beyond. J. Coatings Technology 50 [644] 70-82 (1978).

68. Culler, S.R., Ishida, H. and Koenig, J.L., Structure of Silane Coupling Agents Adsorbed on Silicon Powder. J. Coll. Intef. Sci. 106 [2] 334-346 (1985).

69. Morrell, S.H., A Review of the Effects of Surface Treatments on Polymer-filler Interactions. Plast. and Rubb. Proc. and Appln. 1 179-186 (1981).

70. Han, C.D., Sandford, C. and You, H.J., Effects of Titanate Coupling Agents on the Rheological and Mechanical Properties of Filled Polyolefins. Polym. Eng. Sci. 18 [11] 849-854 (1978).

71. Han, C.D., Van den Weghe, T., Shete, P. and Haw, J.R., Effects of Coupling Agents on the Rheological Properties, Processability and Mechanical Properties of Filled Polypropylene. Polym. Eng. Sci. 21 [4] 196-204 (1981).

72. Han, C.D., Lu, H.L. and Mijovic, J., Effects of Coupling Agents on the Rheological Behaviour and Mechanical Properties of Filled Nylon 6. S.P.E. Antec 28 82-83 (1982).

73. Bigg, D.M., Rheological Analysis of Highly Loaded Polymeric Composites filled with Non-Agglomerating Spherical Filler Particles. Polym. Eng. Sci. 22 [8] 512-518 (1982).

74. Zhang, J.G., Edirisinghe, M.J. and Evans, J.R.G., The use of Silane Coupling Agents in Ceramic Injection Moulding Formations. J. Mater. Sci. 23 2115-2120 (1988).

75. Hunt, K.H., Evans, J.R.G. and Woodthorpe, J., The Role of Coupling Agents in Zirconia-Polypropylene Suspensions for Ceramic Injection Moulding. Polym. Eng. Sci. 28 1572-1577 1988.

76. Hyatt J.W., U.S. Patent 202,441, April 16, 1878; assigned to Celluloid Manufacturing Co.

77. Peshek, J.R., Machinery for Injection Moulding of Ceramic Shapes. In Advances in Ceramics, vol 9.; Mangels, J. (Ed) 234-238 (1984).

78. Rottenholber, P., Langer, M., Storm, R.S. and Frechette, F., Design Fabrication and Testing of an Experimental Alpha Silicon Carbide Turbocharger Rotor. Society of Automotive Engineers USA Publication No. 810523 (1981).

79. Peltsman, I. and Peltsman, M., Improvements in Machinery for Hot Moulding of Ceramics Under Low Pressure. Ceram. Eng. and Sci. Proc. 3 865-868 (1982).

80. Bevis, M.J., Gaspar, E., Allan P. and Hornsby P.R., U.K. Patent Application 8511152; Date of filing May 2 1984; assigned to Brunel University. Int. Patent WO86/06321; assigned to Brunel University.

81. Farris, R.J., Prediction of the Viscosity of Multimodel Suspension from Unimodel Viscosity Data. Trans. Soc. Rheol. 12 281-301 (1968).

82. Rubin, I.I., Injection Moulding Theory and Practice. Wiley (1972).

83. Criens, R.M. and Moste, H.G., On the Influence of Knit Lines on the Mechanical Behaviour of Injection Moulded Structured Elements. S.P.E. Antec. 28 22-24 (1982).

84. Mangels, J.A. and Trela, W., Ceramic Component by Injection Moulding. In Advances in Ceramics vol 9; Mangels, J. (Ed) 220-223 (1984).

85. Edirisinghe, M.J. and Evans, J.R.G., Properties of Ceramic Injection Moulding Formulations II. J. Mater. Sci. 22 2267-2273 (1987).

86. Sugano, T., Materials for Injection Moulding Silicon Carbide. Proceedings of the 1st Symposium on R & D for Basic Technology for Future Industry Fine Ceramics Project Japan Tech. Assoc. pp67-84 Tokyo (1983).

87. Wu, S., Surface and Interfacial Tension of Polymer Melts. J. Coll. Interf. Sci. 31 [2] 153-161 (1969).

88. Hunt, K.N., Evans, J.R.G. and Woodthorpe, J., The Influence of Mixing Route on the Properties of Ceramic Injection Moulding Blends. Br. Ceram. Trans. J. 87 17-21 1988.

89. Zhang, J.G., Edirisinghe, M.J. and Evans, J.R.G., A Catalogue of Ceramic Injection Moulding Defects and their Causes. Industrial Ceramics 9 72-82 (1989).

90. Mangels, J.A. and Williams, R.M., Injection Moulding Ceramics to High Green Densities. Amer. Ceram Soc. Bull. 62 601-6 (1983).

91. Markoff, C.J. Mutsuddy, B.C. and Lennon, J.W., Method for Determining Critical Ceramic Powder Volume Concentration in the Plastic Forming of Ceramic Mixes. In Advances in Ceramics, vol. 9; Mangels, J. (Ed) 246-250 (1984).

92. Chong, J.S., Christiansen, E.B. and Baer, A.D., Rheology of Concentrated Suspensions, J. Appl. Polym. Sci. 15 2007-2021 (1971).

93. Zhang, T. and Evans, J.R.G., Predicting the Viscosity of Ceramic Injection Mouldings Suspensions. To be published.

94. Gupta, R.K. and Seshadri, S.G., Maximum Loading Levels in Filled Liquid Systems. J. Rheol. 30 503-508 1986.

95. Dabak, I. and Yucel, O. Shear Viscosity Behaviour of Highly Concentrated Suspensions at Low and High Shear Rates. Rheol. Acta 25 527-533 (1986).

96. Bierwagen, G.P., Saunders, T.E., Studies of the Effects of Particle Size Distribution on the Packing Efficiency of Particles. Powder Technology 10 111-119 (1974).

97. Mutsuddy, B.C., Injection Moulding Research Paves Way to Ceramic Engine Parts. Industrial Research and Development 25 76-80 (1983).

98. Cox, H.W., Mentzer, C.C. and Custer, R.C., The Flow of Thermoplastic Melts - Experimental and Predicted. S.P.E. Antec 29 694-697 (1983).

99. Weir, F.E., Doyle, M.E. and Norton, D.G., Mouldability of Plastics Based on Melt Rheology. S.P.E. Transactions 3 32-36 (1963).

100. Edirisinghe, M.J. and Evans, J.R.G., The Rheology of Ceramic Injection Moulding Blends. Br. Ceram. Trans. J. 86 18-22 (1987).

101. Saito, K., Tanaka, T. and Hibino T., U.K. Patent 1,426,317, Feb. 25, 1976; assigned to Tokyo Shibaura Electric Co. Ltd.

102. Weiner, E., U.S. Patent 2,593,507, Apr 22 1952; assigned to Thomson Products Inc.

103. Litman, A.M., Schott, N.R. and Tozlowski, S.W., Rheological properties of highly filled polyolefin/ceramic systems suited to injection moulding. Soc. Plant. Eng. Tech. 22 549-551 (1976).

104. Katayama, K. Watanabe, T., Matoba, K. and Katoh, N., Development of Nissan High Response Ceramic Turbocharger Rotor. SAE Tech. Pap. 861128 (1986).

105. Isayev, A.I. and Wong, K.K., Effect of Processing Conditions on the Residual Stresses in the Injection Moulding of Amorphous Polymers. S.P.E. Antec 28 295-297 (1982).

106. Coxon, L.D. and White, J.R., Residual Stresses and Ageing in Injection Moulded Polypropylene Polym. Eng. Sci. 20 [3] 230-236 (1980).

107. White, J.R., On the Layer Removal Analysis of Residual Stress. J. Mat. Sci. 20 2377-2387 (1985).

108. Sandilands, G.J. and White, J.R., Effect of Injection Pressure and Crazing on Internal Stresses in Injection Moulded Polystyrene. Polymer 21 338-343 (1980).

109. Mills, N.J., Computation of Residual Stresses in Extruded and Injection Moulded Products. Plast. and Rubb. Proc. and Appl. 3 [2] 181-188 (1983).

110. Kubat, J. and Rigdahl, M., Influence of High Injection Pressures on the Internal Stress Level in Injection Moulded Specimens. Polymer 16 925-929 (1975).

111. Thomas, M.S. and Evans, J.R.G., Non-uniform Shrinkage in Ceramic Injection Moulding. Br. Ceram. Trans. J. 87 22-26 1988.

112. Hunt, K.M., Evans, J.R.G. and Woodthorpe, J., Computer Modelling of the Origin of Defects in Ceramic Injection Moulding. To be published.

113. Hull, D., An Introduction to Composite Materials, CUP Chp. 3 (1981).

114. Packham, D.E., Mechanics of Failure of Adhesive Bonds Between Metals and Polyethylene and other Polyolefins. In Developments in Adhesives 2; Kinlock, A.J. (Ed), Appl. Sci. Publ. (1981).

115. Morrison, R.V., U.S. Patent 4,412,805, Nov. 1, 1983; assigned to Discovision Assoc.

116. Hunt, K.N., Evans, J.R.G. and Woodthorpe, J. The use of a Heated Sprue in Ceramic Injection Moulding. To be published.

117. Thomas, I., Injection Moulding of Plastics, Rheinhold pp77-89 (1947).

118. Demag Kunststofftechnik, U.K. Patent 1,553,924, Oct. 10, 1979.

119. Menges, G., Koenig, D., Luettgers, R., Sarholz, R. and Schuermann, E., Follow up Pressure Pulsation in the Manufacture of Thick Walled High Strength Plastic Components. Plastverabeiter 31 185-193 (1980).

120. Allan, P.S. and Bevis, M., The Production of Void Free Thick-section Injection Flow Mouldings I. Plast. and Rubb. Proc. and Appl. 3 85-91 (1983).

121. Idem, Producing Void-Free Thick Section Thermoplastic and Fibre Reinforced Thermoplastic Mouldings. Plas. Rubb. Inst. 9 32-36 (1984).

122. Allan, P.S., Bevis, M.J., Edirisinghe, M.J., Evans, J.R.G. and Hornsby, P.R., Avoidance of Defects in Injection Moulded Technical Ceramics. J. Mat. Sci. Lett. 6 2267-2273 (1987).

123. Edirisinghe, M.J. and Evans, J.R.G., An Oscillating Pressure Unit for Ceramic Injection Moulding A. Materials and Design 8 284-288 (1987).

124. Idem, An Oscillating Pressure Unit for Ceramic Injection Moulding B. ibid 9 85-93 (1988).

125. Zhang, J.G., Edirisinghe, M.J., and Evans, J.R.G., The Use of Modulated Pressure in Ceramic Injection Moulding. J. Euro. Ceramics Soc. 5 63-72 (1989).

126. Zhang, J.G., Edirisinghe, M.J. and Evans, J.R.G., The Control of Sprue Solidification Time in Ceramic Injection Moulding. J. Mater. Sci. 24 840-848 1989.

127. Pelstman, I. and Peltsman, M., Low Pressure Moulding of Ceramic Materials. Interceram 4 56 (1984).

128. Wei, T.S. and German, R.M., Injection Moulded Tungsten Heavy Alloy. Int. J. Powder Met. 24 327-335 1988.

129. Wiech, R.E., European Patent 0032403, Jul. 22 1981.

130. Twentyman, M.E., High Temperature Metallizing. J. Mat. Sci. 10 765-776 (1975).

131. de Bruyne, N.A. The Extent of Contact Between Glue and Adherend. Aero Research Tech. Notes. Bull No. 168 1-12 (1956).

132. Priel, Z. and Silberberg, A. The Thickness of Adsorbed Polymer Layers at a Liquid-solid Interface as a function of Bulk Concentration. J. Polym. Sci. A2. Polym. Phys. 16 1917-1925 1978.

133. Strivens, M.A., U.K. Patent 808,583, Feb. 4, 1959; assigned to Std. Telephones and Cables Ltd.

134. Wiech, R.E., U.S. Patent 4,197,118, Apr. 1980; assigned to Parmatech Corporation.

135. Wright, J.K., Evans, J.R.G. and Edirisinghe. Degradation of Polyolefin Blends Used for Ceramic Injection Moulding. J. Amer. Ceram. Soc. In press.

136. Bendix Aviation Corporation; U.K. Patent 706,728; Apr. 7, 1954.

137. Pett, R.A., Rao, V.D.N and Qaderi, S.B.A, U.S. Patent 4,265,794, May 5, 1981; assigned to Ford Motor Co.

138. Wiech, R.E., European Patent 0032404, Jul. 22 1981.

139. Hermann, E.R., U.S. Patent 3,234,308, Feb. 8, 1966; assigned to Corning Glass Works.

140. Rivers, R.D, U.S. Patent 4,113,480, Sept. 12, 1978; assigned to Cabot Corp.

141. Johnsson, A., Carlstrom, E., Hermansson, L. and Carlsson, R., Rate Controlled Thermal Extraction of Organic Binders from Injection Moulded Ceramic Bodies. Proc. Brit. Ceram. Soc. 33 139-147 (1983).

142. Tsuchiya, Y. and Sumi, K., Thermal Decomposition Products of Polyethylene. J. Polym. Sci. A-I 6 415-424 (1968).

143. Stivala, S.S., Kimira, J. and Reich, L., Kinetics of Degradation Reactions. In Degradation and Stabilization of Polymers 1; Jellinek, H.H.G. (Ed), Elsevier, New York (1983).

144. Kiang, J.K.Y., Uden, P.C. and Chien, J.C.W., Polymer Reactions VII: Thermal Pyrolysis of Polypropylene Polymer Degradation and Stability 2 113-127 (1980).

145. Seiger, M. and Gitter, R.J., Thermal Decomposition and Volatilization of Polyolefins. J. Polym. Sci. Polym. Chem. 15 1393-1402 (1977).

146. Tsuchiya, Y. and Sumi, K., Thermal Decomposition Products of Polypropylene. J. Polym. Sci. A-I 7 1599-1607 (1969).

147. Norling, P.M. and Tobolsky, A.V., Fundamental Reactions in Oxidative Chemistry. In Thermal Solubility of Polymers; Conley, R.T. (Ed), Vol. 1, p. 113, Marcel Dekker, New York (1970).

148. Woodthorpe, J., Edirisinghe, M.J., Evans, J.R.G., Properties of Ceramic Injection Moulding Formulations III Polymer Removed. J. Mater. Sci. 24 1038-1048 (1989).

149. Mutsuddy, B.C., Oxidative Removal of Organic Binders from Injection Moulded Ceramics, Proc. Int. Conf. on Non-Oxide Tech. and Eng. Ceram. NIHE Limerick, 10-12 July (1985).

150. Gilissen, R. and Smolders, A., Binder Removal from Injection Moulded Bodies. In High. Tech. Ceramics; Vincenzini, P. (Ed) Elsevier, Amsterdam pp. 591-594 (1987).

151. Kiran, E. and Gillham, J.K., Pyrolysis-Molecular Weight-Chromatography-Vapour Phase Infra Red Spectrophotometry: An On Line System for Analysis of Polymers. In Developments in Polymer Degradation 2; Grassie, N. (Ed), Appl. Sci. Publ. London (1979).

152. Saito, K., Tanaka, T. and Hibino, T. U.S. Pat. 4000110. Assigned to Tokyo Shibaura Electric Co. Ltd. 28 Dec. (1976).

153. Stedman, S.J., Evans, J.R.G. and Woodthorpe, J., A Method for Selecting Organic Materials for Ceramic Injection Moulding. Ceramics International. In press.

154. Masia, S., Calvert, P.D., Rhine, W.E. and Bowen, H.K. Effects of Oxides on Binder Burnout During Ceramic Processing. J. Mater. Sci. 24 1907-1912 (1989).

155. Zhang, J.G., Edirisinge, M.J. and Evans, J.R.G. Initial Heating Rate for Binder Removal from Ceramic Mouldings. Materials Letters 7 15-18 (1988).

156. Wright, J.K. Thompson, R.M. and Evans, J.R.G., On the Fabrication of Ceramic Windings. J. Mater. Sci. In press.

7

Electrophoretic Deposition as a Processing Route for Ceramics

S.N. Heavens

Chloride Silent Power Limited, Davy Road, Astmoor,
Runcorn, Cheshire WA7 1PZ, UK.

1. INTRODUCTION

The phenomenon of electrophoresis was discovered in 1807 by the Russian physicist F.F. Reuss, who observed that when an electric current was passed through a suspension of clay in water the clay particles migrated towards the anode. It is interesting that, 170 years later, the first industrial application of electrophoresis in ceramic forming should be based on the same system – aqueous clay suspensions – as Reuss' original glass tube experiment. During the intervening period (Table 1) electrophoresis has been studied and applied in a wide variety of systems, especially as a technique for the coating of metal components. Metals, oxides, phosphors, inorganic and organic paints, rubber, dielectrics and glasses have been deposited by this technique using both aqueous and non-aqueous media. Some have found large-scale application in manufacturing, notably rubber products, and the application of paint in the automotive industry. In the ceramic manufacturing industry, however, the number of applications has until recently been rather limited. Two cases will be described in detail: the "Elephant" clay strip forming device manufactured by Handle GmbH & Co. KG, and the production of beta"-alumina tubes for the sodium/sulphur battery under development at Chloride Silent Power Ltd.

255

1808	Reuss	Discovery of electrophoresis (1)
1879	von Helmholtz	Theory of electrical double layer (2)
1908	Cockerill	Rubber separation form latex (3)
1927	Sheppard	Commercial production of rubber goods (4)
1939	de Boer et al	Use of organic suspensions Ceramic oxide coatings (5)
1948	Berkman	Forming of porcelain crucibles (6)
1969	Andrews et al	Forming of alumina shapes (7)
1971	Fally et al	Forming of beta-alumina and zirconia (8)
1975	Chronberg	Clay strip forming production machine (9)

Table 1: Historical development of electrophoretic forming.

2. APPLICATIONS OF ELECTRODEPOSITION

Ceramic Forming
1. Porcelain cups and sanitaryware[6,10-17]
2. Continuous forming of clay strip[9,18-22]
3. Alumina radomes[7]
4. Beta/beta"-alumina electrolyte tubes[8,23-27]

Ceramic and Glass Coatings
1. Heat-resistant oxide coatings for tube cathodes[5,28-33]
2. Ceramic/metal coatings for wear/corrosion resistance[34-40]
3. Carbide-coated cutting tools[40]
4. Vitreous enamelling[41]
5. Electronic devices:
    Thin-film dielectric coatings for capacitors[42]
    Thick-film substrates for hybrid circuits[43]
    Glass passivation of semiconductor surfaces[44-46]
6. Phosphor coatings:
    Screens for cathode-ray tubes[47-49]
    Thin-film solar cells[50]

Other Applications
1. Rubber products from latex[4,51]
2. Organic and oxide paints for sheet metal[52-54]
3. Polymer coatings[55]
4. Dehydration of suspensions and emulsions[56]

3. FUNDAMENTALS OF ELECTRODEPOSITION

3.1 Definitions

The term "electrodeposition" is often used somewhat ambiguously to refer to either electroplating or electrophoretic deposition, although it more usually refers to the former. In Table 2 the distinction between the two processes is indicated. There may, in any case, be a certain amount of confusion between the two in cases in which the chemistry of the process is unclear or does not fall neatly into either of the two categories.   In deposition from aqueous clay suspensions, for example, it has been argued[19] that the rate-limiting process is charge transfer at the electrode and not the mass transfer of particles. In electrophoretic deposition the particles do not immediately lose their charge on being deposited, which can be shown from the observation that reversal of the electric field will strip off the deposited layer[57].

	Electroplating	Electrophoretic deposition
Moving species	ions	solid particles
Charge transfer on deposition	ion reduction	none
Required conductance of liquid medium	high	low
Preferred liquid	water	organic
Deposition rate	≈ 0.1 μm/minute	≈ 1 mm/minute

Table 2: Characteristics of electrodeposition techniques.

3.2 Origins of Charging

When phases come into contact, some redistribution of positive and negative charge invariably occurs, leading to the formation of an electric double layer and to a potential difference between the phases. Double layers can be formed by one or more of the following mechanisms:

1.   selective adsorption of ions onto the solid particle from the liquid phase,

2.   dissociation of ions from the solid phase into the liquid,

3.   adsorption or orientation of dipolar molecules at the particle surface,

4.   electron transfer between solid and liquid phases due to differences in work function.

In the case of ceramic or glass particles in water or organic liquids, the last mechanism is inapplicable but the first two invariably occur. The sign of the net charge on the particle will depend not only on whether the ions involved are positive or negative, but also on whether mechanism 1 or 2 is dominant.   Also, a positively-charged particle may even behave like a negative one i.e. be attracted to a positive electrode, if an excess of negative ions is attracted to the vicinity of the particle.   An example of this is beta-alumina in dichloromethane[23], in which addition of trichloracetic acid results in negative charging; similarly for acetone/nitrocellulose suspensions[49], in which the sulfuric or phosphoric acid necessary for charging results in anodic, not cathodic, deposition. Consequently it is difficult to predict in an unknown system whether anodic, cathodic, or no deposition will occur.   In water most solid particles acquire a negative charge but in organic liquids charging may be either positive or negative.   Once again beta-alumina serves as an example.   When beta-alumina particles are dispersed in an alcohol they acquire a net a positive charge which is thought to result from the adsorption of protons from dissolved water[24]:

$$Na\text{-}\beta + H_2O \ \rightarrow \ H^+Na\text{-}\beta + OH^- \hspace{4cm} 1$$

If the dispersion is milled, the charge is first neutralized and subsequently reversed[58].   In this event the charging mechanism is thought to result from friction- or fracture-induced dissociation of sodium ions[24] or protons[58] from the beta-alumina particles:

$$Na-\beta \rightarrow \beta^- + Na^+ \qquad\qquad 2$$

Or:

$$Na-\beta + H_2O \rightarrow H-\beta + NaOH \qquad\qquad 3$$

$$H-\beta \rightarrow \beta^- + H^+ \qquad\qquad 4$$

Consequently the sign of the charge may depend on the amount of water in the suspension medium and on the extent of milling. If, by chance, these effects are equal, the charge is neutralized and no deposition will take place. For this reason it is often preferable to control the charge state by suitable additives.

In an aqueous clay suspension the particles generally acquire a negative charge but this does not mean that the charging mechanism is unique. Kaolinite acquires its charge by the adsorption of hydroxyl ions, while muscovite mica is charged by the dissociation of potassium ions[59].

3.3 Conditions for Electrophoresis

Electrophoresis is the phenomenon of motion of particles in a colloidal solution or suspension in an electric field, and generally occurs when the distance over which the double layer charge falls to zero is large compared to the particle size. In this condition the particles will move relative to the liquid phase when the field is applied. Colloidal particles, which are 1 μm or less in diameter, tend to remain in suspension for long periods due to Brownian motion. Particles larger than 1 μm require continuous hydrodynamic agitation to remain in suspension. An important property of the suspension is its stability. Suspension stability is characterised by settling rate and tendency to undergo or avoid flocculation[60]. Stable suspensions show no tendency to flocculate, and settle slowly, forming dense, strongly adhering deposits at the bottom of the container. Flocculating suspensions settle rapidly and form low density, weakly adhering deposits. According to some theories[57] electrophoretic deposition is analogous to gravitational sedimentation; the charge on the particles is immaterial so far as the properties of the deposit are concerned. The function of the electric field is solely to drive the particles towards the electrode and to exert pressure on the deposited layer in the same way as a gravitational force. The analogy can only be pursued qualitatively. Measured particle velocities under electrophoresis are much lower than those calculated from Stokes' law. Owing to interionic interaction between the liquid carried along with the particle (Stern layer) and the surrounding liquid, the ions of opposite charge in the double layer, moving in the opposite direction to the particle, exert a retarding force. An additional retarding force is caused by the continual break-up and re-forming of the double layer around the moving particle.

If the suspension is too stable, the repulsive forces between the particles will not be overcome by the electric field, and deposition will not occur. According to some models for electrophoretic deposition the suspension should be unstable in the vicinity of the electrode[61]. Local instability could be caused by the formation of ions from electrolysis or discharge of the particles; these ions then cause flocculation close to the electrode surface.

Optimum particle size for electrophoretic deposition is normally in the range 1-20 μm. For electrophoresis to occur with larger particles, either a very strong surface charge must be obtained, or the double layer region must increase in size. Consequently the ionic concentration in the liquid must remain low, a condition favored in liquids of low dielectric constant. Powers[24] investigated beta-alumina suspensions in numerous organic media and determined the incidence of deposition as a function of the dielectric constant of the liquid and the conductivity of the suspension. A sharp increase in conductivity with dielectric constant was noted; this observation apparently refers to the liquids in their pure state. It should also be noted that impurities, in particular water, affect the conductivity, and that the conductivity of a milled suspension is very different to that of the pure liquid[27], as a consequence of the dissociative or adsorptive charging modes. Powers[24] obtained deposits only with liquids for which the dielectric constant was in the range 12-25. With too low a dielectric constant, deposition fails because of insufficient dissociating power, whilst with a high dielectric constant, the high ionic concentration in the liquid reduces the size of the double layer region and consequently the electrophoretic mobility.

3.4 Electrophoretic Yield

The rate of deposition of material can be determined theoretically from Hamaker's equation[28]:

$$M = \int_0^t aAC\mu E.dt \qquad\qquad 5$$

where M = mass deposited in time t

C = particle concentration in the suspension (kg/m^3)

E = electric field (V/m)

A = electrode area (m^2)

μ = electrophoretic mobility (m^2/Vs)

a = coefficient representing the fraction of particles near the electrode that get deposited.

While A, C and E are externally controlled variables, μ is determined by the properties of the suspension. In simplified double-layer theory the interface is treated as a capacitance and μ is given by the Smoluchowski equation:

$$\mu = \epsilon\phi_z/4\pi\eta \qquad\qquad 6$$

where ϵ = dielectric constant of the liquid

η = viscosity of the liquid

ψ_z = electrokinetic (zeta) potential

The physical significance of the zeta potential is discussed in the following section. The suspension could be characterised by particle charge density, which can in principle be determined from the electrophoretic mobility, but which requires certain assumptions regarding the particle size and shape distribution and conductivity effects. The zeta potential is the most commonly used parameter for characterising a suspension, and can be determined from measurements of particle velocity or mobility in an applied field using commercially available electrophoresis cells.   In practice electrophoretic mobilities are not easy to measure accurately, and since the Smoluchowski equation is based on a model of doubtful validity, the view sometimes expressed that "zeta potentials are difficult to measure and impossible to interpret" has a ring of truth but is probably unduly pessimistic. The Smoluchowski relation is valid provided that the double layer region is small relative to the particle radius[62], for example with large particles in aqueous suspension. For small particles in non-ionizing solvents it is more appropriate to use the Huckel equation:

$$\mu = \epsilon\phi_z/6\pi\eta \qquad\qquad 7$$

which is strictly valid for spherical particles only.

3.5 Zeta Potential

The zeta potential is identified with the value of the double-layer potential at the outer edge of the Stern layer, which is the layer of fluid that remains attached to the particle during relative motion of the particles and liquid during electrophoresis (Figure 1). The region extending outwards from the Stern layer is termed the diffuse region, in which electrostatic forces are opposed by those of thermal agitation, and the potential decays approximately exponentially. In Figure 1 ϕ_z represents the electrokinetic (zeta) potential, and ϕ_{DL} the Nernst (double-layer) potential, which is the potential at the surface of the particle relative to a remote point in the liquid.

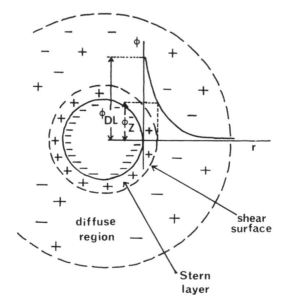

Figure 1: Model of charged spherical particles in a liquid indicating potential distribution $\phi(r)$.

Numerous physical and chemical factors can influence the magnitude and sign of the zeta potential. It may depend on the presence or absence of added electrolytes, or on the concentration of the suspension itself. In liquids that contain no surfactant ions the zeta potential falls with increasing concentration. In the presence of surface-active or multivalent ions the sign of the zeta potential may be reversed. The topography of the solid/liquid interface is also significant.

Choudhary et al[63] investigated the variation of zeta potential of aqueous oxide suspensions in which the pH was controlled by addition of triethylamine, and found that zeta potential increased with pH up to a certain value; this value was, however, not constant and probably dependent on physical factors. Frey and Rackl[64] investigated the electrophoretic mobility of numerous clay suspensions and were unable to deduce any significant influences related to pH, conductivity, viscosity or particle size distribution.

From a fundamental point of view there are therefore a number of difficulties. There is no satisfactory theory that accounts for all observations on electrophoretic deposition. It is desirable to find suitable physical/chemical parameters that characterise a suspension sufficiently in order that its ability to deposit can be predicted. Most investigators use zeta potential or electrophoretic mobility, but these do not uniquely determine the ability of a suspension to deposit. For example, in suspensions of aluminium in alcohol the addition of an electrolyte causes no significant change to the zeta potential, but deposits can only be obtained in the presence of the electrolyte[65]. The stability of the suspension is evidently its most significant property, but this is a somewhat empirical property not closely related to fundamental parameters.

3.6 Practical Considerations

In view of the sensitivity of the electrophoretic mobility to factors such as chemical environment and particle surface topography, and the need for suspensions of marginal stability, it might be thought that a process based on electrophoretic deposition would be inherently difficult to control. This situation is not helped by the shortcomings in fundamental understanding of electrophoretic deposition, and it is almost impossible to predict whether suspensions will deposit electrophoretically. A summary of known working systems is given in the following section. With any system it is of course absolutely necessary to avoid contamination by any impurity that can adversely influence the electrokinetic properties of the suspension; a stringent requirement, but one that should perhaps be regarded as a strength rather than a weakness of the process.

Furthermore, in the concentrated suspensions that are generally used in electrodeposition in order to minimise forming times, variations in electrophoretic mobility from one particle to another tend to become averaged out and there is no segregation as occurs in biological electrophoresis. It is quite possible for particles of very different zeta potential to codeposit uniformly. Thus there is little or no segregation of particles of differing chemical composition, density or particle size during electrodeposition[12,66]. The probable reason is that the dense flux of strongly charged particles will entrain the weakly charged ones.

A practical problem that is likely to prove more relevant concerns the degree of adherence of the deposit to the electrode. For ceramic coatings a strong adherence is desired; for ceramic shapes, on the other hand, it is essential that the shape can be easily removed from the mold or mandrel. In the latter case, one still requires a suspension that deposits with good adherence - but not too good. The degree of adherence is influenced by the nature of the electrode surface, drying shrinkage, the presence of electrolytes, binders and so forth and it is not possible to draw any general conclusions, except that for ceramic shape forming the use of binders is not usually favored; although they may improve the green strength of the material, the problem of excessive adherence to the electrode is increased.

4. EXAMPLES OF WORKING SYSTEMS

4.1 Aqueous Suspensions

Electrophoretic deposition from an aqueous suspension has generally involved clays, oxide ceramic powders, paint or polymers (Table 3). All are characterised by anodic deposition, although it is possible to deposit paint cathodically[67].

SOLID PHASE	ADDITIVES	REFERENCE
Al_2O_3	polyacrylic acid, triethylamine	68
CdS	ethanol, NH_4OH	50
Cr_2O_3, Fe_2O_3, NiO	polyacrylic acid, triethylamine	69
Cr_2O_3,Fe_2O_3,NiO,TiO_2,MnO_2	polyacrylic acid, methylamine	52
TiO2	polyacrylic acid, triethylamine	70
Kaolinite	Na_2CO_3, NaOH, Na_2SiO_3, $Na_2P_4O_7$	11
Kaolinite, mica, talc, pyrophillite	$Na_2P_4O_7$	59
Kaolinite, feldspar, quartz	$Na_2P_4O_7$	64
Kaolinite, illite	Na_2CO_3, Na_2SiO_3	16
Porcelain	water glass	13
Quartz	none	71
Al, Al-Cr, Al-Si	amine-stabilized acrylic resin	36
Latex	none	72
Latex	lipines, cholesterols, choleic acid, ammonium sorbital borate, diglycol stearate	51
Paint (styrene-acrylic)	2-amino-2-methyl-1-propanol	54
Polymers (PVC, PTFE, nylon, polyacrylates, polyester)	not disclosed	55

Table 3: Electrophoretic deposition using aqueous suspensions.

The use of water-based suspensions causes a number of problems in electrophoretic forming. Electrolysis of water occurs at low voltages, and gas evolution at the electrodes is inevitable at field strengths high enough to give reasonably short deposit times. This causes bubbles to be trapped within the deposit unless special procedures are adopted, such as the use of absorbing or porous electrode materials, or high speed chamber flows. Current densities are high, leading to Joule heating of the suspension, and electrochemical attack of the metallic electrodes leads to progressive degradation and to contamination of the deposit. To avoid some of these difficulties the use of isopressed graphite/clay or graphite/cement electrodes has been suggested. Unfortunately these molds are rather soft, or require firing, with consequent loss of dimensional control[14,17].

4.2 Non-aqueous Suspensions

In the Appendix a survey is made of the literature on laboratory electrophoretic deposition systems involving non-aqueous media, including ceramic, glass, metallic and mixed powders. The list is unlikely to be exhaustive. Patents are not listed but can be found from the papers referred to. Furthermore, there are numerous examples of suspensions that have been described but which do not exhibit useful deposition characteristics; these have not been included. No attempt has been made here to identify the role of the additives, which may according to circumstances function as dispersants, deflocculants, electrolytes or binders. Experimental parameters such as zeta potential are often not reported, but in some cases electrophoretic mobility or coulombic yield have been measured or can be estimated from the data. Fuller details will be found in the references.

Where materials are labelled with an asterisk, a large number of powders were successfully deposited using the suspension medium described. Mizuguchi et al[49] included alumina; barium, strontium and calcium carbonates; magnesia, zinc oxide, titanium dioxide, silica, indium oxide, lanthanum boride, tungsten carbide, cadmium sulfide and several metals and phosphors. The list of materials described by Gutierrez et al[35] included several metals; carbides of molybdenum, zirconium, tungsten, thorium, uranium, neptunium and plutonium; zirconium hydride, tantalum oxide and uranium dioxide. In addition, many metallic and oxide powder suspensions in alcohols, acetone and dinitromethane were studied by Brown and Salt[65].

In general, organic liquids are superior to water as a suspension medium for electrophoretic forming. While the generally lower dielectric constant in organic liquids limits the charge on the particles as a result of the lower dissociating power, much higher field strengths can be used since the problems of electrolytic gas evolution, Joule heating and electrochemical attack of the electrodes are greatly reduced or non-existent. Against this must be set the disadvantages of the cost, toxicity and flammability of organic liquids, and judicious selection of a suitable medium and the practice of solvent reclamation will be needed in order to minimise these problems.

4.3 Design of Electrophoretic Apparatus

The design and construction of an electrophoretic deposition facility is straightforward, an attraction of the process being its flexibility in regard to green shape geometry.   Flat plates, crucible shapes, tubes of circular or rectangular cross-section of any size can be deposited by means of appropriate design of the mandrel or forming electrode.   The maximum wall thickness will probably be limited by the electrical resistance of the deposit, but deposits up to 10 mm in thickness have been achieved[7,22].

Although the equipment requirements for electrophoretic deposition are simple (Figure 2), a number of design features need to be taken into consideration.   The materials used in construction of the chamber must be selected carefully to avoid neutralization of the charge on the suspended particles.   Being of marginal stability the suspension will require continuous agitation to avoid gravitational settling if deposition times are more than a few seconds.   Agitation can be achieved by magnetic stirring or by peristaltic pumping.   The chamber geometry may need to be carefully designed to ensure a uniform electric field between the electrodes.   Chamber design criteria have been described for obtaining an electric field

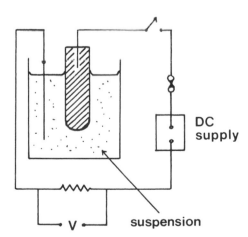

Figure 2: Apparatus for electrophoretic deposition.

distribution that yields uniform wall thickness in cup-shaped deposits[15] and in plates and disks[71]. The thickness profile can be influenced not only by the field distribution but also by the flow pattern of the suspension in the chamber.

5. APPLICATIONS DEVELOPED TO PILOT PLANT SCALE

5.1 Continuous Clay Strip Forming

A machine based on the principle of electrophoretic deposition has been devised for fabricating thin continuous strips of clay, suitable for cutting into tiles or plates of arbitrary shape or size. The device, marketed under the trade name "Elephant", is shown schematically in Figure 3. Two adjacent contrarotating zinc-coated cylinders serve as the anode. Between the cylinders is located a fixed cathode in the shape of a wedge, so that the interelectrode spacing is 12 mm. An aqueous slip of 50% solids content is pumped into the space between the electrodes, and the negatively-charged clay particles deposit on both rotating cylinders. The two deposits thicken and join to form a single strip which is guided away by means of a conveyor belt. This strip contains 10-18% by weight of water and is in a plastic state that can be machined prior to drying and firing. Excess slip flows towards the upper rim of the wedge and is returned to the mixer for recycling.

The width of the strip is determined by the dimensions of the cylinders, and thickness of the strip is a function of the speed of rotation (typically 5 revolutions per hour). Strips of thickness ranging from 2 to 20 mm can be produced. The purpose of forming the strip from two deposits is to avoid the warping of the tiles during drying and firing that occurs in a single deposit as a result of varying water content through the deposit thickness. By joining two deposits the internal stresses during drying are balanced out. The join between the two deposits after firing is invisible under the scanning electron microscope[22].

Bubbles in the deposit from hydrogen evolution due to electrolysis are avoided by ensuring that the slip is pumped rapidly. Excess slip not deposited is recycled and mixed with incoming fresh slip. Oxygen is absorbed by the zinc anodes, the zinc oxide being removed continuously by brushing.

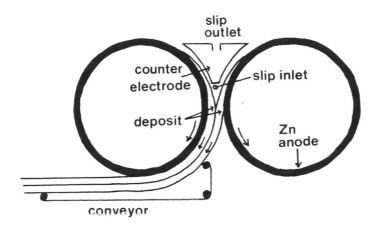

Figure 3: Schematic diagram of 'Elephant' clay strip production machine.

The anodes need to be periodically recoated with zinc, approximately once per year, which consumes 0.5 kg of zinc per tonne of clay strip processed.

A variety of clay minerals can be deposited with the equipment, the most suitable[21,22] being illite-type materials or kaolinite with a low fines content.   Quartz, feldspar and fireclay have also been successfully deposited.

At the time of writing, two pilot scale machines and two production machines have been in operation. The production machines contain cylinders of diameter 1.5 m and width 700 mm, with an output of some 250 kg/h drawn strip, or approximately 500 m² per day. These machines have been used for the production of thin ceramic tiles. The processing time from slip to fabricated tile amounts to 2-3 h, and the complete installation requires no more than two persons for operation. Energy consumption is around 30 kWh/tonne in wet processing and 45 kWh/tonne in drying and firing.

In the absence of a comparative study with conventional extrusion processes it is difficult to assess the potential of the Elephant process.

No measurements of the strength of the fired material have been reported. The principal advantages claimed for the Elephant process are its great flexibility with low capital investment costs, and low energy consumption and maintenance costs.   When the Elephant machine became available commercially in 1980 its cost was approximately US$120,000, a complete tile fabrication system including slip preparation plant, drying and tunnel kiln costing around US$800,000[22].   The system allows the fabrication of special products at short notice, and a wide range of ceramic plate dimensions is possible, from mosaic tiles to sheets of unlimited length.   So far the ceramic tile industry has been slow to realise the possibilities afforded by this machine.

5.2 Beta"-alumina Tube

The sodium/sulphur high energy density battery, which is expected to become available commercially during the 1990's, uses beta"-alumina as its solid electrolyte.   From the early days of its development the need became apparent for a means of fabricating closed-end thin-walled tubes to exacting standards.   If such tubes are manufactured by, for example, isostatic pressing, flaws are easily generated during filling of the mold and the quality of the closed-end region is often inferior, because it is difficult to produce beta"-alumina powders that possess good flow properties and crush readily under pressure, even when spray-dried.   The electrophoretic deposition process lends itself naturally to the formation of uniformly thin-walled tubes with flat or dome-shaped ends.   Electrodeposited beta"-alumina has good sinterability, and isostatic pressing of the green shape is not necessary.   Some developers[8] have nonetheless preferred to isopress the green shapes.

Powers[24] used commercial beta aluminas, Monofrax H fused brick and Alcoa XB-2 powder, as raw materials.   These materials can be synthesized, with the addition of dopants for stabilizing the more conductive beta"-alumina phase if desired, from the calcination of commercial aluminum salts or oxides with sodium carbonate or similar material.   Beta-alumina and beta"-alumina (magnesia- and/or lithia-stabilized) have both been satisfactorily deposited.

During development of the process at General Electric the liquid used was pentan-1-ol (primary amyl alcohol), mainly because of its dielectric properties but also in view of its relatively non-flammable and non-toxic nature and commercial availability. Potential differences of up to 500 volts could be used without undesirable electrochemical reactions occurring.

Powers[24] charged beta-alumina suspensions by milling; both ball milling and vibratory milling could be used, while jet milling of the powder prior to dispersion in the liquid was found to be unsatisfactory. Grinding media of alumina or zirconia can be used, with zirconia being preferred as its wear rate is much lower than that of alumina. Binders were found to be undesirable since, although they could be used to improve green shape strength, they caused the deposit to adhere to the mandrel. The mandrels should be very slightly tapered to facilitate removal but no special surface finish or polish is required.

Electrophoretic deposition is most conveniently carried out by making the chamber wall the counter-electrode. The deposition process itself takes typically three minutes with an electric field of $/ \times 10^4$ V/m. Immediately following deposition the shape is placed in a low humidity drying oven to evaporate the liquid. With conventional drying techniques a few hours' drying is necessary before the shape can be removed from the mandrel.

Other complications from the point of view of large-scale production are (i) the need to pre-convert raw materials into ß/ß"-alumina powder prior to preparation of the suspension, and (ii) the need to used a non aqueous medium. Nevertheless, provided that costs can be maintained sufficiently low and that processing yields are high, there are substantial compensating advantages: pre-conversion of the powder, followed by the extended milling process, ensures that the green shape will be chemically and physically homogeneous, and the processing defects commonly encountered in mechanically pressed beta-alumina shapes - typically voids, inclusions and microcracks - are completely absent in electrodeposited material. Such processing defects are very damaging to the durability of the beta-alumina electrolyte in sodium/sulphur cell operating conditions. With electrodeposited beta-alumina the mechanical strength is governed not by processing flaws but by the microstructural control obtained during firing, and the absence of processing flaws results in a relatively high Weibull modulus in both mechanical strength and cell life characteristics.

Another characteristic of electrodeposited green shapes is their low green density, with close to 50% volume porosity, leading to the unusually high linear shrinkage of 20% on firing. In conventional philosophy this high shrinkage is a disadvantage, however for shapes that are not excessively large - electrolyte tubes are typically 33 mm in diameter, with a wall thickness of 1.5 mm - problems of distortion and cracking on firing do not appear to be severe.

Through appropriate design of the deposition mandrel and chamber, and control of the voltage/current and time of deposition, it is possible to exercise control of the green shape thickness to within 2% without recourse to electrical probes[77] or to the use of load cells to monitor deposit weight.

An example of the microstructure of electrodeposited beta"-alumina is shown in Figure 4.   Fired densities are in excess of 99% of the theoretical X-ray crystallographic density, with a mean pore size of 2-3 μm. The matrix grain size is also 2-3 μm, with a small number of large grains up to around 50 μm. The mean strength of the material, as measured by C-ring diametral compression testing[78], is 250 MPa with a Weibull modulus of 11; this compares favorably with that of isostatically pressed material.

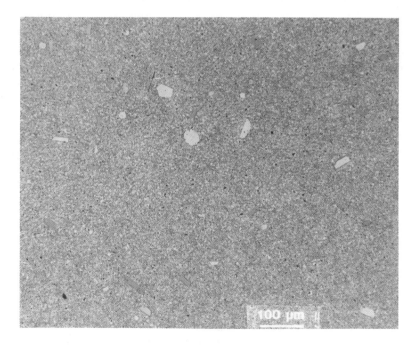

Figure 4: Microstructure of electrodeposited beta" alumina.

Enhancement of the mechanical properties of beta"-alumina, without adversely affecting the electrical characteristics, can be obtained by introducing zirconia as a second phase in order to effect transformation toughening[79]. The milling operation integral to the electrophoretic deposition process ensures a well-dispersed zirconia phase in beta"-alumina/zirconia composites, and substantial increases in both strength and fracture toughness have been obtained, to around 350 MPa and 4 MPam$^{\frac{1}{2}}$ respectively, with a Weibull modulus of 16. These results are obtained at levels of zirconia addition limited to around 5-6% by volume, which is low enough to avoid significant increase in the ionic resistivity of the ceramic. Much of the strength increase is due not so much to transformation toughening as to improved grain size control. Figure 5 shows an optical micrograph of a toughened beta"-alumina ceramic formed by electrophoretic deposition, and Figure 6 shows a scanning electron micrograph which gives an impression of pore size and zirconia distribution. These ceramics have shown impressive performance and durability in sodium/sulfur cell operation.

In a recent study of zirconia-toughened alumina processed by two different routes[80], a considerable difference in fracture toughness behavior was observed according to whether a mechanical or a wet chemical

Figure 5: Microstructure of electrodeposited beta" alumina/zirconia composite.

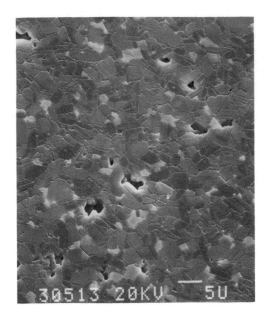

Figure 6: Scanning electron micrograph of electrodeposited beta" alumina/zirconia composite. The white particles are zirconia.

dispersion technique was used. In the first route the material was attrition milled, spray-dried and hot pressed; in the second it was slip cast and sintered conventionally. Because different levels of zirconia addition and different degrees of stabilization were involved, it is difficult to draw a direct comparison between the two processes from the limited data. For the same reason, little attempt has been made here to draw comparisons of electrodeposited beta"-alumina and beta"-alumina/zirconia with the same materials processed by different routes. Some data on the latter are available and can be found in the literature[78,79,81-92].

6. CONCLUSION

Although electrophoresis has been known for nearly two centuries, and is widely used in industrial coating applications, its use as a ceramic forming technique has been attempted only during the last 40 years in the USSR and during the last 20 years in the West. A traditional ceramist might be deterred from electrophoretic forming by such features as the complex physics and chemistry of the charging process, and the high firing shrinkage. But there are a number of advantages to be considered:

1. High coulombic yield is obtained, and typical forming times are in the range 1-5 minutes.

2. There is considerable flexibility in regard to shape forming, and electrodeposition is particularly appropriate for closed-end tubes and crucibles, and continuous strip forming.

3. Electrophoretic forming is not sensitive to suspension rheology, and precise control of deposit thickness can be achieved.

4. Although electrodeposited shapes have a green density that is lower than usual, the porosity is finely distributed and excellent sinterability is obtained. Fired shapes exhibit good microstructural homogeneity, and the processing flaws often difficult to avoid in slip-cast or isopressed shapes are absent in electro-deposited material. Gas bubble formation is one of the main processing flaws in electrophoretic forming but can be avoided by good system design.

REFERENCES

1. Reuss, F.F., Notice sur un nouvel effet de l'electricite galvanique. Mem. Soc. Imp. Natur. Moscou 2 327-37 (1809).

2. Helmholtz, H., Studien uber electrische Grenzschichten. Ann. Phys. Chem. 7 337-82 (1879).

3. Gutierrez, C.P. and Mosley, J.R., in Encyclopaedia of Electrochemistry; Hampel, C.A. (Ed), pp 542-544, Reinhold Publishing Corporation, New York (1964).

4. Sheppard, S.E., The electrical deposition of rubber. Trans. Am. Electrochem. Soc. 52 47-82 (1927).

5. De Boer, J.H., Hamaker, H.C. and Verwey, E.J.W., Electro-deposition of a thin layer of powdered substances. Rec. Trav. Chim. 58 662-65 (1939).

6. Berkman, A.S., Tr. Keram. Inst. 20 6 (1948).

7. Andrews, J.M., Collins, A.H., Cornish, D.C. and Dracass, J., The forming of ceramic bodies by electrophoretic deposition. Proc. Br. Ceram. Soc. 12 211-29 (1969).

8. Fally, J., Lasne, C., LeCars, Y. and Margotin, P., Study of a beta-alumina electrolyte for sodium-sulfur battery. J. Electro-chem. Soc. 120 1296-98 (1973).

9. Chronberg, M.S., Procedes et machines pour la preparation de materiaux par electrophorese. Ind. Ceram. 706 318-21 (1977).

10. Kainarskii, I.S. and Malinovskii, K.B., Electrophoresis as a method of forming thin ceramic articles. Steklo Keram. 4 26-30 (1958).

11. Aveline, M., Faconnage par electrophorese. Ind. Ceram. 581 28-31 (1966).

12. Boncoeur, M. and Carpentier, S., Le formage par depot electro-phoretique. Ind. Ceram. 648 79-81 (1972).

13. Entelis, F.S. and Sheinina, M.E., Electrophoretic forming of porcelain articles. Glass Ceram. (USSR) 36 634-37 (1979).

14. Ryan, W. and Massoud, E., Electrophoretic deposition could speed up ceramic casting. Interceram. 28 117-19 (1979).

15. Entelis, F.S. and Sheinina, M.E., Design of cathodes for electro-phoretic forming of porcelain cups. Glass Ceram. (USSR) 36 683-85 (1979).

16. Vander Poorten, H., Caracterisation de l'electrodeposition et des electrodepots de pates ceramiques. Silic. Ind. 9 159-72 (1981).

17. Ryan, W., Massoud, E. and Perera, C.T.S.B., Fabrication by electrophoresis. Br. Ceram. Trans. J. 80 46-47 (1981).

18. Schmidt, E.W., ELEPHANT - Experiment oder Zukunft!? ZI Int. 4 217-20 (1978).

19. Chronberg, S., Possibilites de l'electrophorese pour le faconnage des produits ceramiques. Ind. Ceram. 718 423-26 (1978).

20. Chronberg, M.S. and Handle, F., Processes and equipment for the production of materials by electrophoresis ELEPHANT. Interceram. 27 33-34 (1978).

21. Frey, E., Formgebung im Schlicker durch Elektrophorese. Silik. Z. 18 399-400 (1979).

22. Handle, F., Elektrophoretische Verformung von keramischen Materialen. Keram. Z. 32 185-88 (1980).

23. Kennedy, J.H. and Foissy, A., Fabrication of beta-alumina tubes by electrophoretic deposition from suspensions in dichloromethane. J. Electrochem. Soc. 122 482-86 (1975).

24. Powers, R.W., The electrophoretic forming of beta-alumina ceramic. J. Electrochem. Soc. 122 490-500 (1975).

25. Kosinskii, S., Posnik, T. and Rog, G., Elektroforetyczne formowanie rurek z proskow β-Al_2O_3. Sklo Ceram. 32 102-3 (1981).

26. Foissy, A.A and Robert, G., Electrophoretic forming of beta-alumina from dichloromethane suspensions. Ceram. Bull. 61 251-55 (1982).

27. Heavens, S.N., in Novel Ceramic Fabrication Processes and Applications; Davidge R.W. (Ed), Brit Ceram Proc No. 38, pp 119-26, Institute of Ceramics, (1986).

28. Hamaker, H.C., Formation of a deposit by electrophoresis. Trans. Faraday Soc. 36 287-95 (1940).

29. Benjamin, M. and Osborn, A.B., The deposition of oxide coatings by cataphoresis. Trans. Faraday Soc. 36 287-95 (1940).

30. Avgustinik, A.I., Vigdergauz, V.S. and Zhuravlev, G.I., Electrophoretic deposition of ceramic masses from suspensions and calculation of deposit yields. J. Appl. Chem. (USSR) 35 2090-93 (1962).

31. Panov, V.I. and Petrov, V.A., Electrophoretic deposition of suspensions of alkaline-earth metal oxides. Sov. J. Colloid. Sci. 37 359-62 (1975).

32. Krishna Rao, D.U. and Subbarao, E.C., Electrophoretic deposition of magnesia. Ceram. Bull. 58 467-69 (1979).

33. Kinney, G.F. and Festa, J.V., Some aspects of electrophoretic deposition. Sylvania Technologist 10 48-52 (1957).

34. Lamb, V.A. and Reid, W.E.,Jr., Electrophoretic deposition of metals, metalloids and refractory oxides. Plating 47 291-96 (1960).

35. Gutierrez, C.P., Mosley, J.R. and Wallace, T.C., Electrophoretic deposition: a versatile coating method. J. Electrochem. Soc. 109 923-27 (1962).

36. Fisch, H.A., Electrophoretic deposition of aluminide coatings from aqueous suspensions. J. Electrochem. Soc. 119 57-64 (1972).

37. Arai, T. and Yoda, R., Coating of nichrome alloy on molybdenum by electrophoresis. Trans. Nat. Res. Inst. Metals 23 227-35 (1981).

38. Ramasamy, A., Totlani, M.K. and Ramanathan, P.S., Electrophoretic deposition of NiO from suspensions in n-butanol. J. Electrochem. Soc. India 32 113-20 (1983).

39. Furman, V.V., Gaiduchenko, G.K., Vlastuk, R.Z. and Deimontovich, V.B., Structure and properties of electrophoretic chromium carbide coatings with a eutectic composition nickel-boron binder. Sov. Powder Metall. & Met. Ceram. 22 942-45 (1983).

40. Ortner, M., Recent developments in electrophoretic coatings. Plating 51 885-89 (1964).

41. Hoffman, H., Theory and practice of electrocoating of porcelain enamel. Ceram. Bull. 57 605-8 (1978).

42. Surowiak, Z., On the technology of deposition of polycrystalline thin films of ferro- and antiferroelectrics on metallic substrates. Acta Phys. Polon. A43 543-50 (1973).

43. Sussman, A. and Ward, T.J., Electrophoretic deposition of coatings from glass-isopropanol slurries. RCA Rev. 42 178-97 (1981).

44. Shimbo, M., Tanzawa, K., Miyakawa, M. and Emoto, T., Electrophoretic deposition of glass powder for passivation of high voltage transistors. J. Electrochem. Soc. 132 393-98 (1985).

45. Trap, H.J.L., L'emploi des verres au germano-silicate dans la passivation des dispositifs a semi-conducteurs. Verres Refract. 32 17-23 (1978).

46. Miwa, K., Kanno, Kawamura, S. and Shibuya, T., Glass passivation of silicon devices by electrophoresis. Denki Kagaku 40 478-84 (1972).

47. Grosso, P.F., Rutherford, R.E.,Jr. and Sargent, D.E., Electrophoretic deposition of luminescent materials. J. Electrochem. Soc. 117 1456-59 (1968).

48. Livesey, R.G. and Lyford, E., Electrophoretic deposition of luminescent powder. J. Sci. Instrum. (J. Phys. E) 1 948 (1968).

49. Mizuguchi, J., Sumi, K. and Muchi, T., A highly stable nonaqueous suspension for the electrophoretic deposition of powdered substances. J. Electrochem. Soc. 130 1819-25 (1983).

50. Williams, E.W., Jones, K., Griffiths, A.J., Roghley, D.J., Bell, J.M., Steven, J.H., Huson, M.J., Rhodes, M., Costlich, T. and Cobley, U.T., The electrophoresis of thin film CdS/Cu$_2$S solar cells. Solar Cells 1 357-66 (1979/80).

51. Murphy, E.A., Electro-decantation for concentrating and purifying latex. Trans. Inst. Rubber Ind. 18 173-80 (1942).

52. Caley, W.F. and Flengas, S.N., Electrophoretic properties of the oxides NiO, Fe$_2$O$_3$, Cr$_2$O$_3$, MnO$_2$ and TiO$_2$ in aqueous suspension. Can. Metall. Quart. 15 375-82 (1976).

53. Bushey, A., Electrodeposition of thin primer coatings on aluminum. J. Coatings Technol. 48 51-54 (1976).

54. Standish, J.V. and Boerio, F.J., Anodic deposition of paint for coil coating galvanized steel. J.Coatings Technol. 52 29-39 (1980).

55. Damm, E.P.,Jr. and Feigenbaum, M.A., Coating polymers by electro-phoretic deposition. SPE Journal 27 58-61 (1971).

56. Itkis, Yu.A. and Svinukhov, V.M., A new machine for dehydrating a kaolin suspension. Glass Ceram. (USSR) 32 406-7 (1975).

57. Troelstra, S.A., Applying coatings by electrophoresis. Philips Tech. Rev. 12 293-303 (1951).

58. Kennedy, J.H. and Foissy, A., Measurement of mobility and zeta potential of beta-alumina suspensions in various solvents. J. Am. Ceram. Soc. 60 33-36 (1977).

59. Kiefer, C., Origine de la charge des suspensions des mineraux phylliteux et sa relation avec la structure reticulaire super-ficielle de ces mineraux. Bull. Soc. Franc. Ceram. 2 24-30 (1949).

60. Hamaker, H.C. and Verwey, E.J.W., The role of the forces between the particles in electrodeposition and other phenomena. Trans. Faraday Soc. 36 180-85 (1940).

61. Frens, G., Huisman, H.F., Vondeling, J.K. and van der Waarde, K.M., Suspension technology. Philips Tech. Rev. 36 264-70 (1976).

62. Smith, A.L., in Dispersions of Powders in Liquids; Parfitt, G.D. (Ed), pp. 99-147, Applied Science Publishers, (1981).

63. Choudhary, J.Y., Ray, H.S. and Rai, K.N., Measurement of zetapotential and its relevance in kinetics of electrophoretic deposition. Trans. Indian Inst. Met. 36 363-67 (1983).

64. Frey, E. and Rackl, M., Elektrophoretische Ablagerung keramischer Rohstoffe. Keram. Z. 34 154-56 (1982).

65. Brown, D.R. and Salt, F.W., The mechanism of electrophoretic deposition. J. Appl. Chem. 15 40-48 (1965).

66. Shyne, J.J. and Scheible, H.G. in Modern Electroplating; Lowenheim, F.A. (Ed), pp. 714-730, John Wiley & Son, (1963).

67. Pierce, P.E., The physical chemistry of the cathodic electrodeposition process. J. Coatings Technol. 53 52-67 (1981).

68. Choudhary, J.Y., Ray, H.S. and Rai, K.N., Electrophoretic deposition of alumina from aqueous suspensions. Trans. Br. Ceram. Soc. 81 189-93 (1982).

69. Choudhary, J.Y., Ray, H.S. and Rai, K.N. in Sintered Metal-Ceramic Composites; Upadhyaya, G.S. (Ed), pp. 147-8, Elsevier Science Publishers B.V., (1984).

70. Das, R.K., Ray, H.S. and Chander, S., Electrophoretic deposition of titanium dioxide from aqueous suspensions. Trans. Indian Inst. Metals 32 364-68 (1979).

71. Tsarev, V.F., Effect of shaping parameters on the properties of quartz ceramics. Glass Ceram. (USSR) 36 637-41 (1979).

72. Fink, C.G. and Feinleib, M., Electrodeposition and electrochemistry of the deposition of synthetic resins. Trans. Electrochem Soc. 94 309-40 (1948).

73. Apininskaya, L.M. and Furman, V.V., Electrophoretic deposition of coatings (a literature review). Sov. J. Powder Metall. 122 136-39 (1973).

74. Sacks, M.D. and Khalidar, C.S., Milling and suspension behaviour of Al_2O_3 in methanol and methyl isobutyl ketone. J. Am. Ceram.Soc. 66 488-94 (1983).

75. Koelmans, H. and Overbeek, J.Th.G., Stability and electrophoretic deposition of suspensions in non-aqueous media. Disc. Faraday Soc. 18 52-63 (1954).

76. Koelmans, H., Suspensions in non-aqueous media. Philips Res. Rep. 10 161-93 (1955).

77. Cornish, D.C., A probe electrode technique for controlling the thickness of electrophoretic ceramic deposits. J. Sci. Instrum. (J. Phys. E) 2 123-4 (1969).

78. Tan, S.R., May, G.J., McLaren, J.R., Tappin, G. and Davidge, R.W., Strength controlling flaws in beta-alumina. Br. Ceram. Trans. J. $\underline{79}$ 120-27 (1980).

79. Viswanathan, L., Ikuma, Y. and Virkar, A.V., Transformation toughening of ß"-alumina by incorporation of zirconia. J. Mat. Sci. $\underline{18}$ 109-13 (1983).

80. Schafer, J., Homerin, P., Thevenot, F., Orange, G., Fantozzi, G. and Cambier, F., Zusammenhang zwischen Fertigungsart und Festigkeit von ZrO_2-verstarktem Al_2O_3. Keram. Z. $\underline{38}$ 320-22 (1986).

81. Gordon, R.S., McEntire, B.J., Miller, M.L. and Virkar, A.V., Processing and characterization of polycrystalline ß"-alumina ceramic electrolytes. Mat. Sci. Res. $\underline{11}$ 405-20 (1978).

82. Miller, M.L., MeEntire, B.J., Miller, G.R. and Gordon, R.S., A prepilot process for the fabrication of polycrystalline ß"-alumina electrolyte tubing. Ceram. Bull. $\underline{58}$ 522-26 (1979).

83. Johnson, D.W.Jr., Granstaff, S.M.Jr. and Rhodes, W.W., Preparation of ß"-Al2O3 pressing powders by spray-drying. Ceram. Bull. $\underline{58}$ 849-55 (1979).

84. Davidge, R.W., Tappin, G., McLaren, J.R. and May, G.J., Strength and delayed fracture of beta alumina. Ceram. Bull. $\underline{58}$ 771-74 (1979).

85. May, G.J., Tan, S.R. and Jones, I.W., Hot isostatic pressing of beta-alumina. J. Mat. Sci. $\underline{15}$ 2311-16 (1980).

86. Kvachkov, R., Yanakiev, A, Poulieff, C.N., Balkanov, I., Yankulov, P.D. and Budevski, E., Effect of the starting Al_2O_3 and of the method of preparation on the characteristics of Li-stabilized ß"-Al2O3 ceramics. J. Mat. Sci. $\underline{16}$ 2710-16 (1981).

87. Lange, F.F., Davis, B.I. and Raleigh, D.O., Transformation strenghening of ß"-Al_2O_3 with tetragonal ZrO_2. J. Am. Ceram. Soc. $\underline{66}$ C50-52 (1983).

88. Green, D.J. and Metcalf, M.G., Properties of slip-cast transformation-toughened ß"-Al_2O_3/ZrO_2 composites. Ceram. Bull. $\underline{63}$ 803-07 (1984).

89. Evans, J.R.G., Stevens, R. and Tan, S.R., Thermal shock of ß-alumina with zirconia additions. J. Mat. Sci. $\underline{19}$ 4068-76 (1984).

90. Pett, R.A., Theodore, A.N., Tennenhouse, G.J. and Runkle, F.D., Plate-type beta"-alumina electrolytes for an advanced sodium-sulfur cell design. Ceram. Bull. $\underline{64}$ 589-92 (1985).

91. Binner, J.G.P. and Stevens, R., Improvement in the mechanical properties of polycrystalline beta-alumina via the use of zirconia particles containing stabilizing oxide additions. J. Mat. Sci. $\underline{20}$ 3119-24 (1985).

92. Green, D.J., Transformation toughening and grain size control in ß"-Al_2O_3/ZrO_2 composites. J. Mat. Sci. $\underline{20}$ 2639-46 (1985).

APPENDIX

Electrophoretic deposition using non-aqueous suspensions

SOLID	LIQUID	ADDITIVES	CHARGE	REFERENCE
Al	acetone	butylamine		73
Al	ethanol	$Al(NO_3)_3$		73
Al	ethanol	NH_4OH		65
Al_2O_3*	acetone	cellulose nitrate, $N(CH_3)_4OH$, H_2SO_4	−	49
Al_2O_3	ethanol	water, shellac		7
Al_2O_3	methanol	none	+	28,74
Al_2O_3	butanone	cellulose nitrate, dichloracetic acid		7
Al_2O_3	xylene	oleic/stearic acid, sodium dioctyl-sulfosuccinate, sorbitan palmitate	+	75
β-Al_2O_3	dichloromethane	acetic acid	+	58
β-Al_2O_3	dichloromethane	trichloracetic acid	−	23,26,58
β-Al_2O_3	nitromethane	acetic/benzoic acid	+	23,58
β-Al_2O_3	C2-6 alcohols, benzyl alcohol, acetone, pentan-3-one, methanol/1,4-dioxan	none	−	24
β-Al_2O_3	pentan-1-ol	none	−	24,25,58
β-Al_2O_3	pyridine	none	+	24
alundum	pentan-1-ol	polyvinylbutyral, ammonium thiocyanate		73
$BaCO_3$	ethanol	none	+	5,65
$BaCO_3$	methanol	none	+	28
$BaCO_3$, $SrCO_3$	acetone	cellulose nitrate, ethane-1,2-diol	+	29
BaO glass	propan-1-ol	water, KCl, BaO, CO_2	+	43
BaO, $BaSrCaO_3$	pentan-2-ol	none		31

$BaSO_4$	xylene	sorbitan mono-oleate	+	75
$BaSr(CO_3)_2$	ethanol/acetone	none	+	28
$BaSrCa(CO_3)_3$	propanol	polyvinylbutyral		73
$BaTiO_3$	dimethoxyethane	Penetrol-60	-	34
$BaTiO_3$	1,4-dimethoxy-diethylether	Na dioctylsulfosuccinate	+	34
$BaTiO_3$	1,4-dimethoxy-diethylether	Penetrol-60	-	34,66
$BaTiO_3$	nitropropane	none	+	34
$BaTiO_3$	pyridine	none	+	34
$BaTiO_3$, $PbTiO_3$, $PbZrO_3$	C1-5 alcohols, methyl acetate, acetone	HCl, cellulose nitrate, methyl polymethacrylate, collodion		42
$CaCO_3$	acetone	none	+	65
CaF_2	ethanol	none	+	28
$2CaO.MgO$ $.2SiO$	propan-2-ol acetone	water, $Mg(NO3)2$	+	47
Fe	acetone, CH3NO2	none	+	65
Fe_2O_3	xylene	oleic/stearic/caproic acid, sodium dioctyl-sulfosuccinate	+	75
$MgCO_3$	methanol	KI	+	74,75
MgO	ethanol	none	+	65
MgO	1-nitropropane	none	-	65
MgO	pentan-2-ol, benzyl alcohol	none	+	32
MgO, $MgCO_3$	methanol	none		28,75
MgO, NbC	ethanol	HCl		30
Ni	ethanol	none	+	65
Ni-Cr	propan-2-ol/nitromethane	zein, $Co(NO_3)_2$	+	37
NiO	butan-1-ol	trichloroethylene		38
NiO	1-nitropropane	ethylcellulose	+	34
NiO/Al_2O_3	propan-1-ol	none		66

NiO/Ni-Cr	propan-2-ol/ nitromethane	zein		66
PbO	methanol, ethanol	none	+	65
PbO-Al$_2$O$_3$- -SiO$_2$ glass	methanol	AlCl$_3$		45
SiO$_2$	acetone	LiCl	-	75, 76
talc	acetone, ethanol	none	-	65
TiH$_2$	1,4-dimethoxy- diethylether	ethylcellulose, tannic acid, Penetrol-60	-	34
TiH$_2$	1-nitropropane	none	+	34
TiH$_2$	kerosene	Igepal CA	+	34
TiO$_2$	ethanol	none	+	65
UO$_2$*	propan-2-ol, nitromethane	NH$_4$OH, zein, benzoic acid		35
U$_3$O$_8$	propan-2-ol	nitric/benzoic acid		73
W	acetone	none		73
WC, TaC	propan-2-ol, nitromethane	zein, Co(NO$_3$)$_2$		40
Zn	ethanol, CH$_3$NO$_2$	none	+	65
ZnO-B$_2$O$_3$ -SiO$_2$ glass	propan-2-ol, acetone	HF, NH$_4$OH		46
ZnO-B$_2$O$_3$ -SiO$_2$ glass	propan-2-ol	Mg(NO$_3$)$_2$, Y(NO$_3$)$_3$		44
ZnSiO$_3$:Mn	propan-2-ol	Al(NO$_3$)$_3$, La(NO$_3$)$_3$		48

ACKNOWLEDGEMENTS
I thank M. McNamee for the photomicrographs and strength data, and Chloride Silent Power Ltd. for permission to publish beta"-alumina data.

8

Microwave Processing of Ceramics

A.C. Metaxas [†] and J.G.P. Binner [*]

† Electricity Utilisation Studies Centre, Department of Engineering, Trumpington Street, University of Cambridge, Cambridge, CB2 1PZ, UK

* Department of Materials Engineering and Materials Design, University of Nottingham, University Park, Nottingham, NG7 2RD, UK

1. INTRODUCTION

It has been well accepted that the seventies was the decade which highlighted the problems associated with diminishing fossil fuels and the need to look carefully at alternative sources of energy for carrying out many processes in industry. The volatile nature of oil fuel costs coupled with gas unavailability in some regions has compelled many industrialists to consider novel electrical methods for supplying their process heat needs. Microwave heating is just one such area where considerable progress has been made in the last decade, culminating in numerous installations throughout many manufacturing and processing industries.

The acceptance of microwave heating as an established technique has not come about without considerable problems. In the early sixties such techniques were pushed forward into industry too fast without adequate research and development back-up, resulting in numerous costly mistakes and loss of confidence. The major problem has been the user-supplier relationship in that each was largely unaware of the details and limitations of the equipment and process, respectively, under consideration. Nowadays, however, more time is spent clearly defining the requirement, the specific design and capability of the equipment and making cost comparisons with alternative equipment, before any firm commitments are made. It is

285

imperative that expert independent advice is sought, particularly in the early stages during preliminary feasibility tests, in order to avoid misinterpretation of available data.

To date, the application of microwave energy to the processing of ceramic materials has been uncoordinated and slow. There are many reasons for this, not least being the fast development and diverse nature of materials which come under the heading of ceramics. Many of these new ceramics and composites have potentially exciting properties but can be difficult to process by conventional means. Recently, this has led to considerably more attention being focused on the processing possibilities associated with the use of microwaves. This chapter summarises the theory and practice of microwave heating as applied to the processing of ceramic materials.

2. FUNDAMENTALS OF MICROWAVE HEATING

2.1 Dielectric Loss Mechanisms

It has long been established that a dielectric material, such as many types of ceramics, can be heated with energy in the form of high frequency electromagnetic waves. The frequency range used for microwave heating lies between 400 MHz and 40 GHz, however the allowed frequencies are restricted to distinct bands which have been allocated for Industrial, Scientific and Medical (ISM) use, as shown in Table 1. The principal frequencies are centred at 433 MHz, 915 MHz (896 MHz in the UK) and 2450 MHz since specific industrial equipment can be readily purchased.

In the microwave frequency range there are primarily two physical mechanisms through which energy can be transferred to a ceramic material. Firstly there is the flow of conductive currents, resulting in an ohmic type of loss mechanism where the ceramic conductivity, σ, plays a prominent role. Secondly, the existence of permanent dipoles in the ceramic can give rise to a loss mechanism referred to as re-orientation or dipolar loss.

Frequency MHz	Frequency tolerance ±	Area permitted
433.92	0.2%	Austria, Netherlands, Portugal, Switzerland, West Germany, Yugoslavia.
896	10 MHz	Great Britain.
915	13 MHz	North and South America.
2375	50 MHz	Albania, Bulgaria, Czechoslovakia, Hungary, Rumania, USSR.
2450	50 MHz	Worldwide except where 2375 MHz is used.
3390	0.6%	Netherlands.
5800	75 MHz	Worldwide.
6780	0.6%	Netherlands.
24150	125 MHz	Worldwide.
40680		Great Britain.

Table 1: Frequency allocation for industrial, scientific and medical purposes in the range 400 MHz to 40 GHz.

It is well established that at radio frequencies (1-100 MHz) the dominant mechanism is due to conductive currents flowing within the materials, this in turn is due to the movement of ionic constituents. This mechanism is enhanced by the presence of impurity contents such as salts, etc. Since this chapter deals only with microwave heating, the conductive loss mechanism only substantially affects the overall loss at the lower end of the frequency range considered in Table 1, where both dipolar relaxation and conductivity loss mechanisms could be equally dominant.

In the range 1 to 10 GHz energy absorption is primarily due to the existence of permanent dipole molecules which tend to reorientate under the influence of a microwave electric field E, as shown in Figure 1. The origin of this heating effect lies in the inability of the polarisation to follow extremely rapid reversals of the electric field. At such high frequencies the resulting polarisation phasor P (basically a collection of charges) lags the applied electric field ensuring that the resulting current density has a component which is in phase with the electric field as shown in Figure 2

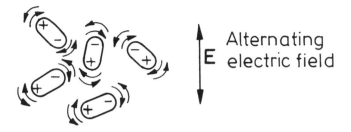

Figure 1: Dipolar reorientation

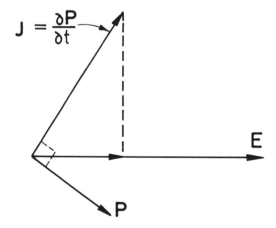

Figure 2: Phasor diagram between external electric field, polarisation and current phasors.

and therefore power is dissipated in the ceramic material.

The dipolar or re-orientation loss mechanism was originally examined in dipolar electrolytes by Debye[1] who deduced the well known relation for the complex relative dielectric constant:

$$\epsilon^* = \epsilon' - j\epsilon'' = \epsilon_\infty + \left[\frac{\epsilon_s - \epsilon_\infty}{1 + \omega^2\tau^2}\right] - \left[\frac{j(\epsilon_s - \epsilon_\infty)}{1 + \omega^2\tau^2}\right] \qquad 1$$

where ϵ_s and ϵ_∞ are the relative dielectric constants at d.c. and very high frequencies respectively and τ is the relaxation time of the system.   This latter controls the build-up and decay of the polarisation as the external field is applied and removed respectively.   The absolute dielectric constant is given by $\epsilon = \epsilon_0\epsilon^*$.   By separating the real and imaginary parts of equation 1 the parameters ϵ' and ϵ'' can be obtained.   They are shown plotted as a function of $\omega\tau$ in Figure 3.

A simple interpretation of the Debye plots is as follows.   At very low frequencies the dipoles have ample time to follow the variations of the

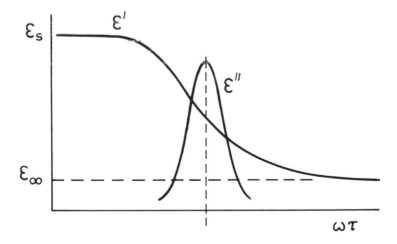

Figure 3: Debye relaxation.

applied electric field and the dielectric constant is at its maximum value
ϵ_s. We can say that the bound charge density attains its maximum value and
the energy from the external source is stored in the material.   As the
frequency increases the dipoles are unable to fully restore their original
positions during field reversals and as a consequence the dipolar
polarisation lags the applied field. As the frequency increases further, the
re-orientation polarisation fails to follow the applied field and contributes
less and less to the total polarisation.   The fall of the effective
polarisation manifests itself as a fall of the dielectric constant ϵ' and a
rise of the loss factor ϵ''. Power is now drawn from the external source and
dissipated as heat in the ceramic material.

This theory does not apply very well to dipolar solids because Debye's
interpretation of the above relaxation was given in terms of dipolar rotation
against frictional forces in the material.   It is difficult to imagine,
particularly in solid dielectrics, the dipoles as spheres in a medium where
viscosity is the dominant mechanism. Moreover, when many atoms and molecules
are bonded together to form a dielectric, the dipoles will be influenced by
the forces of all the neighbouring particles and these must be taken into
account in the theory. Thus the concept of an activation energy U_a (ie, the
energy required to dislodge a dipole from its equipropable position) must be
introduced.   What is needed is a link between the activation energy and
relaxation time constant.   Such a link is found in the double well theory
which deals with the transitions of dipoles between two equipropable
positions separated by a potential energy, as depicted in Figure 4.   The
external field distorts the potential energy from its original symmetrical
form.   The theory considers the number of dipoles decaying from state 1 to
state 2, it calculates the resulting polarisation and hence the total current
density which, in turn, gives the two terms ϵ' and ϵ''. These are shown to
obey a Debye type relaxation with a time constant given by:

$$\tau_o = \frac{\exp\left[\dfrac{U_a}{kT}\right]}{\nu} \cdot \frac{\left[\epsilon_s + 2\right]}{\left[\epsilon_\infty + 2\right]} \qquad\qquad 2$$

where $1/\nu$ is the time for a single collision in the potential well.

Equation 2 introduces a very important relation between the activation

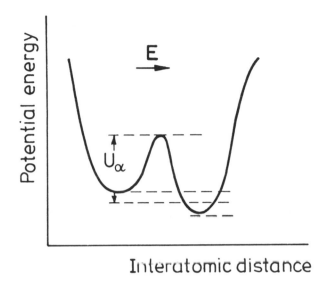

Figure 4: Potential energy diagram for two positions of a permanent dipole.

energy and τ since it is now possible to extend the simple Debye theory and bring it closer to what is observed in practice in many real dielectrics. This is because in many solid dielectrics dipoles can exhibit more than one activation energy, corresponding to transitions from different wells and giving rise to multiple relaxation times. The concept of a spread of relaxation times can now be easily envisaged in an inhomogeneous solid dielectric, where hindered dipole redistribution results in multirelaxational spectra as shown qualitatively in Figure 5. In general, in solids the molecular dipoles have more than two equipropable positions so that $\epsilon''(\omega)$ is much flatter and broader than is predicted from the simple Debye theory.

A third relaxation type mechanism, which once again has its strongest influence in the radio frequency range, is a loss due to space charges in interfaces between different dielectric materials, referred to as Maxwell-Wagner loss. It may be assumed that at the low frequencies where these losses are important, the conductivity loss mechanism referred to above will be the dominant one for ceramic materials and therefore losses due to Maxwell-Wagner polarisation will not be considered any further.

There are two other principal polarisation mechanisms, namely electronic

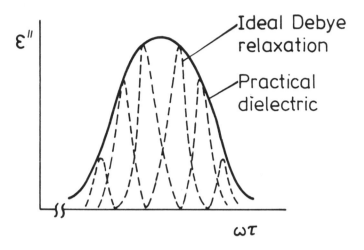

Figure 5: Multirelaxational Debye relaxation.

and atomic, which are induced by the action of the electric field. Their influence occurs at optical and infrared frequencies and as such will also not be considered further.

2.2 Total Current Density

The total current density flow in a dielectric material is given by Maxwell's equation:

$$\oint \left[\frac{B}{\mu} \right] . dl = \int_S J_c . dS + \int_S \left[\frac{dE}{dt} \right] . dS \qquad 3$$

where J_c is the current density due to conduction (ohmic) effects,
B is the magnetic flux density, and
μ is the permeability of the medium.

The conduction current density is given by σE, where σ is the conductivity of the ceramic material.

For periodic waveforms, such as usually used for high frequency heating, equation 3 attains the following form:

$$\times \underline{H} = \underline{J}_c + jwe\underline{E}$$ 4

where $\underline{H}$ is the magnetic field.

Substituting for $\epsilon = \epsilon_o \epsilon^*$ using equation 1 in equation 4 we obtain:

$$\times \underline{H} = \sigma\underline{E} + jwe_o\epsilon'\underline{E} + we_o\epsilon''\underline{E} = we_o\epsilon''_e\underline{E} + jwe_o\epsilon'\underline{E}$$ 5

where ϵ''_e is the effective loss factor comprising dipolar as well as conductivity losses and is given by:

$$\epsilon''_e = \epsilon'' + (\sigma/we_o)$$ 6

Note that the total losses can quite easily be expressed in terms of an effective conductivity, σ_e defined as:

$$\sigma_e = we_o\epsilon''_e = we_o\epsilon'' + \sigma$$ 6a

where the first term relates the dipolar loss factor to an equivalent conductivity $we_o\epsilon''$, whilst the second term is due to the ionic effects.

2.3 Equivalent circuit presentation

It has just been shown that the total current flow in a medium under the influence of a high frequency electric field is given by a conductive current and a displacement current, the latter being strongly influenced by the dielectric constant and comprising a real and imaginary component as shown in Figure 6. The ratio of the effective loss factor to the relative dielectric constant is given by:

$$\tan \delta_e = \epsilon''_e/\epsilon'$$ 7

and is termed the effective loss tangent. The relation between $\tan \delta_e$ and the dipolar loss tangent, $\tan \delta$, is given by:

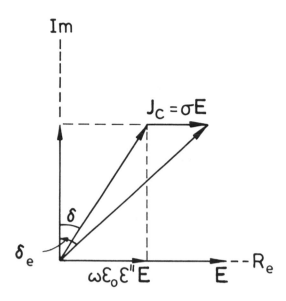

Figure 6: Current density phasors.

$$\tan \delta_e = \tan \delta + \sigma/\omega\epsilon_o\epsilon' \qquad 8$$

An equivalent circuit representation for a lossy dielectric is that depicted in Figure 7. The capacitance of a pure dielectric is denoted by C and R_d accounts for relaxational losses, i.e. the term $\omega\epsilon_o\epsilon''$ and R_c accounts for any additional losses due to dielectric conductivity effects. By applying a sinusoidal voltage $\underline{V} = \underline{V}e^{j\omega t}$ across the circuit we obtain:

$$\underline{I} = C\frac{d\underline{V}}{dt} + \frac{\underline{V}}{R_d} + \frac{\underline{V}}{R_c} \qquad 9$$

The relaxational losses represented by R_d can be written as:

$$R_d = \frac{(1/\omega\epsilon_o\epsilon'').1}{A} \qquad 10$$

while the conductivity losses represented by R_c can be approximated to:

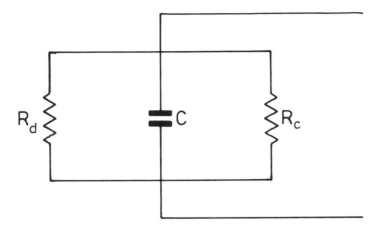

Figure 7: Equivalent circuit of a lossy ceramic.

$$R_c = 1/\sigma A \qquad\qquad 11$$

where 1 and A are the equivalent length and cross sectional area of the ceramic.

Substituting for R_d and R_c using the above expressions in equation 9, we obtain, with $d\underline{V}/dt = j\omega\underline{V}$:

$$\underline{I} = j\omega C\underline{V} + \frac{\underline{V}\omega\epsilon_o\epsilon''A}{1} + \frac{\underline{V}A\sigma}{1} \qquad\qquad 12$$

However, since the capacitance C of the equivalent pure ceramic is given by:

$$C = \frac{\epsilon_o\epsilon'A}{1} \qquad\qquad 13$$

Substitution in eqn 12 gives the total current density $\underline{J}$, with $\underline{E} = (\underline{V}/1)$:

$$\underline{J} = \omega\epsilon_o\epsilon''\underline{E} + j\omega\epsilon_o\epsilon'\underline{E} = j\omega\epsilon_o K\underline{E}$$

which is an expression identical to equation 5. The parameter K is given by:

$$K = \frac{\left[\epsilon - \dfrac{j\sigma}{\omega} \right]}{\epsilon_o} = \epsilon' - j\epsilon_e'' = \epsilon' - \frac{j\sigma_e}{\omega\epsilon_o} \qquad 14$$

For a ceramic with no conductive loss, $\sigma = 0$ and $R_c \to \infty$ which leaves R_d in parallel with C as the equivalent medium circuit presentation. For example, for barium titanate the complex dielectric constant at 1 GHz is given by $\epsilon^* = 950 - j0.15$ (see Figure 8a and b). Assuming that the losses are solely due to dipolar relaxation, we can represent such a ceramic of 1 mm in thickness and 0.01 m^2 in cross sectional area by a parallel R/C circuit of parameters given by equations 10 and 13 respectively. Substitution in these equations of $1 = 1$ mm, $\epsilon_o = 8.8$ pFm^{-1}, $\epsilon' = 950$, $\epsilon'' = 0.15$ and $f = 1$ GHz yields $R_d = 12.06 \; \Omega$ and $C = 0.084 \; \mu F$.

Similar equivalent circuits can be found for any ceramic from knowledge of the dielectric properties, these being discussed in the following section. Other approximations such as, for example, a dominant conductive loss mechanism and a negligible dipolar loss process, ie $\epsilon'' \to 0$, can be readily considered through the general circuit of Figure 7.

Although in this section we have attempted to separate the ϵ_e'' into its constituent ϵ'' or $\sigma/\omega\epsilon_o$ components, the experimental data presented in the following section do not at first sight distinguish the various component losses and thus the generic terms ϵ_e'' or tan δ_e are used. Closer scrutiny of the experimental data is needed to be able to differentiate between dipolar and conductivity losses.

3. DIELECTRIC PROPERTIES

3.1 General data

Proper design of microwave industrial systems requires a knowledge of the dielectric properties of the ceramics to be processed. As was stated above, the property which describes the behaviour of the dielectric under the

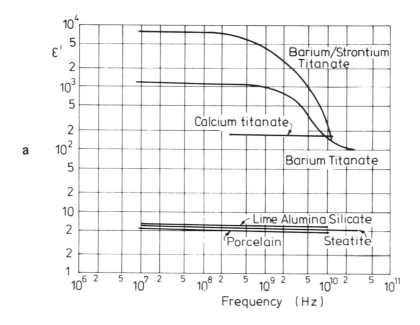

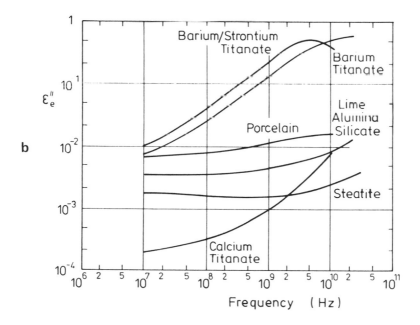

Figure 8: Dielectric properties of various ceramics at 25°C. (After ref 3).

influence of an electromagnetic field is the effective loss factor ϵ''_e. Both ϵ' and ϵ''_e are frequency, temperature, density and electric field direction dependent and a considerable amount of data have been amassed on such variations.

Von Hippel[2] compiled the first tabulation of measurements for numerous materials over a wide range of frequencies (10 Hz - 10 GHz). The materials characterised were both organic and inorganic in nature; both liquid and solid. Although the bulk of the results were obtained at room temperature, some higher temperature measurements up to a few hundred degrees centigrade, were also performed. Subsequently, dielectric data on high temperature solid materials, including oxides, nitrides, silicates, rocks and minerals were obtained by Inglesias and Westphal[3]. Couderc et al[4] have also described a dynamic method for measuring the dielectric properties of various ceramics.

Since then a series of 4 reports have been produced at the same MIT Laboratory for Insulation Research where Von Hippel obtained his results[5]. These reports contain much valuable information on the frequency and temperature dependence of the dielectric properties of a large number of ceramic materials (both oxide and some non-oxide), together with various minerals, many organic compounds and some foodstuffs. The frequency range was increased for some materials up to 25 GHz and the temperature increased into the 500-1000°C region.

More recently still the Rockwell International Science Center has commenced characterisation programmes intended to provide dielectric data on materials at millimetre-wave frequencies, ie above 30 GHz[6]. It is also intended that high temperature (>1000°C) property measurements will be more systematic in nature than previous studies. Georgia Tech Research Institute appears to be independently pursuing a similar approach[7].

Table 2 shows the dielectric properties of a range of ceramic materials under various conditions and near the two frequencies for which industrial equipment can be readily purchased. It is evident that the effective losses of various ceramics depend upon the material density and the temperature, frequency and field orientation, giving a range of effective loss factors from above 70 for silicon carbide to 3×10^{-4} for boron nitride. This latter ceramic can be considered as transparent to microwave energy and may be used as an insulating material in microwave industrial equipment or as a microwave window in waveguides. In fact, any material with an effective loss factor

Material	q	T	p	E-field direction	frequency 10^9 Hz		frequency 3×10^9 Hz	
	gcm^{-3}	°C	ohmcm		ϵ'	ϵ''_e	ϵ'	ϵ''_e
Silicon carbide		25	35		107	73.4	60	34.8
BeO + SiC				‖	23-75	2.3-30	20-60	3-27
Zinc oxide		25	100	⊥	11	17	8.5	10
Alumina (Al-500)	3.665	25			9	0.056	9	0.063
Alumina (Al-18)	3.676 3.675	25 500			9.1 9.7	0.0025 0.02		
Zirconia		25		⊥‖	3	0.04		
Boron nitride		25 25			3.5 4.9	3×10^{-4} 1×10^{-3}		
Steatite		25			6.9	3.5×10^{-4}	6.9	3.5×10^{-4}
Magnesia		500			7.1	1×10^{-3}	7.08	7.5×10^{-4}

Table 2: Dielectric properties of a range of ceramics.

q: density   T: temperature   p: resistivity

below about 0.01 is considered as a low loss material and would normally not be considered for microwave processing without careful examination of the merits of such treatment. This is primarily because at such low loss factors very high electric fields need to be set up in the ceramic in order to dissipate a reasonable amount of power within it (see equation 25).

The data shown in Table 2 have been extracted from experimental results of ϵ' and ϵ_e'' as a function of the frequency with typical variations shown in Figures 8 and 9 for various ceramics. In Figure 8 pronounced dipolar relaxations occur for the titanate ceramics such as barium or barium/strontium titanate in the frequency range 10 MHz - 100 GHz. The effective loss factors of these two ceramics at 2.45 GHz are 0.2 and 0.3 respectively indicating that both these materials will readily absorb microwave radiation at this industrial allocated frequency. Lime alumina silicate, steatite and calcium titanate on the other hand have loss factors below 0.02 and as such are not obvious candidates for microwave heating. Figure 10 shows the material properties of a family of ferrites, some types can have extremely high loss factors.

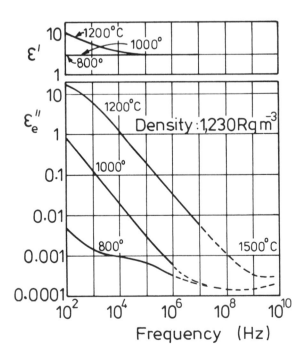

Figure 9: Dielectric properties of pyrolytic boron nitride.   (After ref 3).

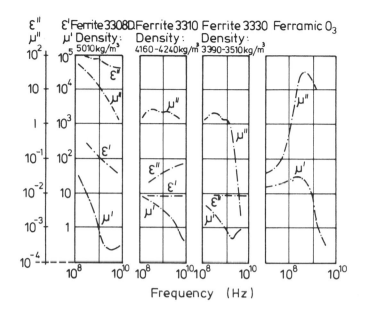

Figure 10: Properties of some ferrites in the microwave regime.   (After ref 3).

3.2 Temperature effects

The importance of temperature in the microwave treatment of ceramics cannot be over emphasized.   Figure 9, for example, highlights this for pyrolytic boron nitride where at 100 MHz the loss factor increases by an order of magnitude as the temperature changes from 800°C to 1200°C.   Figure 11, however, shows that for hot pressed boron nitride near 5 GHz the loss factor increases only by a factor of 2.

In general, the effective loss factor of many ceramics increases with temperature, with the rate of increase depending upon the type of ceramic and the operating frequency. The data displayed in Figure 12, measured at 10 GHz, show that in the range 20 - 100°C the increases in tan δ_e for each ceramic are modest although at much higher temperatures the increases may be much faster.   Such a trend in tan δ is indicated in Figure 13 for hot pressed aluminium nitride at temperatures above 500°C and at 8.5 GHz. Note, however, the sharp increase in tan δ_e at the radio frequency of 10 MHz.   Two different aluminas, shown in Figures 14 and 15 respectively, show similar trends in

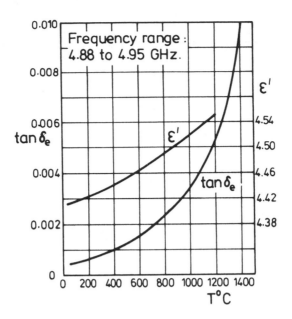

Figure 11: The dielectric constant and the tan δ_e as a function of the temperature for hot pressed boron nitride, grade HBN.   (After ref 3).

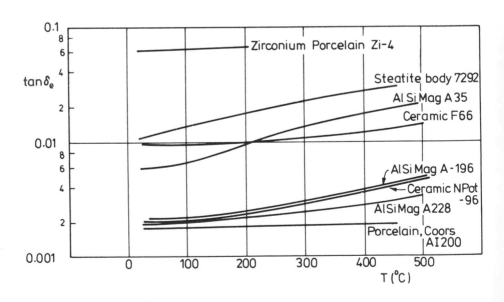

Figure 12: Variation of the tan δ_e with temperature for steatite, titania and titanate ceramics and porcelain at 10 GHz.   (Extracted from ref 2 by permission of the MIT Press).

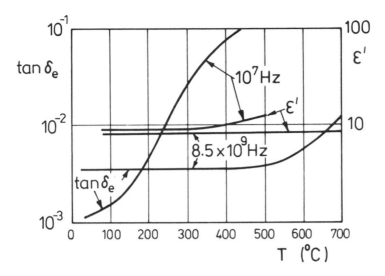

Figure 13: The dielectric constant and the effective tan δ_e for hot pressed aluminium nitride as a function of the temperature. (After ref 3).

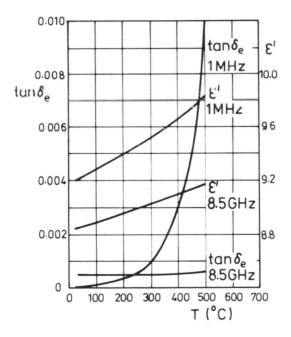

Figure 14: The dielectric constant and the effective tan δ_e for alumina 206 (95% Al_2O_3) as a function of the temperature. (After ref 3).

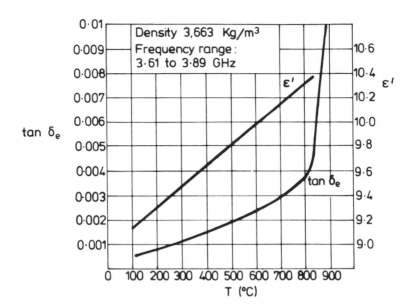

Figure 15: The dielectric constant and the tan δ_e for alumina as a function of temperature. (After ref 3).

the way in which the tan δ_e increases with temperature beyond a certain critical temperature, T_c. A change in tan δ_e of about an order of magnitude between 800°C and 900°C can be observed in Figure 15.

The effective losses in fused silica remain substantially constant as shown in Figure 16. Fused silica is in fact used in microwave applicator equipment as a support material on account of its extremely low microwave absorbtion properties. McMillan and Partridge[8] have also measured the properties of a family of glass ceramics at 9.37 GHz and shown that the tan δ's were below 50.10^{-4} for temperatures up to 400°C. Small additions of CaO in the glass ceramics, however, affected the measured ϵ' and tan δ values.

3.3 Thermal runaway

Of great importance to dielectric heating applications is the 'runaway effect' or the uncontrolled rise in temperature in a material brought about as a result of a positive slope, $d\epsilon_e''/dT$, of the ϵ_e'' vs T response. A typical example is the case of hot pressed boron nitride; its dielectric properties

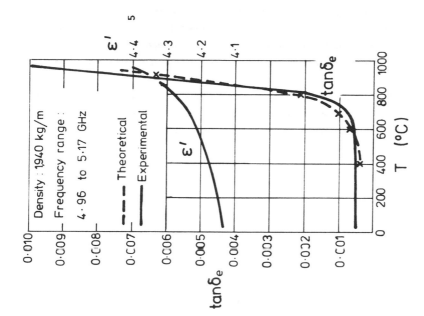

Figure 17: Temperature variation of the dielectric constant and the tan δ_e for hot pressed boron nitride, grade HD 0056. (After Inglesias and Westphall, MIT, 1967).

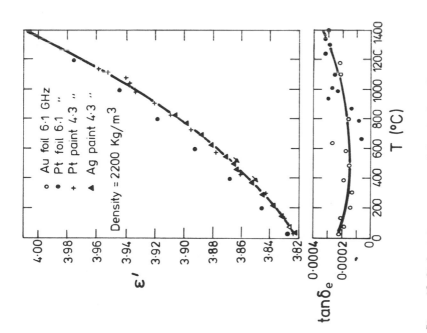

Figure 16: Dielectric properties of fused silica (Corning 7940). (After Inglesias and Westphall, MIT, 1967).

are shown in Figure 17. After an initial absorbtion of the microwave energy, the temperature rise causes the $\tan \delta_e$ to increase which, in turn, results in a further temperature increase and so on. Damage to the material is highly probable unless steps are taken to avoid such a cumulative effect.

It is often required to express mathematically the effective loss factor or $\tan \delta_e$ in terms of the temperature. In specific temperature ranges such an expression can take the form of a straight line:

$$\epsilon_e'' = aT + b \qquad\qquad 15$$

or a parabola:

$$\epsilon_e'' = c \times T^2 \qquad\qquad 16$$

where a, b and c are constants. Alternatively, a good fit of the experimental data can often be afforded by the empirical formula:

$$\tan \delta_e = \tan \delta_o + \frac{A \times T^2}{B-T} \qquad\qquad 17$$

where the constants $\tan \delta_o$, A and B are chosen to best fit the data. For example, by fitting equation 17 to the data shown in Figure 17, the following empirical relation is found to hold for boron nitride, grade HD 0056, in the temperature regime 400°C - 900°C:

$$\tan \delta_e = (2 \times 10^{-4}) + \frac{(49 \times 10^{-8} T^2)}{(965-T)} \qquad\qquad 17a$$

where T is in °C.

3.4 Purity and microstructure

Ho[6] has measured the dielectric properties of a wide variety of silicon nitride ceramics (CVD, sintered, hot pressed, differing composition and sintering aids) and of polycrystalline alumina versus single crystal sapphire. His results have shown that the temperature dependence of the dielectric constant is intrinsic to the crystalline lattice properties of the material whilst other parameters, such as impurities (of a few percent) and microstructure, are of secondary importance.

In contrast, the dielectric losses in ceramics are directly related to microstructural and grain boundary properties. Thus polycrystalline materials, such as alumina, can have an order of magnitude higher loss than their single crystal counterparts. It is this extrinsic nature of the absorbtion mechanism which Ho believes accounts, at least in part, for the rapid rise in loss tangent of polycrystalline ceramics. Further, the mechanism giving rise to the observed loss at low and moderate temperatures appeared to be quite different from that at higher temperatures.

Several researchers have indicated that the degree of microwave absorption by a material does not correlate with the bulk purity of the sample or with the presence and extent of secondary crystalline phases. Ho suggests that absorption appears to be associated with impurities incorporated into the primary microcrystalline grains and grain boundaries, with fine-grained materials (which have greater surface area) proving to be more susceptible to impurity content.

Based on their observations Ho and Harker[9] constructed a theory attributing the low and intermediate temperature loss mechanism to localised doping of the primary and secondary crystalline phases to create isolated microscopic regions of finite electrical conductivity localised within the grains and along individual grain junctions and grain boundaries. A distribution of electrical conductivity values was shown to explain the observed frequency dependence between sapphire and alumina.

The high temperature mechanism, which gives rise to the rapid increase in loss tangent, has been attributed to increasing electrical conductivity as the softening temperatures of residual glassy phases are approached[6]. Experimental results[10] have shown good agreement with these theories using alumina doped with low concentrations of titania and soda.

Recent work by Varadan and Varadan[11] on microwave interactions with polymeric and organic materials has shown that helical shaped inclusions, provided that they are all one-handed, show significantly superior absorbtion characteristics compared with similar, spherical inclusions. This has interesting implications for work in ceramic composite systems.

3.5 Wet Ceramics

Concerted emphasis has been given to the variation of the dielectric properties with moisture content; the ensuing data being used for the effective optimisation of applicator designs intended for industrial drying applications. Typical variations of ϵ_e'' with moisture content (on dry basis) and electric field orientation established within the material is given in Figure 18.

The addition of water to many materials enhances the dielectric properties of the mixture because water in its natural state is a highly polar substance with large numbers of permanent dipoles in its structure. When the water is bound to another material the combined properties of the mixture are radically different to those of free water, as shown in Figure 19.

Considering the qualitative representation of the loss factor versus moisture content shown in Figure 18, we can deduce the following: the slope of the ϵ_e'' versus M curve is critical to industrial applications where

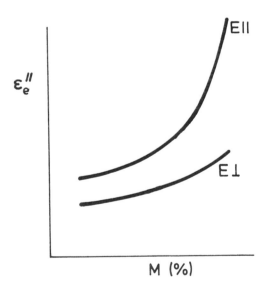

Figure 18: Qualitative representation of ϵ''_e versus moisture content and electric field orientation. (After ref 14 by permission of Peter Perigrinus, Ltd).

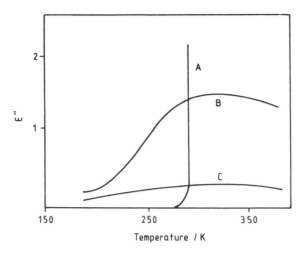

A   Free water
B   26% water on silica gel
C   8% water on alumina

Figure 19: Temperature dependence of the dielectric losses of free and absorbed water at microwave frequencies.   (After ref 15).

moisture levelling of a material in planar form is the prime requirement. This is because moisture levelling is very effective for moisture contents above a critical value, M_c, since the higher ϵ_e'' ensures that the wetter parts of the planar material absorb more power than the dryer parts and thus tends to level off the initial uneven moisture distribution.   Conversely, moisture levelling for moisture contents below the critical value is much less pronounced since the ϵ_e'' is practically independent of the moisture content and consequently different parts of the material absorb virtually the same amount of energy from the high frequency source irrespective of their degree of wetness.   In Figure 18, the curve with electric field parallel to the plane of the paper, $E\|$, is far more effective for moisture levelling than the curve with the field perpendicular, $E\perp$.

The dielectric properties of bricks at 3 GHz are shown in Figure 20 where the loss factor is about 2 for 30% water content.   Such a loss is significant in terms of microwave processing in that only modest electric fields would be required to establish quite an adequate power density within the bricks[12].   The use of microwaves in the drying of ceramics will be discussed in more detail in Section 7.2.

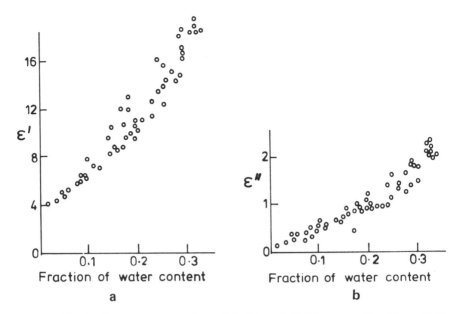

Figure 20: Dielectric properties of bricks at 3 GHz as a function of the water content.   (Extracted from ref 12 by permission of Chapman and Hall Ltd.)

4 DEVELOPMENT OF THE BASIC THEORETICAL CONCEPTS

4.1 Electric field distribution

The usual analysis of dielectric heating makes use of the wave equation, which is derived through Maxwell equations for $\underline{H}$ and $\underline{E}$.   For electric and magnetic fields with time harmonic variations, it is usual to represent these with the real parts of the expressions $\underline{E} = \underline{E}e^{j\omega t}$ and $\underline{H} = \underline{H}e^{j\omega t}$ respectively. The wave equation reduces to the following form in one dimension for a semi-infinite ceramic dielectric slab as shown in Figure 21[13]:

$$\frac{\delta^2 E_z}{\delta y^2}(y) = -\mu\epsilon_o K E_z(y) \tag{18}$$

where μ is the magnetic permeability and $E_z(y)$ refers to the electric field in the z-direction as it varies in the y-direction.   The assumption is made

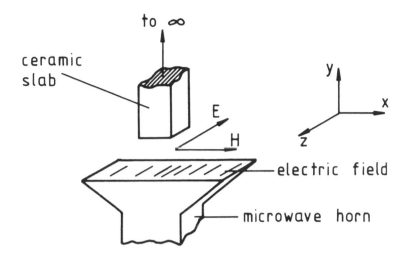

Figure 21: Interaction of microwave energy with a semi-infinite ceramic slab (After ref 13).

that within the slab, E_z does not vary in the x-direction.   The slab tends to infinity in the y-direction.   Although this is a special case of a microwave horn radiating towards a ceramic slab material, similar arguments follow for other microwave applicator/ceramic material configurations.   The propagation constant of the microwave energy traversing the ceramic material is given by[14]:

$$\gamma = j\omega\sqrt{K\epsilon_o\mu} = \alpha + j\beta \qquad\qquad 19$$

where α and β are the attenuation and phase constants respectively and, as shall be seen below, these two parameters are in turn very strongly dependent on the dielectric properties of the material under consideration.

Solution of equation (18) including time variations, assuming that when $y \rightarrow \infty$, E must be finite, yields:

$$E_z(y) = E_o e^{-\gamma y} e^{j\omega t} \qquad\qquad 20$$

and by taking the real part, the electric field expression simplifies to:

$$E_z(y) = E_o e^{-\alpha y} \cos (\omega t - \beta y) \qquad\qquad 21$$

where E_o is the maximum value of the electric field intensity at the ceramic/air interface, that is the value of E_z at $y \to 0$. This is a very significant result in that it allows us to estimate the extent to which the electric field decays in traversing the material.

4.2 Power Dissipation Within The Ceramic

It is often required to estimate the amount of power that can be safely dissipated in a ceramic given that the effective loss factor is known. This can be obtained from considering the Poynting vector $\underline{E} \times \underline{H}$ which leads to the following expression for the power dissipated per unit volume P_v:

$$P_v = \frac{1}{2} R_e(\underline{E} \cdot \underline{J}^*) = \frac{1}{2} R_e(\sigma + \omega \epsilon_o \epsilon'')\underline{E} \cdot \underline{E}^* - j\omega \epsilon_o \epsilon' \underline{E} \cdot \underline{E}^*$$

$$= \frac{1}{2}(\sigma + \omega \epsilon_o \epsilon'')\underline{E} \cdot \underline{E}^* = \frac{1}{2}(\sigma + \omega \epsilon_o \epsilon'')E^2 \qquad (Wm^{-3}) \qquad 22$$

The electric field is given by equation 20. In microwave heating the electric field is assumed not to decay too fast within the material, therefore it is usual to make the following approximation:

$$E = E_o e^{-\alpha y} = E_o\left(1 - \alpha y + \frac{(\alpha y)^2}{2!} \cdots\right) \approx E_o \qquad 23$$

Substituting equation 23 into equation 22 yields the power per unit volume developed in the ceramic dielectric:

$$P_v = \frac{1}{2}(\sigma + \omega \epsilon_o \epsilon'')|E|^2 = \omega \epsilon_o \epsilon_e'' E_{rms}^2 = \sigma_e E_{rms}^2 \qquad 24$$

$$P_v = 0.556 \times 10^{-10} \epsilon_e'' f E_{rms}^2 \qquad (Wm^{-3}) \qquad 24a$$

where f is in Hz, E_{rms} is in Vm^{-1} and $E_o = \sqrt{2E_{rms}}$. If the approximation of equation 23 cannot be made then the power density dissipation has to be evaluated by integrating throughout the volume and using expression 20 for the electric field.

It is fairly evident that in a material which exhibits the same dielectric properties at say, 900 MHz and at 2450 MHz, the electric field at the lower frequency has to be much higher in order to develop similar power densities at the two frequencies. For example, for a power dissipation of 10 MWm^{-3} and $\epsilon_e'' = 0.1$ the required electric fields at 900 MHz and at 2450 MHz are 44.5 kVm^{-1} and 27 kVm^{-1} respectively.

4.3 Attenuation Constant And Skin Depth

As the electromagnetic energy penetrates into the interior of the material it attenuates to an extent depending upon the effective loss factor ϵ_e''. The inverse of the attenuation constant is defined as the skin depth, δ;

$$\delta = 1/a \qquad\qquad 25$$

which is the depth at which the magnitude of the electric field drops to 1/e of the value at the surface. Expanding and equating the real and imaginary parts of equation 19 gives, when $\epsilon_e'' \gg 1$:

$$a = \omega \sqrt{(\mu\epsilon_o\epsilon_e'')/2} \qquad\qquad 26$$

Such approximation applies to very wet ceramics ($\epsilon'' = 0$) where $\sigma = \omega\epsilon_o\epsilon_e''$ and the dominant loss mechanism is dielectric conductivity. Therefore substitution and inversion leads to the skin penetration depth:

$$\delta = 1/a = \sqrt{(2/\omega\epsilon\mu)} \qquad\qquad 27$$

For a low loss ceramic, the usual approximation is now $\epsilon_e'' \ll 1$, which leads to the equivalent depth for a low loss dielectric:

$$\delta = 1/a = (2/\omega\epsilon'')\sqrt{(\epsilon'/\mu\epsilon_o)} \qquad\qquad 28$$

For a lossy ceramic where both dipolar and conductivity losses are present, no such approximation is applicable and the full expression involving the attenuation constant α must be used.

4.4 Power Penetration Depth

It has been shown that the power dissipated in the dielectric medium is controlled by the term $|E|^2$. Therefore:

$$P = P_o e^{-2\alpha y} = P_o e^{-2y/\delta} \qquad\qquad 29$$

where P_o is the power at the surface. At $y = \delta$:

$$P_\delta = \frac{P_o}{e^2} = 0.14\ P_o \qquad\qquad 30$$

giving 86% dissipation. We can define a power penetration depth D_p at which the power drops to $1/e$ from its value at the surface. This gives:

$$P = P_o e^{-y/D_p} \qquad\qquad 31$$

where:

$$D_p = 1/2\alpha = \delta/2 \qquad\qquad 32$$

It is important to differentiate between power penetration depth and electric field or skin depth as shown in Figure 22. At the frequencies allocated for industrial use in the microwave regime, the penetration depths could be very small indeed and the size of the ceramic to be treated, particularly when it is fairly lossy, could be many times larger than D_p, resulting in unacceptable temperature non-uniformities.

Table 3 illustrates the dependence of D_p on the material properties for various ceramics. The D_p ranges from 169 m for pyrolytic boron nitride, which is a very low loss ceramic, to 12 cm for hot pressed aluminium nitride, which at 700°C and 8500 MHz is a highly "lossy" ceramic material. The allowable size of the component therefore depends critically on the

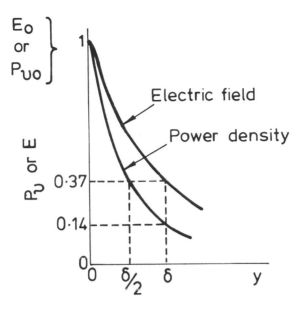

Figure 22: Skin and power penetration depths.

materials' dielectric properties, emphasizing the need to establish the latter before any specific industrial equipment is built.

Large variations of the D_p can also occur in a material due to abrupt changes of the tan δ_e with temperature. Hot pressed boron nitride is prone to such an effect as illustrated in Table 4 where the D_p reduces from 7.62 m to 45 cm as the temperature increases from 600°C to 950°C.

5. APPLICATORS

5.1 Introduction

The term applicator is a generic one and is used in microwave heating to refer to a device into which the material is inserted for processing. The domestic oven can be regarded as a form of applicator for cooking foodstuffs.

Industrial microwave applicators for processing a variety of materials

Material	T	f	ϵ'	ϵ''_e	tan δ_e	D_p
	°C	MHz				m
Pyrolytic boron nitride	800	2450	3	2×10^{-4}		169
Calcium titanate	25	2450	180	2×10^{-3}		131
Steatite	25	2450	6	2×10^{-3}		23.9
Lime alumina silicate	25	2450	7	6×10^{-3}		8.6
Porcelain	25	2450	5	1.5×10^{-2}		2.9
Barium/strontium titanate	25	2450	2000	0.5		1.74
Barium titanate	25	2450	700	0.3		1.72
Hot pressed aluminium nitride	500	8500	9		4×10^{-3}	0.47
	700	8500	9		1.5×10^{-2}	0.12

Table 3: Power penetration depths, D_p.

T	ϵ'	tan δ_e	D_p
°C			m
600	4.22	6×10^{-4}	7.62
700	4.25	8×10^{-4}	5.70
800	4.28	20×10^{-4}	2.27
900	4.35	70×10^{-4}	0.64
950	4.4	100×10^{-4}	0.45

Table 4: Temperature effects of D_p in hot pressed boron nitride, $T_c \approx 700°C$.

including ceramics, fall broadly into three categories: travelling wave, single and multimode applicators. The most common form of industrial applicator is the multimode type which in principle is an extension of the domestic microwave oven but built for large scale material processing. However, for ceramic processing all three types could find specific uses and therefore a brief description of each will be given below.

5.2 Travelling wave applicator

The principle of operation of the travelling wave applicator is illustrated in Figure 23, where a magnetron or klystron microwave source transfers energy to a water load via a waveguide section. At microwave frequencies used industrially for heating, the energy transfer is accomplished in special channels called waveguides, the dimensions of which depend upon the operating frequency. At the 896/915 MHz band, for example, waveguide type WG4 is used with dimensions a = 248 mm and b = 123 mm (see Figure 23), whilst at 2450 MHz two waveguides are used, WG9A (86x43 mm) and WG10 (72x34 mm).

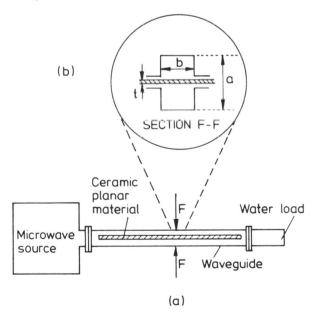

Figure 23: A travelling wave applicator: a) a source feeding a waveguide and a water load; b) cross-section of the waveguide showing the position of the planar ceramic material. (After ref 14 by permission of Peter Peregrinus, Ltd).

The planar material to be processed is inserted in a slot at the centre of the broad dimension along the waveguide length and absorbs an amount of energy from the travelling microwave field dependent upon the effective loss factor of the insertion. Any remaining power is absorbed by the matched water load, with very little energy reflected back towards the source.

The electric field in the waveguide decays exponentially with a factor $e^{-\alpha y}$ where α is the attenuation per unit length and is strongly dependent on the effective loss factor. The arrangement in Figure 23 poses a problem when processing planar ceramics because it can lead to high non-uniformity of heating or insignificant heating depending upon the value of $t\epsilon_e''$ where t is the thickness of the material.

To overcome this effect the material is passed through a number of waveguide lengths (passes), which leads to the development of the meander applicator illustrated in Figure 24. This device consists of a number of rectangular waveguides mounted with their broad faces side by side and electrically connected in series, each waveguide having slots at the centre of the broad face through which the ceramic workload travels at right angles.

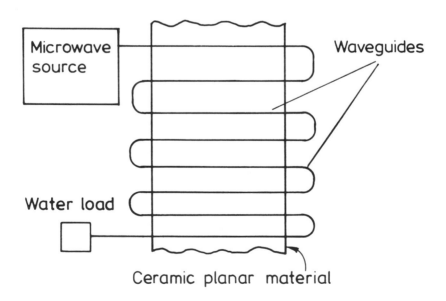

Figure 24: A meander travelling wave applicator. (Adapted from ref 14 by permission of Peter Peregrinus, Ltd).

As the electric field decays, the power at a point y away from the source (Figure 23) is given by:

$$P(y) = Pe^{-2\alpha y} \qquad\qquad 33$$

where P is the output power from the source incident at the input of the rectangular waveguide. Differentiation of equation 33 yields:

$$\frac{dP}{dy}(y) = -2\alpha Pe^{-2\alpha y} = 2\alpha P(y) \qquad\qquad 34$$

The fractional loss of power per metre is therefore given by 2α. The attenuation coefficient, α, relates to the partially loaded rectangular waveguide as shown in Figure 23b. Assuming no wall or other losses, α is given by[14]:

$$\alpha = \frac{17.37\pi\epsilon_e'' t\lambda_g}{a\lambda_o^2} \qquad (dBm^{-1}) \qquad\qquad 35$$

where λ_o and λ_g are the free space and waveguide wavelengths respectively. Substituting the values of the principal frequencies 896 MHz, 915 MHz and 2450 MHz with the approximate waveguide sizes yields for a ceramic slab of 1 mm thickness:

At 896 MHz; $2\alpha l = 0.891 \times 21 \times \epsilon_e''$ (dB)   36a
At 915 MHz; $2\alpha l = 0.894 \times 21 \times \epsilon_e''$ (dB)   36b
At 2450 MHz; $2\alpha l = 7.35 \times 21 \times \epsilon_e''$ (dB)   36c

where l is the width of the process material in metres. For example, for a ceramic with $\epsilon_e'' = 1$ and $l = \frac{1}{2}$ m, equation 36b gives $2\alpha l = 0.894$ dB. This represents a reduction of about 19% of the input power in one pass and six passes would be required to dissipate 70% of the available power. Alternatively if ϵ_e'' is only 0.1 the fractional loss is now 0.089 dB per pass which results in just 2% dissipation of the input power. Such meanders would be unacceptably large and thus resonant applicators would be more suitable for ceramics with low ϵ_e'' values.

5.3 Single Mode Resonant Applicators

Resonant applicators consist of metallic enclosures into which a launched microwave signal of the correct electromagnetic field polarisation undergoes multiple reflections. The superposition of the incident and reflected waves gives rise to a standing wave pattern which for some simple structures is very well defined in space. Knowing precisely how the electromagnetic field is distributed within the resonant applicator enables the ceramic material to be placed in the position of maximum electric field for optimum transfer of the electromagnetic energy within it.

A most versatile single mode resonant applicator is shown in Figure 25. It consists of a straight piece of rectangular waveguide connected to a flange with a coupling iris on one side and a non contacting short circuit plunger on the other side. Such a transverse electric type of applicator, where the electric field is transverse to the direction of propagation, results in a cosinusoidal electric field distribution as illustrated in Figure 25c. The ceramic material to be processed is inserted in the applicator through a slot in the broad dimension of the waveguide. As can be seen, the ceramic experiences the maximum static electric field set up

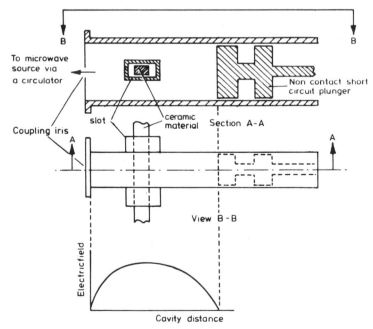

Figure 25: A single mode TE_{10n} resonant applicator. (Adapted from ref 14 by permission of Peter Peregrinus Ltd).

within the applicator. This ranges within $1 < E < 15$ kVcm^{-1} with a moderate power level from the source, say a few kW's, and a large rate of heating can be achieved.

For optimum performance both the position of the plunger, which determines the operating frequency, and the size of the iris, which establishes how much of the energy is transferred to the applicator, are dependent upon the dielectric properties of the ceramic material under consideration. Careful optimisation procedures must be carried out at low power with the ceramic material in place before high power tests can start. The versatility of such an applicator is unquestioned because a variable plunger and a flange which has on it a variable aperture, enable this applicator to treat a wide range of ceramics of different effective loss factors simply by choosing the right dimensions[14].

Another useful but less versatile single mode resonant applicator is shown in Figure 26. It consists of a circular waveguide operating in the transverse magnetic or TM$_{010}$ mode and is shortened at both ends. This fixes the overall dimensions which restrictions the operation to a given narrow frequency band for a given ceramic.

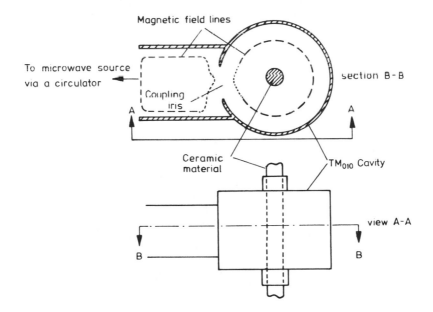

Figure 26: A single mode TM$_{010}$ resonant applicator. (Adapted from ref 14 by permission of Peter Peregrinus, Ltd).

Coupling of the energy inside the applicator from the connecting waveguide and source is afforded through the magnetic fields near the aperture as shown dotted in Figure 26. Such an applicator establishes a zero order Bessel type electric field distribution, $E_oJ_o(kr)$, with a peak value at the centre of the applicator where the ceramic dielectric is located[14].

5.4 Multimode Applicators

These applicators are widely used in the domestic sector and have also found many applications in industry. Multimode applicators are very versatile in that they can accept a wide range of material loads of different effective loss factors and sizes.

The multimode microwave applicator consists of a metallic container and a coupling mechanism such as a simple iris or a number of irises forming an array, to couple the microwave energy from the magnetron source to the metallic enclosure. In a given frequency range such an applicator will support a number of resonant modes. In general a mode refers to a specific way in which the electromagnetic field establishes itself within the applicator. For an empty applicator each of these modes exhibit sharp resonant responses as illustrated in Figure 27 for f_1 and f_2. However, for

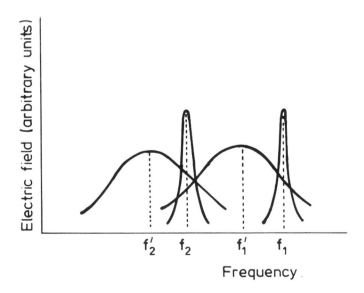

Figure 27: Typical modes in a multimode applicator. (After ref 14).

an applicator partially filled with a ceramic material which couples reasonably to the microwave electromagnetic field, the resonant responses of the modes will overlap in frequency to give a continuous coupling with the ceramic load. This broadening of the modes is also shown qualitatively in Figure 27 where a frequency shift is seen to occur as well.

In the multimode applicator the electric field distribution is given by the sum of all the modes excited at a particular frequency, each mode resulting in a heating effect which is proportional to $(1 - \cos \theta)$, θ being related to a linear dimension within the metal enclosure. The heating effect of each mode therefore shows a basic trigonometric variation along the principal coordinate axes, which results in a non-uniformity of heating in the ceramic workload within the applicator and much consideration has been given to minimise such an effect.

To improve the uniformity of heating two basic methods are used. A turntable which rotates at a constant speed can be used to move the load successively through the nodes and antinodes. Alternatively, a mode stirrer can be used. This is a structure such as a metallic multiblade fan which perturbs the electromagnetic field continuously by its rotation inside the multimode oven. It is quite common to find both of these techniques in operation.

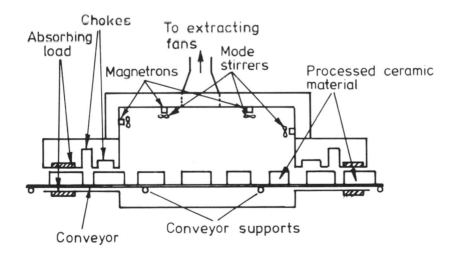

Figure 28: A mulitmode resonant applicator. (After ref 14 by permission of Peter Peregrinus, Ltd).

A typical on line multimode oven applicator for industrial processing of ceramic loads is shown in Figure 28. Protective devices such as absorbing or reflective loads are normally fitted at the input and output ports to prevent leakage of electromagnetic energy to the surrounding space. The ceramic workloads are carried through the applicator on conveyor belts suitably designed to operate within the oven enclosure. Four magnetrons are shown to feed power to the applicator; however industrial systems with many tens of magnetrons feeding one applicator have been designed. With multiple generators the opportunity exists to distribute the power so as to achieve a better excitation of the modes and better uniformity of heating than can be achieved with a single feed, by distributing the feed points around the walls of the oven and by feeding at different polarisations.

6. HEAT TRANSFER

Having described the electrical parameters which control the current density and power dissipation within the ceramic material and having outlined various applicator structures, attention is now switched to the equation which controls the heat transfer. In the present context it will suffice to quote the following continuity equations for heat:

$$\frac{\delta T}{\delta t} = a_\tau {}^2T + \frac{P_v}{c\gamma} \qquad (Wm^{-3}) \qquad 37a$$

$$\frac{\delta T}{\delta t} = a_\tau {}^2T + \frac{H_{fg}}{c}.\frac{\delta M}{\delta t} + \frac{P_v}{c\gamma} \qquad (Wm^{-3}) \qquad 37b$$

where P_v is the localised power density, a_t is the temperature diffusivity, M is the moisture content in % of dry mass, H_{fg} is the enthalpy of evaporation, γ and c are the density and specific heat respectively. For equation (37b) two assumptions are inferred, that any moisture is predominantly in the liquid phase and that convection heat transfer is negligable. In microwave heating, a number of distinct processes can be analysed using the equation for the heat flow accordingly simplified.

6.1 Heating rates

For a ceramic material with no liquid phase (equation 37a) and with no spatial distribution of temperature within it ($\Delta T = 0$), as in the simple case where the temperature rises at a uniform rate throughout the body of the ceramic, the continuity equation for heat gives:

$$\frac{dT}{dt} = \frac{P_v}{c\gamma} \qquad\qquad 38$$

Substitution of P using equation 25 gives:

$$\frac{dT}{dt} = \frac{0.556 \times 10^{-10} \epsilon_e'' f E_{rms}^2}{c\gamma} \qquad (°Cs^{-1}) \qquad\qquad 39$$

where the electric field is given in Vm^{-1}, the density in kgm^{-3} and the specific heat is given in $Jkg^{-1}°C^{-1}$. Therefore, for a material heated by microwave fields the rate of rise of temperature at a given frequency depends on the product ($\epsilon_e''.E_{rms}^2$). This is a function of the temperature because of the variation of ϵ_e'' with T.

If the variation is simply parabolic with the form $\epsilon_e'' = aT^2$, substitution in equation 39 gives:

$$\frac{dT}{dt} = \frac{0.556 \times 10^{-10} aT^2 f E_{rms}^2}{c\gamma} = gT^2 \qquad\qquad 40$$

where g is constant. Integration between limits T_1 and T_2 where $T_1 < T_2$ gives:

$$gt = [\,(1/T_1) - (1/T_2)\,]$$

6.2 Temperature Distribution

The temperature distribution in ceramic materials heated by microwave energy can be obtained by solving the continuity equation for heat (equation

37a).   By substituting P using equation 24 and noting any functional relationship between ϵ_e'' and T, various solutions of T as a function of y (considering only one dimension) and t can be obtained.

This ability to predict temperature distributions in media being heated by microwave energy has been the focus of many investigations[15-19].   De Wagter[17] used a computer to simulate the temperature distributions generated by microwave absorption in multilayered media.   Qualitative agreement was achieved between the simulation and subsequent experimental results.   The lack of complete fit between the two sets of results was attributed to a restricted ability to measure the experimental temperature profiles accurately.   A key feature of the simulation was the capability of allowing for temperature dependence of the dielectric properties through use of third order polynomials.

For continuous media, Figure 29 represents the initial distribution of absorbed power; however the temperature profile will change with time as shown by Figure 30[20].   This can lead to an inverse temperature profile, that is a higher internal temperature.   Watters et al[19] have confirmed the above argument by analyzing the electromagnetic heating of an infinite half

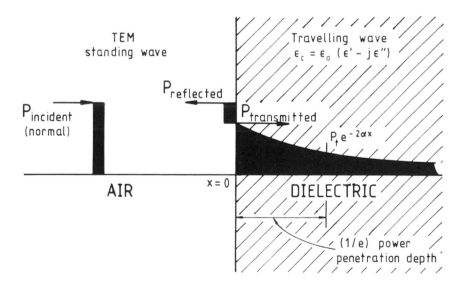

Figure 29: Power transmission across air-dielectric interface. (Adapted from ref 15).

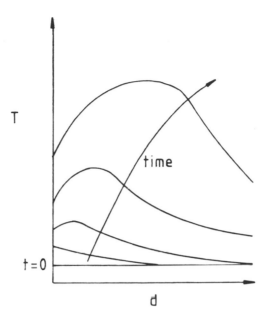

Figure 30: Proposed variation in temperature profile with time. (After ref 20).

space in terms of the heat equation, subject to heat loss boundary conditions. Once the necessary equations had been formulated Watters and co-workers were able to determine the effect of a number of variables. They determined that increasing electrical conductivity reduced the inverse temperature gradient and thus reduced temperature inhomogeneity; increasing the thermal conductivity also decreased the temperature inhomogeneity. Changes in the incident microwave power level primarily affected temperature values rather than the distribution.

Recently, Meek[21] attempted to develop a model for the sintering of a dielectric in a microwave field. By assuming that partial sintering had taken place and that the sample consisted of fully dense and 50% dense material, Meek compared the power absorbtion, half-power depth and heating rates in the two regions. His conclusions were that all three would be considerably higher in the 50% dense regions. However, power absorbtion and depth of penetration must be inversely related, invalidating the theory.

A key feature of the model was the assumption that the loss tangent of the lower density regions was approximately equal to that of the higher

density regions. By further asssuming the electric fields were perpendicular
to the media interface, this led to the equation:

$$\frac{P_p}{P_d} \propto \frac{\epsilon'_d}{\epsilon'_p}$$ 41

where the subscripts p and d refer to the porous and dense regions
respectively.   In the limiting case where the porous region is virtually
100% air, $\epsilon'_p \approx 1$ and thus approximately ϵ'_d more power would be absorbed by
this region than by the dense material.   On this argument the low density
fibre-based insulation used in the experiments would absorb power more
efficiently than the samples.

Whilst the exact nature of the relationship between loss tangent and
density is uncertain, by allowing that $\tan \delta_p \neq \tan \delta_d$ we find:

$$\frac{P_p}{P_d} \propto \frac{\epsilon'_d}{\epsilon'_p} \cdot \frac{\tan \delta_p}{\tan \delta_d}$$ 42

and thus for an ultra low density region, $P_p \approx 0$. A similar result has been
obtained by Tinga[22].

Varadan et al[23] have used a computer to simulate microwave sintering
via multiple scattering theory involving correlation function, shape and
size effects. The increase in the imaginary part of the dielectric constant
was primarily ascribed to the collapse of pore concentrations and the
enlargement of pore sizes with increasing temperature. Thus, according to
the model, there is a critical pore concentration at which the heating rate
of the material due to microwave absorbtion is dramatically raised and green
samples begin to densify much more effectively than by conventional furnace
heating. The critical concentration, c_r, depends exclusively on the thermal
runaway temperature.

Microwave sintering of strontium titanate and α-alumina was used to
experimentally verify the importance of porosity level and distribution by
comparing heating rates for samples of varying porosity. The results clearly
indicated that the model is essentially correct; the fastest heating rate (ie
the maximum power absorbtion) being obtained for samples with approximately
50% porosity. Power absorbtion fell sharply for samples with lower or higher

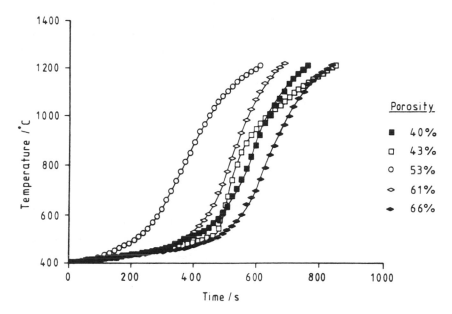

Figure 31: Temperature-time curves for strontium titanate samples with different initial porosities. (After ref 23).

green densities, Figure 31.

An additional conclusion which may be drawn from these results is the importance of homogeneity with respect to the density of green samples. A lack of such homogeneity may account for some of the non-uniform heating experienced by researchers attempting to microwave sinter green ceramic pieces (see section 7.5).

6.3 Temperature Measurement

The accurate measurement of sample temperature during heating by a microwave field is seen as one of the major problems to be overcome for the successful exploitation of microwave ceramic processing. Infra-red pyrometry has been used extensively to date, particularly when a single mode applicator forms the basis of the sintering system. However, this technique suffers from the disadvantage that only the surface temperature of the samples is recorded.

The major alternative to pyrometry involves the use of electrically shielded and insulated thermocouples. Ideally they must touch the sample since the latter forms its own heat source, yet this still only provides surface temperature. Electric shielding is usually achieved by inserting the thermocouple in a hollow metal tube, whilst electrical insulation (to prevent discharge from the specimen) is provided by a thin ceramic coating. If the latter is too thick then the finite (and generally low) thermal conductivity of suitable coating materials will result in inaccurate measurements being obtained. King[24] has discussed the insulated antenna approach in some detail.

Tinga[22] has suggested monitoring thermocouple responses during brief periods when the electro-magnetic field is switched off. This would still necessitate the use of electrically shielded and insulated thermocouples for the periods of microwave heating and whilst it is envisaged that this approach may find some success it is likely to be very system dependent. In some situations the decrease in sample temperature when the power is switched off may be excessive.

Thermistors have been used[25] in another attempt to avoid heating of thermocouple wires. Unfortunately, these devices are liable to be limited in the temperature ranges within which they can be used to obtain results[22]. Bosisio et al[26] developed an alternative approach in which the material properties were monitored by a calibrated signal at a second, lower frequency. Changes in this second signal could then be used to provide an indication of temperature. Such a system may well find some applications, though it is liable to be expensive.

A further technique was developed by Araneta et al[27] in which the dielectric properties of the material were continuously estimated from comparison of the experimentally determined admittance of the system with the theoretical value. Such a procedure is limited in its applicability to the problem of temperature control and the range of suitable sample geometries is also restricted.

More recently a new technique has started to be implemented, involving the use of optical fibre thermometry[f1]. This involves the use of a single crystal sapphire rod coated at its tip with a thin film of a precious metal

---

f1 Accufibre Inc, Beaverton, Oregon, USA

to form a blackbody cavity.   When the sample is heated it emits radiant
energy which is transmitted via the sapphire crystal to a low temperature
optical fibre and thence to a remote optical detector.   This latter converts
the radiation to an electrical signal which can be used both to measure
temperature and in a feedback system to control the temperature, the
measurements being immune to microwave interference.

7. APPLICATIONS

7.1 Introduction

Microwave energy has been used in industry in a number of distinct ways
during the last two decades.   The advantages of systems involving microwave
techniques, whether alone or in combination, over purely conventional fuel
fired systems have been well documented[14].   These range from faster
processing times and better uniformity of heating to space savings and
improved quality of product.

Microwave energy offers several advantages to the industrial processing
of many ceramics.   Firstly, the inverse temperature profile which can be
generated (figure 30) results in much faster drying and moisture levelling
if the ceramics are wet (or alternatively the microwaves can be used to
control the amount of water content in the original wet materials) and may
also have significant potential applications when microwaves are used to
sinter ceramics.   In addition it may prove possible to generate uniform
temperature profiles by combining surface heating techniques with microwave
heating.

Secondly, the coupling mechanism between microwave radiation and many
ceramic materials gives rise to a high energy transfer efficiency, leading
to greater control of microstructural development via good on-off control.
It has also been claimed that coupling can be localised (e.g. at interfaces),
allowing joining and sintering to occur with significantly faster kinetics
than by conventional radiant heating and also providing the potential for
controlling interfaces in composites systems.   The combination of all these
potential advantages offers the possibility of creating new structures with
novel properties.

Application	Intention / Comments
Drying	to accelerate the removal of moisture from green bodies.
Slip casting	to increase the casting rate; also includes accelerating the drying of moulds and ware.
Sol-gel processing	to reduce the number of processing steps involved in producing a sintered body from sol-gel technology.
Calcination	to accelerate the process; so far only superconducting systems and some electro-ceramics appear to have been studied.
Sintering	a wide range of ceramic and ceramic composite systems are currently being studied with a view to accelerating the process (and improving economics), and to generating novel microstructures.
Joining	to decrease time required to join ceramics and, possibly, to allow a greater range of ceramics to be joined. Both oxides and nonoxides are currently under investigation.
Plasma-assisted sintering	to use microwaves to generate plasmas to assist oxide sintering.
CVD	to use microwaves to decompose gaseous species which recombine to form thin films or powders.
Other applications	includes process control, clinkering of cement products, heating of optical fibre preforms, etc.

Table 5: Applications for microwave energy in ceramic processing.

Apart from drying and sintering, there are many other possible uses of microwaves in the field of ceramic processing; some of these are summarised in Table 5.   The rest of this review will concentrate on examining the potential for microwave energy in some of these applications.

7.2 Drying

7.2.1 Theory
The removal of moisture from green ware is an important process in the manufacture of both traditional and technical ceramic components.   The economics of the situation usually demand that the process is performed in as short a time interval as possible; however attempts to accelerate convective drying (the conventional drying method) can lead to a number of problems. To understand the advantages microwave energy can offer the drying of ceramics it is necessary to briefly review the mechanism by which moisture removal occurs.

The drying process can be conveniently divided into two periods; a constant rate period and a falling rate period (Figure 32).   During the

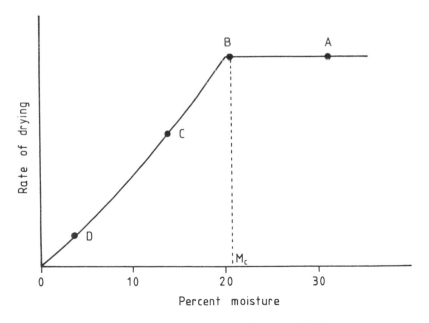

Figure 32: The two stage drying process.   (After ref 28).

former, the wet material contains so much water that the surface of the body consists of a continuous film of water. This evaporates at a rate comparable to that of a free water surface and is thus dependent only on the ambient conditions (temperature, velocity and humidity of the air) and the surface area and temperature of the water film. Once the critical moisture content (M_c) is reached, however, the continuous film is broken and the falling period starts.

This stage involves two processes; migration of moisture within the material to the surface and removal of the moisture from the surface. The former can occur by three mechanisms; vapour flow, liquid flow and diffusion. The latter is usually only significant when the moisture content drops below the saturation point. Liquid flow dominates with convective drying, where the rate of moisture movement is given by[28]:

$$\frac{dM}{dt} = k.\frac{(c_d - c_w)\ p}{1}.\frac{p}{\eta}$$

43

where c_d and c_w are the water concentrations at the dry and wet faces of the body respectively, p is the permeability of the body, 1 is the path length (approximately equal to half the body thickness), η is the viscosity of water and k is a constant.

From this relation it is evident that to achieve a high rate of water flow a large moisture gradient or a low water viscosity (i.e. a high drying temperature) would be required for a body of given permeablity and size.

Figure 33 shows characteristic curves for the conventional drying of a wet body. Moisture is removed initially from the external surface of the body which produces the required moisture gradient for outward moisture flow. However, as a wet body looses moisture it shrinks and since moisture is lost non-uniformly throughout the body, shrinkage is non-uniform. If the drying rate is excessive, surface cracking can occur as the dry surface is placed in tension by the still moist interior.

To overcome these problems it is normal to use either a slow heating rate or to raise the humidity around the ware prior to heating to prevent moisture loss. Once the body has acheived a uniform temperature the humidity is lowered and drying begins. In either case, drying can be slow and expensive. The use of microwave energy for drying, therefore, can offer a number of advantages.

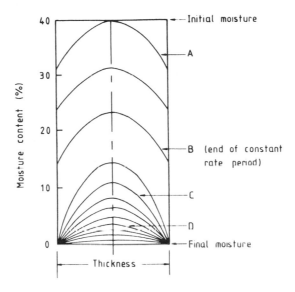

Figure 33: Conventional drying profile. (After ref 28).

7.2.2 Microwave drying

It was seen in Section 6.2 that the microwave irradiation of a body of finite size can lead to an inverse temperature profile (Figure 30). Several researchers have examined the effect of this on the drying of porous media[16,18,29].

Overall, much faster drying rates were found compared with conventional surface heating, together with a lack of a significant moisture concentration gradient during the process. This may be explained by the exponential dependence of the diffusivity of moisture on temperature. The diffusional flow rate for a given moisture gradient will be much higher near the centre of a body than near the surface leading to a strong levelling process. Thus microwave dried bodies have a completely different moisture distribution pattern with time than conventionally dried ware, as illustrated below.

Both Wei et al[18] and Perkin[16] have attempted to model the situation arising from the use of microwaves. Wei and colleagues first developed a mathematical model to explain the heat and mass transfer phenomena occurring in convectionally heated wet, porous materials[30]; water-filled sandstone was used to compare theory with practical results. Subsequent work developed this approach further for microwave heated systems by incorporating Lambert's law for the attenuation of microwave power in dielectric media[18]. Once again the fluid flow and heat transfer phenomena were predicted for a water-laden sandstone. Figures 34a and b show comparisons of the experimental and calculated values of temperature and evaporation rate versus time for the convectional and microwave heated samples respectively.

It can be seen from these curves that the model appears to fit the microwave case better than the convection heating situation. The lack of fit in the latter case is attributed by the authors to difficulty in obtaining the surface temperature values since near the surface the temperature gradient in the drying medium was large, about 75 Kcm^{-1}.

Figure 35 shows the theoretically calculated moisture profiles for the microwave case (after ref. 18). They have been redrawn to allow direct comparison with Figure 33, illustrating the much greater uniformity of microwave heating for moisture removal. This feature is emphasised by the results obtained by Perkin in his theoretical analysis of heat and mass transfer characteristics of boiling point drying using dielectric heating. The mass transfer equation developed is:

$$\frac{dM}{dt} = \frac{-P_{bp}}{L_s} \cdot \frac{M - M_c}{M_{in} - M_c} \qquad\qquad 44$$

The equation shows that the rate of internal evaporation at a point is directly proportional to the amount of liquid present. Consequently, with an initial non-uniform moisture profile the fastest rate evaporation occurs where the solid is wettest. In general, provided the overall result of any changes in the loss factor and electric field is for the absorbed power P to decrease with decreasing moisture content, this levelling effect will be obtained.

Whilst both the models of Wei et al[18] and Perkin[16] are useful indicators of the potential offered by the use of microwaves, their limited application to ceramic processing should be noted. Boiling of the moisture in a green body[16] will lead to disruption of the ware, whilst the work of Wei et al, by using wet sandstone, eliminates the complication of sample shrinkage.

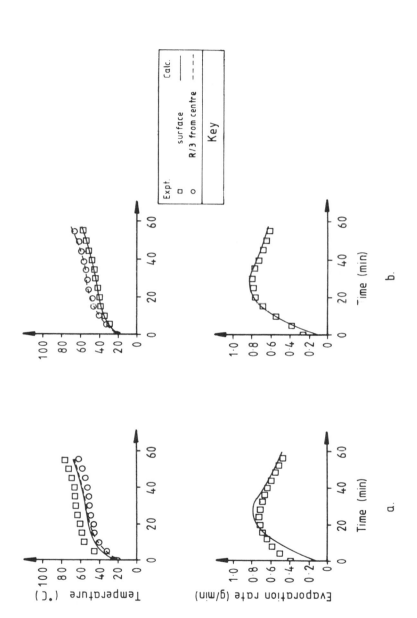

Figure 34: Plots of temperature profile and evaporation rate for sandstone; a) conventional drying, b) microwave drying. (After refs 30 & 18 respectively).

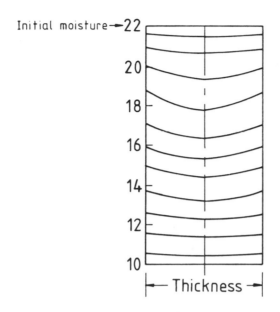

Initial moisture→22

Figure 35: Microwave drying profiles (from ref 18), redrawn to allow direct comparison with Figure 33.

Considerable research is now underway into the drying of green ceramic bodies. As one example, current research being performed at Ceram Research in the UK is focussed on two different sectors of the ceramic whitewares industry[31].   In the first, microwave drying of clay items during the manufacture of sanitaryware is being researched to replace current shop drying methods. Clay items are normally dried overnight in a heated workshop which means large stocks of work in progress and long processing times for individual items. Feasibility studies have indicated that clay items can be dried by microwave energy techniques in times of approximately 1-1½ hours. Industrial development of this concept offers the opportunity of more rapid processing and eventual flow line manufacturing methods.

The second sector is the tableware industry, for which microwave drying is being considered as a means of speeding up release of the green piece from the former, conventionally a plaster mould.   The concept is applicable to both the casting and plastic making processes. A further feature of the research is the combination of vacuum with microwave energy.   By this approach the temperature of the evaporating water never rises much above 40°C thereby removing any risk of damaging the plaster. Initial work has shown that release times can be reduced from a normal 20 minutes to approximately one minute.

Recently, Japanese research has also been directed towards combining microwave drying with the use of vacuums or low pressure systems[32]. This approach is claimed to improve further drying uniformity in addition to allowing lower temperatures to be used.

The major benefit from these studies is seen to lie in the development of more compact, flexible and efficient fabricating systems where the need for large mould rounds and dryers covering considerable floor areas would be removed.

7.2.3 Drying of powders

In addition to the drying of green bodies, microwaves have been used to dry powders. Over twenty years ago Allan[33] produced a feasibility study for the use of microwave heated moving-bed driers. In this paper Allan attempted to provide a method for determining whether extensive pilot plant investigations could be econonmically justifiable in situations where limited data is available.

More recently a conventional, domestic microwave oven has been used to dry advanced ceramic powders produced by a wet chemistry route[34]. The use of microwaves was found to result in much faster drying and also to produce a powder which consisted of soft agglomerates. These were more easily broken down during pressing than the conventionally dried powders and so yielded higher sintered densities. Although the final strengths were primarily dependent on other factors, there was an indication of slightly higher strengths and reduced variability with the microwave dried powders. These results were attributed to the vapour pressure produced within the agglomerates, the effect of which would be to open them up, reducing the degree of bonding. Futhermore, more uniform drying was believed to occur with local moisture concentrations absorbing more energy and thus evaporating more quickly.

Sheinburg et al[35] have explored several evaporative approaches to powder formation from aqueous nitrate solutions in the Gd-Ba-Cu-O system including microwave drying and decomposition. Simple microwave dehydration of the nitrate solutions yielded fine powders, with continued heating causing some decomposition (though not the $Ba(NO_3)_2$). The principle problems with this route were the evolution of nitrogen oxide fumes and a lack of chemical homogeneity owing to the excessive time required for precipitation of the powder during dehydration.

When citric acid was introduced to the nitrate solution and the pH adjusted to 6 via ammonium hydroxide additions, a citrate-based gel was formed. After drying, continued application of microwave energy caused the product to ignite in a controlled manner, a burning interface moving at about 1 cms^{-1} through the material. The resultant friable product was microcrystalline in nature and relatively homogeneous, though it contained some free barium carbonate and copper oxide. Nevertheless, on 'conventional' sintering it yielded the desired '1-2-3' phase at relatively low temperatures.

7.3 Slip Casting

7.3.1 Theory
The use of microwaves in the drying of green ware through the interaction of the electromagnetic energy with the polar water molecule led to experiments involving slip casting.

The driving force for the extraction of liquid during slip casting is the suction pressure created by the porous structure of the mould. During this dewatering of the slip the solid particles are drawn together to allow attractive particle interactions. Once the cast layer is formed a moisture gradient is created and the process becomes one of diffusion. Since diffusion of the liquid through the cast is much slower than through the mould the former becomes the rate controlling step. Thus casting rate is determined to a large extent by permeability of the cast.

A number of investigators have modelled the casting process, for example Aksay and Schilling[36], who derived the equation:

$$\frac{D^2}{t} = \frac{2P_T}{\eta n \left[a_c + \frac{n a_m}{P_m} \right]}$$
45

where D is the thickness of the cast after time t, P_T is the total pressure difference across the cast and mould, η is the viscosity of the slip medium,

q_c and q_m are cast and mould resistance terms respectively, n is a system parameter given by $(1-v_s-p_c)/v_s$, p_m and p_c are the pore fractions in the mould and cast respectively and v_s is the volume fraction of solids in the slip.

As can be seen from this equation, casting rate is directly related to pressure and various methods have been used to increase the pressure difference across the cast[37]. These include pressure casting, centrifugal casting and vacuum assisted casting.

The temperature of the slip is also of significance since viscosity decreases with increase in temperature. Van Wunnik et al[38] showed the slip temperature, and also that of the mould, to be important parameters in the casting process and Hermann and Cutler[39] found that the rate of casting varied linearly with temperature. This appears to be the principle variable affected by the use of microwaves.

7.3.2 Microwave assisted slip casting

Many researchers[40-47] have experimented with the use of microwaves with respect to slip casting. Generally, they have found that microwaves may be beneficially used in four stages of the slip casting process:

i)   during the casting period,

ii) after draining off the excess slip (to harden the cast),

iii) to dry the resultant green body, and

iv) to dry the moulds ready for re-use.

Of these three applications, the latter two are primarily drying excercises and, as such, have already been discussed above. Further discussion of these two stages will therefore be limited in nature.

The earliest report found on microwave slip casting was made by Blin[40] during the XIth International Ceramic Congress in Madrid. He reported on the drying of earthenware tiles and plates and also on the rapid casting of cups and small jugs. This was followed by a subsequent publication[41] in which further details were given.

As a result of this work, Evans[42] investigated the potential for the slip casting of large sanitaryware pieces and found that the casting cycle could be accelerated significantly with no detrimental effects on the clay bodies produced.

The process was patented by Tobin[43] who used the slip casting of sanitary ware for tests. He found that a three to four minute microwave exposure at approximately 750 watts followed by a set time of a further 20 minutes was sufficient to produce a cast with the same green strength and mechanical stability as a similar body produced in one to two hours without the use of microwaves. Despite these impressive savings in time, it was the subsequent drying of the ware and the moulds by microwaves which gave the greatest economies. A further advantage accruing was the ability to dry the mould with great precision increasing production yields significantly.

The Italian company, Mori SpA[44], has also worked in this field. In comparative tests between conventional and microwave slip cast sanitaryware, production rate was found to increase significantly through the ability to use three work shifts in place of the usual single shift and the overall cost of the product was lowered. Once again, the principle saving in time was concerned with the drying of the moulds which allowed a much faster turn around and removed the need for expensive controlled atmosphere rooms or buildings in which the moulds dry for periods of up to 16-24 hours. Mori also claimed that drying of the ware could be reduced from 24 hours in ambient air to 35-60 minutes by exposure to microwave radiation, the latter resulting in less than 1% residual water in the product. The actual casting stage was reduced from an average of 90 minutes per piece to 10-15 minutes using microwaves. Costs were assessed at Lit.2800 per piece using conventional technology and Lit.2500 per piece using microwave assisted slip casting.

An extensive analysis of the use of microwaves in the slip casting of toilet bowls has been performed by Chabinsky and Eves[45]. The increase in casting rate is explained by the movement of moisture through the mould to the outside mould wall. This, in effect, is claimed to keep the inner wall (which is in contact with the slip) dry to the extent of a new mould. A one hour cycle from cast-to-cast was established with no active drying required for the plaster moulds. In addition, the same mould on repetitive casting showed no deterioration implying a possible increase in mould life.

Subsequently, Chabinsky[46] has attempted to quantify the advantages offered by the use of microwaves.   His conclusions, again based on sanitaryware, are set out below:

- increased productivity - 10-40%
- increased repetivity and quality - 30-50%
- reduced floor space - 50-70%
- reduced mold shop and construction - 60-70%
- reduced material handling - 30-40%
- reduced refire - owing to ware uniformity
- improved slip operation - leading to reduced costs
- reduced personel physical problems
- improved computerisation and thus production control

Relatively little research has been performed into the use of microwaves during the slip casting of advanced ceramics or composites. However, recent work on the slip casting of alumina and alumina-SiC composites by one of the authors[47] has indicated that the use of microwaves can result in an increase in casting rate of approximately $2\frac{1}{2}$ times over conventional casting. This increase was considered to arise principally as a result of a decrease in slip viscosity with increasing temperature. The acceleration was achievd with no noticeable degradation of the physical and mechanical properties of the cast body.

7.4 Calcining

Very little work has been reported in the literature on the use of microwave energy to calcine ceramic materials. As will be discussed in more detail in the next section, many ceramics are essentially transparent to microwaves at room temperature. This is often overcome during the sintering of green bodies by the polar nature of the binders present which are sometimes capable of absorbing sufficient energy to allow the body to reach a high temperature.   The ceramic phase may then become lossy enough to continue the heating process.   For calcining, however, binder phases are generally absent and thus the number of materials which may be heated without the use of some form of susceptor is more restricted.

Nevertheless, there are still a much larger number of ceramics which are microwave absorbing at room temperature than is commonly imagined.   In addition to dielectric structural ceramics such as silicon carbide and nitride, the vast majority of the electroceramics have high loss factors at low temperatures. Harrison et al[48] used microwaves to dry, calcine and fire PZT and PLZT powders. Batches of powder were calcined in air, a dense PZT piece being inserted into the powders to improve coupling. The results of the calcination stage indicated that significantly shorter calcination soak times could be used with microwave heating, however at the expense of higher quantities of unreacted lead oxide.

Ahmad et al[49] have used microwave energy to calcine precursor materials to yield superconducting powders. The process involved the conventional mixed oxide route with the powder undergoing 3 calcinations at 900°C followed each time by grinding and pelletising. After a final pelletisation the resultant product was also microwave sintered in a bed of powder at 750°C for 30 minutes and then cooled to 500°C for 1 hour before final cooling. The sintering, annealing and cooling stages being performed in flowing oxygen. The end product had similar physical properties to 'conventionally' produced ceramic using the same conditions, however the microwave processed material had a lower degree of porosity and a larger and more even grain size, resulting in what was described as an overall 'better microstructure'. Ahmad et al[49] concluded that microwave processing was superior to conventional processing for the mixed oxide route at the sintering and annealing stages but less useful for calcination.

7.5 Sintering

7.5.1 Introduction

Despite the widespread view discussed above that ceramic materials are essentially transparent to microwave frequencies, perhaps the greatest amount of research effort in the field of microwave ceramic processing is concerned with firing or sintering. Most of this effort, however, has occurred over the last 5 years with a large number of ceramic systems now having received attention.

Initially work concentrated on ceramics which couple strongly with the incident radiation. More recently, there has been a trend towards studying materials with particular relevance to modern engineering ceramic applications. Some of these latter ceramics have a very low dielectric loss at room temperature, meaning that the material is essentially transparent to

microwave radiation.   However, by 800-1000°C most such ceramics are
sufficiently lossy to continue heating of their own accord (see Figures 14
& 15 for alumina, for example). Considerable attention has therefore been
focussed on finding methods to achieve the initial heating of low-loss
ceramics and these have included the use of susceptors, additives and high
frequencies.   Each of these approaches will be covered in subsequent
subsections.

In the studies described below, both single mode and multimode systems
have been used to achieve heating effects.   Tinga[22] has addressed the
problem of applicator design and concluded that when irregular objects need
to be processed, multiresonant systems are probably preferable to plane or
travelling wave applicators. Whilst single resonance applicators have the
advantage of being able to generate high field strength in the central core
of the applicator, they are restricted in the geometry of sample they can
handle.

7.5.2 Ceramic systems studied.

The earliest reports of work aimed at generating high temperatures in
ceramic materials using microwave radiation found by the authors was
performed by Ford and Pei[50] as long ago as 1967. A wide range of oxide and
non-oxide materials were heated in a standard multimode microwave oven
consisting of two 800 W, 2.45 GHz magnetrons coupled into a cavity with a
field stirrer.   Sample size was found to be important for the low loss
materials, with faster heating rates being obtained with increasing size of
sample[f2].

---

f2   This phenomena, however, is probably due to a combination of two
effects. Firstly, cavity walls are often made of a relatively low electrical
conductivity stainless steels. This results in them absorbing a significant
percentage of the incident radiation when only a small amount of a low
absorbtion material is placed inside the cavity. Up to a point, therefore,
larger samples will result in greater power dissipation in the material.

Secondly, with small samples the surface area is generally high and
therefore heat losses from the material can be great enough to prevent any
rise in temperature.   (This has frequently led to the practice, during
microwave sintering, of totally surrounding samples with insulation in an
attempt to reduce heat losses.) However, as sample size increases, the heat
lost becomes less than that generated and the temperature rises. Microwave
heating of multiple samples is often more effective, therefore, than heating
single samples (see later in text).

Uranium oxide fuel pellets, which had the fastest heating rate in the initial experiments (1100°C in 6 seconds), were used for subsequent work on sintering. Conventionally, urania could be sintered in hydrogen to about 96% of theoretical density at 1700°C using a sintering cycle that could last up to 20 hours. Ford and Pei, however, reported the ability to microwave sinter the material, without cracking, to 87% density using a maximum temperature as low as 1100°C and only a four hour sintering cycle.

The reduction in length of the cycle is easily explained by the much faster heating and cooling rates used in the microwave experiments. On the basis of these results, Ford and Pei suggested that the rates used in the conventional sintering of urania could be substantially increased.

The lower sintering temperature is almost certainly artificial to some extent, in that quite severe thermal gradients probably existed within the sample. The internal temperature may thus have been considerably hotter than the 1100°C measured. This conclusion is supported by work performed at the Oak Ridge National Laboratory several years later[51]. Tests intended to microwave dry samples of hydrated UO_3 gel spheres as part of a nuclear fuel preparation process using a standard 600 W domestic oven resulted in partial melting of the samples. Published melting points for UO_2 are 2878±20°C. It is interesting to note, however, that during heating a local hot spot (typically 1-3 cm in diameter) would develop which continued to heat without any change in the remaining UO_3.

In the mid 1970's Badot and colleagues, at the National Centre for Scientific Research in France, used a rectangular single mode cavity operating in the TE_{01n} mode rather than a multimode oven[52,53]. In what is now widely regarded as pioneering work, three oxide materials were studied with dielectric property measurements over a range of temperatures being used to aid in the design of the experiments. Sintering of alumina was achieved in 10 to 15 minutes and melting of silica rods could be maintained using only a few hundred watts of microwave energy, provided the rod was preheated to near its melting temperature where losses from ionic conduction became significant. Advantages from the process were considered to be precise temperature control, an absence of pollution and the low power requirements; 400-500 W compared with about 10 kW used by other systems to melt the same sample volume. This was believed to be possible through the very high, near 90%, energy conversion efficiency.

Subsequent work[54] involved the sintering of ß-alumina ceramics. These are superionic conductors with a highly anisotropic crystal structure. Using a TE_{01n} single mode cavity and a 2.45 GHz microwave generator capable of producing a continuously variable output between 20-1500 W, they discovered that after sintering the crystallites were oriented such that their conduction planes were parallel to the electric field vector in the resonant cavity. This phenomenon was thought to be due to one of two reasons; either the centre of the samples completely melted and on recrystallisation the grains were oriented, or reorientation of the grains occurred as a result of the presence of a transitory liquid phase during sintering. As a further advantage, the outer surface of the samples being at a lower temperature to that of the centre, the latter was protected by the surrounding environment and soda losses by volatilisation were reduced. Both these results have important implications for other ceramic systems. For example, the possibility of achieving grain orientation in high T_c superconductors and reduced lead loss in lead-based electroceramics.

Boron carbide[55] was found to couple with 2.45 GHz radiation almost four times better than water on a Joule mol^{-1} of energy absorbed basis. More recently, Katz et al[56] sintered boron carbide disks, 1 cm diameter by 1 cm thick, to greater than 95% density in only 12 minutes. This compares extremely well with the 1-2 hours required using hot pressing. No sintering aids were used and the sintering process appeared to follow a similar pattern to that obtained when conventional techniques are used.

Roy and co-workers[57,58] have examined the heating behaviour of alumina, silica and mullite gels. Results indicated that all three gels could be heated rapidly to temperatures in excess of 1500°C in a standard 600 W domestic oven. Whilst the absorbtion mechanism was not understood, the mass of the gel was found to be a prime factor in controlling the heating rate. This point has already been discussed (see previous footnote). Nevertheless, as the authors conclude, with the emergence of sol-gel technology the high susceptibility of gels to microwave radiation provides interesting possibilities.

Harrison et al[48] used microwaves to sinter green PZT and PLZT pellets. The former were sintered in tubes of fired PZT whilst the latter were sintered in closed magnesia crucibles containing loose $PbZrO_3$ powder. The density of the microwave sintered PZT was approximately the same as that of conventionally sintered pieces, although the soak time was reduced from 60 to 12 minutes. Grain sizes were also smaller, leading to a g_{33} coefficient a factor of 2 larger for the microwave processed material. The principle advantage of the microwaves for the PLZT pellets was the ability to sinter the material to high densities in an air atmosphere. The researchers concluded that the microwave route was particularly attractive for PLZT.

Johnson and co-workers at NorthWestern University in the USA, used a circular cross-section applicator, operating in the TE_{111} mode and with the inner surfaces of the walls given a smooth, silver coating to reflect heat[59], to conduct sintering experiments on Si_3N_4, SiC, Al_2O_3 and Al_2O_3-TiC composites[60-64].

The silicon nitride underwent highly localised heating; some areas decomposing to elemental silicon, in others whiskers and hexagonal rods of an undetermined chemical composition grew into the porosity, whilst the balance of the material remained unchanged[60]. Argon, the usual atmosphere employed in conventional sintering of SiC, proved to be unacceptable for the microwave sintering of this material on account of breakdown and arcing even at low power levels[60,64]. This forced the use of a nitrogen atmosphere and whilst temperatures as high as 2200°C could be achieved without plasma formation through increasing the pressure to about 0.9 MPa, no densification was observed. This was attributed to depletion of the boron sintering additive via reaction with the nitrogen atmosphere.

In contrast, the much lower dielectric constant alumina could be sintered to high densities (>99%) with grain sizes less than 1 μm being retained under optimum conditions[61]. This was believed to be a direct result of the fast heating rates achieved (up to 400°C min^{-1}). Extensive research by Brook and co-workers (summarised by ref 65) has shown that rapid heating leads to improved ceramic microstructures through the retention of fine, uniform grain sizes - the concept of 'fast-firing'. This is achieved through the optimisation of the conditions for a high densification rate and low coarsening rate. Whilst this work has been based on rapid 'conventional' sintering procedures, there seems to be no reason why the use of microwaves should not lead to similar results.

A surprising feature of the work at NorthWestern is that despite the use of an applicator designed to maximise temperature uniformity, severe temperature gradients were found both radially and axially[62]. This resulted in cracking being observed in the Al_2O_3-TiC composite samples, with

the higher TiC contents proving to be more susceptible. A uniform hot zone
could be achieved only by rotation and axial oscillation of the specimens.
Figure 36 shows the temperature dependence of density for microwave sintered
and coventionally fast-fired 70% Al$_2$O$_3$ - 30% TiC, the composition which was
found to microwave sinter with least problems.

An interesting observation to come out of the subsequent work[63], is
that for the same net input power, the higher conductivity samples (ie,
higher TiC concentration) showed lower surface temperatures and lower heating
rates (Figures 37a and b).   This was explained in terms of the higher
conductivity samples having a relatively large hot zone and a more even
temperature distribution which lowered the absorbed power density and
therefore the heating rate. Melting was always observed for samples of pure
α-alumina and 10 wt% TiC composite, but never for 30 wt% and 50 wt%
composites.   Al$_2$O$_3$ - 20 wt% TiC composites behaved in an intermediate
fashion, indicating a transition composition between the two types of
behaviour.   These results appear to be at odds with the earlier findings
where increasing the TiC content led to greater non-uniformity in the
temperature distribution[62]. These results now require further explanation.
A distribution of a high dielectric loss phase in a low loss matrix appears

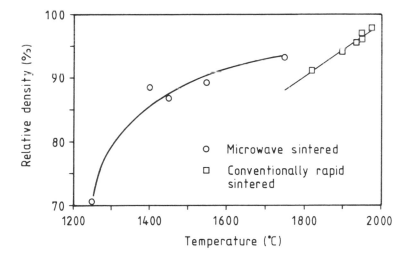

Figure 36: Temperature dependence of density for microwave sintered and
conventionally fast sintered 70% Al$_2$O$_3$ - 30% TiC. (After ref 62).

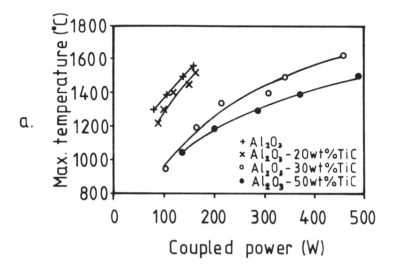

a.

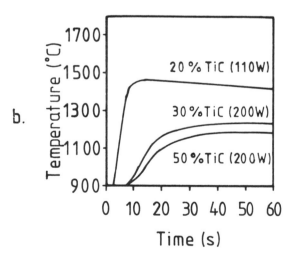

b.

Figure 37: Specimen temperature as functions of; a) coupled (absorbed) power, and b) time. (After ref 62).

to lead to a more uniform temperature distribution (see Section 7.5.5).

In a parallel study[64], cracking during sintering of Al_2O_3 - TiC composites was avoided by use of a two-stage heating process. In the first, presintering step, densities in the order of 82-85% of theoretical were obtained by gradual heating to 1450°C - 1500°C under a relatively low nitrogen heating pressure. The samples were subsequently heated to the full sintering temperature under high N_2 pressures at a high heating rate to minimise grain growth. Thermal cracking in this second stage was minimised by the presintered strength of the samples. Whilst this technique undoubtedly worked, it does seem to negate many of the inherant advantages proposed for microwave sintering.

7.5.3 The use of susceptors

Ferrites have been recognised as efficient absorbers of microwaves for many years and have therefore found applications in microwave devices such as attenuators, isolators and mode suppressors. The heating effect resulting from the absorption of microwaves has generally been counteracted through the use of small components and water or air cooling systems. By the mid 1970's, however, it was realised that the heating effect could be used to advantage[66-68].

In 1981 Krage[67] examined the electromagnetic loss mechanism in barium ferrite and then devised an experimental apparatus, based on a multimode cavity, to achieve sintering. A key feature of the equipment was a microwave-absorbing setter plate (or susceptor) to support the green magnets. One of the problems associated with the microwave sintering of ferrite ceramics is their generally low thermal shock tolerance. On conventional heating the surface is hotter than the centre of a body, leading to the generation of compressive stresses in the surface layers. Since ceramics are much stronger in compression than tension, rapid heating usually causes fewer problems than rapid cooling. However, with microwave heating the generation of an inverse temperature profile can mean that rapid heating poses the same problems as rapid cooling, i.e. the generation of tensile stresses in the surface layers.

Since both the setter plate (a composite based on silicon carbide) and ferrite heated in the microwave field, any heat sink problem from the magnets in contact with the plate was eliminated. In addition, and probably more importantly, the heat given off from the plate heated the interior of the insulated cavity. This will have reduced the heat losses from the surface of the samples and thus lead to a more uniform temperature profile, thus reducing thermal shock problems. As a result of Chabinsky's[69] work on the sintering of alumina spark plugs it is also believed that the use of a large batch size of green samples will have contributed to the uniformity of the temperature profile. Krage found that he could sinter barium ferrite magnets with a heating rate of $9°C$ min^{-1} without danger of magnet cracking and that the physical and magnetic properties of microwave and conventionally sintered material were comparable.

Similar conclusions were reached by Okada et al[68] who used a cylindrical TE_{011} cavity resonator to sinter calcium vanadium garnet ferrite compacts. The internal temperature of the small samples, sintered one at a time, was estimated to be approximately $100°C$ higher than the surface temperature, leading to less uniform properties than those achieved by Krage.

Wilson and Kunz[70] used a disc of silicon carbide as a susceptor during work on sintering yttria partially stabilised zirconia (Y-PSZ). Green bars were sintered in an enlarged region of a waveguide using 2.45 GHz radiation. Visual observation of the zirconia during sintering (by reflection off darkened glass) showed that sintering started by an outward bowing of the upstream edge of the samples, followed by the appearance of small cracks. These healed and other cracks subsequently appeared. Eventually the centre shrank and glowed as it heated and then the shrunken, glowing region gradually moved outward until the whole sample was sintered. These observations appear to support the sintering model proposed in Section 6.2. Densities in excess of 95% were achieved in times considerably less than those required for conventional sintering.

Care must be exercised when designing the shape of the susceptor. Humphrey[71] used a susceptor during work on the sintering of barium titanate ceramics. It consisted of a $\frac{1}{4}$ inch thick closed box made from SiC sheets. Although a measurable electric field apparently existed during firing, it is uncertain what fraction of the sintering achieved was by direct interaction of the microwaves with the barium titanate and what fraction was of a more conventional nature from radiaton by the SiC. It should be noted that SiC has one of the highest loss factors of any ceramic material and therefore one of the lowest penetration depths. The results showed very great similarity between the microwave sintering and conventional sintering using SiC radiant elements.

7.5.4 The use of additives

Meek and co-workers[72,73] have used a variety of additives to achieve initial heating. Sodium nitrate and glycerol were used to accomplish coupling with the microwave field for alumina-silicon carbide whisker and alumina-silicon nitride whisker composites and a pyrex-silicon carbide whisker composite[72]. The alumina-based materials appeared to sinter to full density, presumably as a result of the low viscosity glass phase likely to be formed as a result of the use of a sodium-based additive. The glass-based composite, however, retained a significant amount of gas phase in the form of bubbles.

When sintering zirconia-silicon carbide whisker and zirconia-silicon nitride whisker composites, zirconium nitrate and glycerol were added. The results were a sintered matrix with good bonding between whiskers and matrix. It was also claimed that with the SiC whiskers, alignment between whiskers and the electromagnetic field of the incident microwaves was observed. Whether a multimode cavity would generate a uniaxial field, and if so, whether it would generate sufficient torque to align whiskers, is uncertain however. It is also uncertain whether the samples were initially in the form of a slurry, providing freedom to orientate.

A much wider range of additives were investigated in the subsequent study[73]. Paint grade CaO-stabilised ZrO_2 and Y_2O_3, and fine Y-PSZ and Al_2O_3 powders were die pressed and microwave sintered 'pure' and using additives such as zirconium oxynitrate solution (ZON), niobium, tantalum carbide, silicon carbide, molybdenum disilicide, iron and copper. Resultant porosities appeared to follow no recognisable pattern with respect to additive used or level of additive. Where additives were used a three stage heating process was claimed. Initially the zirconia-based insulation was believed to couple with the microwaves, heating the samples to between 700 and 1000°C. By this temperature the sintering aid material was coupling and raised the ceramic temperature still further until it began to heat of its own accord. When no additive was used the second stage was obviously not present, though it appears to have had negligable effect on the properties of the resultant body. A three stage heating process was also obtained by Tian et al[61], however in this case the stages had to be intrinsic to the samples since no insulation or additives were used. This casts doubt on the explanation provided by Meek and co-workers.

The mechanical and physical properties of microwave sintered paint grade yttria containing 2 wt% zirconia, introduced in the form of zirconium oxynitrate, were investigated further[74] and the results compared with commercially available material of similar composition. Whilst the modulus of elasticity for the microwave processed material varied between 77-206.5 GPa (commercial material, 150.5 GPa), the thermal shock resistance of the former was considerably superior to that of the commercial material. A sudden change in temperature (ΔT) of 200°C decreased the strength from 50-60 MPa to 15 MPa. The microwave sintered material, however, displayed little change in strength with thermal shock testing up to 500°C. If anything, the results indicated that a higher strength was obtained as the degree of thermal shocking increased - even up to 800°C. This was attributed by the authors to the microwave sintered material having disconnected, spherical porosity whilst the commercial material contained connected, non-spherical porosity.

7.5.5 The use of high frequencies

An alternative to either the use of susceptors or additives is to increase the frequency of the incident microwaves. The higher the frequency the greater the power deposited in the sample and thus even low-loss materials can be made to couple. The disadvantage of high frequencies is a decreased depth of penetration making heating a more surface effect.

Kimrey et al[75,76] used a high power gyrotron oscillator that produced hundreds of kilowatts of microwave power and operated at 28 GHz. This was coupled to a large untuned cavity, some 76 cm in diameter by 100 cm long. Experimenting with large (several hundred cubic centimetre) alumina samples resulted in >98% densities being achieved over much of the sample volume. However, samples were cracked, indicating excessive thermal stresses during heating.

Meek et al[77] have experimented with 60 GHz microwave heating. High purity, submicron alumina powders could be sintered to >95% of theoretical density by heating to 1700°C in about 6 minutes, this was followed immediately by cooling. Suprisingly for such a rapid thermal schedule, the average grain size was found to be of the order of 5 μm indicating significant grain growth and yet pockets of incompletely densified material retaining the original powder morphology were also observed. These results are indicative of extremely inhomogeneous coupling of the microwaves with the alumina and might be attributable to the decreased depth of penetration associated with high microwave frequencies.

No such variation in density is mentioned for alumina-silicon carbide whisker composites produced from the same alumina powder. This may indicate a greater degree of heating uniformity as a result of the dispersed high dielectric loss (SiC) phase, which compensates for the decreased depth of penetration (see discussion of refs 62 & 63, section 7.5.2). The maximum density achieved, 77%, is approximately comparable with that achievable through conventional, pressureless sintering.

7.5.6 Non-thermal effects

Many of the references cited contain results indicating that microwave sintering has been achieved in times considerably less than those required for conventional sintering; for example refs 48, 56, 70, 72, & 73. This may be evidence for additional, non-thermal effects on diffusive processes or may result from the use of faster heating rates. Despite substantial evidence for the latter[61-65,78], there appears to be a widespread feeling among researchers working in this field that microwaves may result in improved densification through non-thermal effects such as reduced activation energies. This was particularly noted during discussions at the Materials Research Society Symposium in Reno[f3]. The current authors, however, remain sceptical.

7.6 Joining

The joining of ceramic materials using microwave energy is one of the least researched topics and yet appears to offer considerable potential. Meek and Blake[79] used a conventional, domestic (700 W) microwave oven to make ceramic-glass ceramic seals. Three different sealing glasses were used to join 96 wt% alumina substrates, by making a 'sandwich' out of two substrates and the glass in the form of a slurry. An unspecified coupling agent was added to the glass slurries to enhance the loss factor at low temperatures.

The key result of the work was an observed difference in bonding mechanism between microwave and conventionally heated samples. Extensive inter-diffusion was reported in the former case leading to a diffusion bonding mechanism, whilst the conventionally heated samples displayed very

---

f3 Microwave Processing of Materials, Symposium M, Materials Research Society Spring Meeting, Reno, Nevada, USA. April 5-9, 1988.

little interdiffusion and wetting was determined to be the predominant bonding mechanism.

A more sophisticated approach has been taken by Palaith and co-workers[80,81]. One of the principal differences in the work was the replacement of the multimode cavity of a domestic microwave oven with a single mode cavity. This resulted in a high conversion efficiency, good control over the process and permitted detailed modelling of the microwave radiation - ceramic interaction. The apparatus used is shown in Figure 38. A key feature is the use of a compressor and acoustic probe. The former allows pressure to be applied during joining of rods and bars whilst the latter permits the joint to be monitored throughout the joining process. A further feature is the ability to vary the atmosphere in the cavity (including evacuation), although at the expense of having a fixed, rather than adjustable, iris.

Thus far the joining of two materials has been attempted, these being alumina and mullite, with silicon nitride to follow once the sealing of the cavity has been perfected. The results indicate that use of this technique allows oxide ceramics to be joined in very short time periods, typically 5-15 minutes, and with low power consumptions (100-300 W). Microstructurally the joints resulted in a line of porosity, however no grain growth was observed and in the low purity, porous materials chosen for the study, the joints appear to have had at least as great a strength as the as-received material. This was indicated by the test specimens failing away from the joint region during four point bend strength measurements.

The Toyota Central Research and Development Laboratories in Japan have developed a very similar research programme[82] to that created by Palaith. The Japanese apparatus operates at 6 GHz rather than the 2.45 GHz used by the Americans and does not have the ability to moniter the joint during formation via an acoustic probe. Nevertheless, the results achieved are very similar. 92-96% pure alumina rods could be joined at temperatures of 1850°C under 0.6 MPa of pressure in as little as 3 minutes, the joint displaying very similar strengths to the original material. It was, however, difficult to join rods of >99% purity, even when using a small sheet of lower purity alumina as an adhesive between the ends of the rods. Electron microprobe analysis (EPMA) showed that impurities did not diffuse across the interfaces from the low purity adhesive into the higher purity rods. These results, combined with those of Palaith et al[81], indicate that it is almost certainly the presence

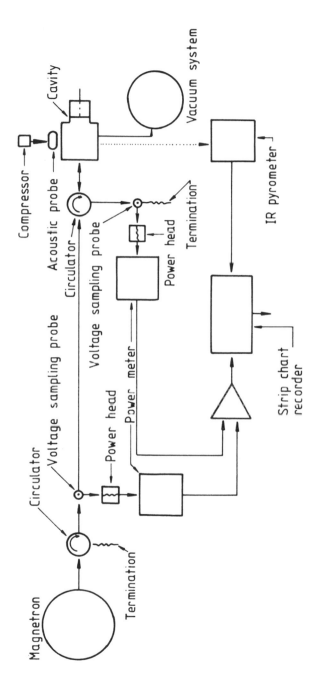

Figure 38: Block diagram of microwave joining apparatus. (After ref 81).

of glassy grain boundary phases which allows the joining to occur.

This conclusion is further supported by Fukushima's work on joining silicon nitride rods. When a low purity 'adhesive' sheet was used the maximum strengths achieved matched those of the adhesive material. Without the use of such an adhesive, joint strengths were limited to only 70-90% of the intrinsic strength of the high purity rods.

An interesting feature of both research programmes is the speed with which the joints occur, comparing very favourably with the 30-120 minutes required with conventional heating element furnaces[83,84]. Palaith et al[81] suggest this may be further evidence for the nonthermal effects believed to be associated with the use of microwaves and discussed briefly in Section 7.5.6. However, an alternative explanation could be that in the conventional case the heating of the glassy grain boundaries is dependent upon the thermal conductivity of the ceramic matrix, whilst the use of microwaves removes this limitation by putting the energy directly into the lossy grain boundaries.

Fukushima[82] also reported an apparent reduction in the degree of scatter in the strength results of the joined material compared with the original rods. No explanation for this was provided though it was thought that it may be attributable to rounding of porosity reducing stress concentrations.

Overall, using microwave energy to join ceramic components appears to have considerable potential, especially for materials containing glassy second phases - and it should be remembered that these materials make up the bulk of current commercial ceramics. Strengths approximately equivalent to those of the original component appear to be achievable and thus the opportunity to use this technology for repair work (perhaps even with portable units) and constructing articles of complicated shape out of several components, seems to be unequalled. Future work should perhaps investigate the dependence on glassy, secondary phases more closely to determine its exact rôle.

7.7 Plasma-based processing

Plasmas are used in a variety of processing applications, including surface deposition techniques and sintering. One of the best known examples of plasmas for deposition is the growth of diamond films by chemical vapour deposition, the plasma assisting in the decomposition of the gaseous species. This is now a very large field in its own right and further discussion really lies beyond the scope of this review, apart from a brief discussion of the advantages of microwave-induced plasmas compared with other excitation techniques.

The use of plasmas to assist in the sintering of ceramic materials was first reported by Bennet et al[85]. A microwave-induced plasma discharge was used to sinter rods and pellets of a range of oxides in several gases at pressures between 130 and 6700 Pa. The results indicated that plasma sintered samples displayed smaller grain sizes and greater densification compared with samples conventionally sintered under similar temperature, time and atmospheric conditions.

More recently these results have been confirmed by the work of Kemer and Johnson[86]. Using green alumina rods they showed that the microwave-induced plasma was capable of producing densities in excess of 99% of theoretical in less than 2 minutes and densities approaching 99.9% in less than 10 minutes. This may be attributable to the very high temperatures (>1900°C) attained, combined with the extremely high heating rates (up to $\approx 100°C\ s^{-1}$) associated with this processing route. These results support the arguments of Brook[65] for enhanced grain boundary and lattice diffusion, leading to densification, without a concomittant increase in surface diffusion which leads to coarsening.

A third factor which was believed might contribute to the rapid sintering rates was the possibility that the plasma chemically interacted with particle surfaces to enhance densification. Whilst this theory was largely discounted by the authors on the basis that rapid sintering rates to high densities were not unique to plasma sintering and plasmas are unlikely to exist within porosity, the presence of plasma etching was suggested as indirect evidence of such non-thermal effects. This whole line of argument has interesting parallels with that proposed for non-thermal effects in microwave processing and briefly discussed in Section 7.5.6.

Plasmas have been generated by both dc devices and radio frequency induction in addition to microwaves. However, the microwave-excited plasmas appear to offer a number of advantages over dc and low frequency plasmas, not just for plasma sintering but for the full range of plasma-based applications. These have been summarised by Ji and Gerling[87];

1. A microwave-induced plasma produces a much higher degree of ionisation and disassociation. This, in turn, provides a field of active species some 10 times higher than obtained with other types of electrically-excited plasma.

2. It is possible to sustain a microwave plasma over a wider range of pressures than other electrically stimulated plasmas, typically from 10^{-5} torr to several atmospheres. This broadens the range of applications.

3. The electron-to-gas temperature ratio, T_e/T_g, is very high with microwave-induced plasmas allowing the carrier gas and substrate to remain moderately cool even in the presence of high electron energy.

4. The absence of internal electrodes removes a source of contamination and makes reaction vessels simpler.

5. The state of the art of microwave engineering at high power levels is relatively well advanced.

8. CONCLUSIONS

The most important property of a ceramic as regards it treatment with microwave energy is its effective loss factor. Essentially the ceramic will absorb microwave energy if it contains polar or ionic components in its structure. It can also sometimes be advantageous to add selective agents which have an affinity to microwave energy, enabling these to heat preferentially and to transfer this energy to the bulk of the low loss ceramic material by conduction.

How readily a ceramic absorbs the available microwave energy depends on the value of the effective loss factor. A simple rule of thumb suggests that in the range $0.05 < \epsilon_e'' < 1$ the ceramic material should heat up with little difficulty, provided there is no tendency of a thermal runaway situation developing.

Below $\epsilon_e'' = 0.01$ the ceramic starts to become a low loss dielectric and as such requires either special applicators to ensure that the electric field set within it is very high, since the power dissipation is proportional to $\epsilon_e'' E^2$, or the use of susceptors. Single mode resonant cavities will give the required high electric fields, however the cavity with the material inserted in it will have a very high Q-factor and that in itself may present difficulties in coupling effectively the energy into it from the microwave source and circulator.

Conversely, effective loss factors much larger than unity may present severe non uniformity of heating in that power penetration depths will be relatively small and so the dimensions of the ceramic to be treated become critical. It must be stressed, however, that irrespective of the value of the effective loss factor, careful consideration must be given in all potential applications to the type of applicator so that its design is tailored to the particular ceramic to be treated.

The use of microwave energy within ceramic processing is now extremely diverse. Ultimately, the deciding factor for microwave processing will be the economics of the situation. A number of studies have been performed in this area[44,88-90] and conflicting results have been obtained. One of the key reasons for the uncertainty is that a high number of initial assumptions have to be made in the current absence of suitable data. Nevertheless, it does seem to be indicated that the use of microwaves for both drying and slip casting may result in financial advantage over conventional processing. This arises largely from increases in productivity, though improved yields can also make a useful contribution.

For sintering, joining and other such applications financial benefit appears to be less guarenteed. It appears that each potential application will need a thorough economic analysis; however, it is suggested that this should be attempted only after the technology has been developed further. It is the authors' view that the major advantage for the use of microwaves in many such operations will have to come from processing benefits; that is, the ability to fabricate materials with properties not previously obtainable. Systems in which conventional processing is difficult, such as the non-oxides, composites and certain electro-ceramics, would therefore appear to offer the greatest opportunities. A further benefit to be gained here is that these materials generally have relatively high dielectric losses.

It is difficult to forecast a realistic timescale for the commercial development of microwave assisted processing techniques. Some, such as drying, are already in operation in a number of countries, whilst other processes, such as slip casting, are expected to have been commercialised to a greater or lesser extent by the early 1990's. For the other applications, considerably more research and development is still required; however, given the interest currently being shown, progress is expected to be rapid.

REFERENCES

1. Debye, P., Polar Molecules. Chemical Catalog, New York, (1929).

2. Von Hipple, A.R., Dielectric Materials and Applications. MIT Press, Cambridge, (1954).

3. Inglesias, J. and Westphal, W.B., Supplementary dielectric constants and loss measurements on high temperature materials. Technical Report 203, MIT, (January 1967).

4. Couderc, D., Giroux, M. and Bosisio, R.G., Dynamic high temperature complex permitivity measurements on samples heated via microwave absorption. J. Microwave Power 8 [1] 69 (1973).

5. Westphal, W.B. and Sils, A., Dielectric constant and loss data. AFML-TR-74-250 Part I, (April 1972); Part II, (December 1975); Part III, (May 1977); Part IV, (December 1980).

6. Ho, W.W., High-temperature dielectric properties of polycrystalline ceramics. Mat Res Soc Symp Proc 124 137-148 (1988).

7. Fuller, J.A., Taylor, T.S., Elfe, T.B. and Hill, G.N., Dielectric properties of ceramics for millimeter-wave tubes. Technical Report AFWAL-TR-84-1005, Air Force Avionics Laboratory, Wright-Patterson AFB, (1984).

8. McMillan, P.W. and Partridge, G., J. Mat. Sci. 7 847-855 (1972).

9. Ho, W.W. and Harker, A.B., Proc. 1st DoD Electromagnetic Window Symposium, MP85-148 1, 2-6-1, Naval Surface Warfare Center, (1985).

10. Ho, W.W. and Morgan, P.E.D., Dielectric loss to detect liquid phase in ceramics at high temperature. J Am Ceram Soc Comm 70 [9] C209-C210 (1987).

11. Varadan, V.K. and Varadan, V.V., Principles of microwave interactions with polymeric and organic materials. Mat Res Soc Symp Proc 124 59-68 (1988).

12. Hasted, J.B., Aqueous Dielectrics. Chapman Hall, London, (1973).

13. Metaxas, A.C., A unified approach to the teaching of electromagnetic heating of industrial materials. Int J Elect Eng Educ 22 101-118 (1985).

14. Metaxas, A.C. and Meredith, R., Industrial microwave heating. Peter Peregrinus, (1983).

15. Stuchly, S.S. and Hamid, M.A.K., Physical parameters in microwave heating processes. J Microwave Power 7 [2] 117-137 (1972).

16. Perkin, R.M., The heat and mass transfer characteristics of boiling point drying using radio frequency and microwave electromagnetic fields. Int J Heat Mass Transfer 23 687-695 (1980).

17. De Wagter, C., Computer simulation predicting temperature distributions generated by microwave absorption in multilayered media. J Microwave Power 19 [2] 97-105 (1983).

18. Wei, C.K., Davis, H.T., Davis, E.A. and Gordon, J., Heat and mass transfer in water-laden sandstone: microwave heating. A I Ch E Journal 31 [5] 842-848 (1985).

19. Watters, D.G., Brodwin, M.E. and Kriegsman, G.A., Dynamic temperature profiles for a uniformly illuminated planar surface. Proceedings of the Materials Research Society, 124 129-136, (1988).

20. Binner, J.G.P. Unpublished work.

21. Meek, T.T., Proposed model for the sintering of a dielectric in a microwave field. J Mat Sci Let, 6 638-640, (1987).

22. Tinga, W.R., Design principles for microwave heating and sintering. Mat Res Soc Symp Proc, 60, 105-116, (1986).

23. Varadan, V.K., Ma, Y., Varadan, V.V. and Lakhtakia, A., Microwave sintering of ceramics. Mat Res Soc Symp Proc, 124 45-58 (1988).

24. King, R.W.P., Proc IEEE, 64 [2] 228-238, (1976).

25. Bowman, R.R., IEEE Trans, MTT 24 43-45, (1976).

26. Bosisio, R.G., Dallaire, R. and Phromothansy, P., J Microwave Power, 12 [4] 309-317, (1977).

27. Araneta, J.C., Brodwin, M. and Kriegsmann, G.E., IEEE Trans, MTT 32 [10] 1328-1335, (1984).

28. Norton, F.H., Elements of Ceramics. Addison-Wesley Publishing Co. pp 114-125, (1974).

29. Lyons, W.D., Hatcher, J.D. and Sunderland, J.E., Drying of a porous medium with internal heat generation. Int J Heat Mass Transfer, 15 897-905, (1972).

30. Wei, C.K., Davis, H.T., Davis, E.A. and Gordon, J., Heat and mass transfer in water-laden sandstone: convective heating. A I Ch E Journal, 31 [8] 1338-1348, (1985).

31. Roberts, W., Head of the Whiteware Division, British Ceramic Research Ltd. Private communication, (1988).

32. Jervis, A., Microwave/vacuum drying systems.   Ceram Ind J, [Oct] 29-30, (1988).

33. Allan, G.B., Microwave moving-bed driers.   J Microwave Power, 3 [1] 21-28, (1968).

34. Binner, J.G.P.   Unpublished work.

35. Sheinburg, H., Phillips, D.S., Ramsey, K.B., Bradbury, J.R. and Foltyn, E.M., Microwave synthesis of superconducting oxide powders.   Presented at the Materials Research Society Symposium, Reno, Nevada, USA. April 5-9, 1988. Not published.

36. Aksay, I.A. and Schilling, C.H., Colloidal filtration route to uniform microstructures. In L.L. Hench and D.R. Ulrich, (Editors), Ultrastructure Processing of Ceramics, Glasses and Composites, pp 439-447.   J. Wiley & Sons, New York, (1984).

37. Cowan, R.E., Slip casting. In F.F.Y Wang, (Editor), Treatise on Material Science and Technology, Vol.9. pp 153-171, (1976).

38. Van Wunnik, J., Dennis, J.S. and Phelps, G.W., The effect of temperature on slips and moulds.   J Can Ceram Soc, 30 106-112, (1961).

39. Hermann, E.R. and Cutler, I.B., The kinetics of slip casting.   Trans Brit Ceram Soc, 61 207-11, (1962).

40. Blin, C., Secharge rapide d'articles ceramiques par haute ou hyperfrequence.   Proceedings of the XIth Int Ceram Congress, Madrid, Spain, pp 223-233.   Sept 22-28, 1968.

41. Blin, C. and Guerga, M., Trocknen von Keramikerzeugnissen durch dielektische Verluste - Ein industrieller Mikrowellen-Trockner. Keramicshe Zeitschrift, 21 [3] 157-159, (1969).

42. Evans, W.A., Rapid casting trials in a microwave drying oven.   T&J Brit Ceram Soc, 72 365-369, (1973).

43. Tobin, L.W. Jr., Ceramic material processing.   US Patent 4 292 262,   Sept 29, 1981.

44. Mori SpA, 'Fast automatic casting of sanitaryware Several short articles by several authors (C. Ferrari, A. Castelfranco, P. Zannini, R. Ronchetti, M. Mori), C.I. News, 6 [1] 7-12, (1986).

45. Chabinsky, I.J. and Eves, E.E., Applications of microwave energy in drying, calcining and firing of ceramics.   Ceram Eng and Sci Proc, 1412-1427 Nov-Dec, (1985).

46. Chabinsky, I.J., A microwave energy enhanced slip casting system. Interceram, 36 [3] 38-39, (1987).

47. Binner, J.G.P., Pyke, S.H. and Hussein, N., Microwave slip casting of advanced ceramics, submitted to J Eur Ceram Soc.

48. Harrison, W.B., Hanson M.R.B and Koepke B.G, Microwave processing and sintering of PZT and PLZT ceramics.   Mat Res Soc Symp Proc, 124 279-288, (1988).

49. Ahmad, I., Chander, G.T. and Clarke, D.E., Processing of superconducting ceramics using microwave energy. Mat Res Soc Symp Proc, 124 239-246, (1988).

50. Ford, J.D. and Pei, D.C.T., High temperature chemical processing via microwave absorption. J Microwave Processing, 2 [2] 61-64, (1967).

51. Haas, P.A., Heating of uranium oxides in a microwave oven. Am Ceram Soc Bull, 58 [9] 873, (1979).

52. Bertaud, A.J. and Badot, J.C., High temperature microwave heating in refractory materials. J Microwave Power, 11 [4] 315-320, (1976).

53. Badot, J.C., Application de l'energie en microondes au frittage et a la fusion d'oxydes refractaires. Unpublished PhD dissertation, L'universitie Pierre et Marie Curie, Paris. (1977).

54. Colomban, Ph. and Badot, J.C., Elaboration de ceramiques superconductrices anisotropes ($Na^+BAl_2O_3$) par chauffage microondes. Mats Res Bul, 13 135-139, (1978).

55. Holcombe, C.E., New microwave coupler material.   Am Ceram Soc Bull, 58 [9] 1388, (1983).

56. Katz, J.D., Blake, R.D., Petrovic, J.J. and Sheinburg, H., Microwave sintering of boron carbide. Mat Res Soc Symp Proc, 124 219-226, (1988).

57. Roy, R., Komarneni, S. and Yang, L.J., Controlled microwave heating and melting of gels. J Am Ceram Soc, 68 [7] 392-395, (1985).

58. Komarneni, S., Breval, E. and Roy, R., Microwave processing of mullite powders. Mat Res Soc Symp, 124 235-238, (1988).

59. Brodwin, M.E. and Johnson, D.L., Microwave sintering of ceramics.   IEEE MTT-S Digest 287-288, (1988).

60. Johnson, D.L. and Brodwin, M.E., Microwave sintering of ceramics.   EM-5890 Research Project 2730-1, Final Report.   Northwestern University, Department of Materials Science and Engineering, (1988).

61. Tian, Y-L., Johnson, D.L. and Brodwin, M.E., Ultrafine microstructure of Al_2O_3 produced by microwave sintering. Proceedings of First International Conference on Ceramic Powder Processing Science, Orlando, Florida, USA, (1987).

62. Tian, Y-L., Johnson, D.L. and Brodwin, M.E., Microwave sintering of Al_2O_3-TiC composites. Proceedings of First International Conference on Ceramic Powder Processing Science, Orlando, Florida, USA, (1987).

63. Tian, Y-L., Dewan, H.S., Brodwin, M.E. and Johnson, D.L., Microwave sintering behaviour of alumina ceramics.   Presented at the 90th Annual American Ceramic Society Meeting, Cincinnati, Ohio, USA, (1988).

64. Tian, Y-L., Brodwin, M.E., Dewan, H.S. and Johnson, D.L., Microwave sintering of ceramics under high gas pressure. Mat Res Soc Symp, 124 213-218, (1988).

65. Brook, R.J., Fabrication principles for the production of ceramics with superior mechanical properties. Proc Brit Ceram Soc, 32 7-24, (1982).

66. Maguire, E.A. and Ready, D.W., Microwave-absorbing ferrite-dielectric composites. J Am Ceram Soc, 59 [9-10] 434-437, (1976).

67. Krage, M., Microwave sintering of ferrites. Am Ceram Soc Bull, 60 [11] 1232-1234, (1981).

68. Okada, F., Tashiro, S. and Suzuki, M., Microwave sintering of ferrites. In F.F.Y. Wang, (Editor), Advances in Ceramics, 15 pp 201-205, (1985).

69. Chabinsky, I.J., Microwave sintering of ceramics. Presented at the Materials Research Society Symposium, Reno, Nevada, USA. April 5-9, 1988. Not published.

70. Wilson, J. and Kunz, S.M., Microwave sintering of partially stabilised zirconia. J Am Ceram Soc Comm, 71 [1] C40-C41, (1988).

71. Humphrey, K.D., Microwave sintering of $BaTiO_3$ ceramics, MS Thesis, University of Misouri-Rolla, (1980).

72. Blake, R.D. and Meek, T.T., Microwave processed composite materials. J Mat Sci Lett, 5 1097-1098, (1986).

73. Meek, T.T., Dykes, N. and Holcombe, C.E., Microwave sintering of some oxides using sintering aids. J Mat Sci Letters, 6 1060-1062, (1987).

74. Holcombe, C.E., Meek, T.T. and Dykes, N., Enhanced thermal shock properties of Y_2O_3-2wt% ZrO_2 heated using 2.45 GHz radiation. Mat Res Soc Symp, 124 227-234, (1988).

75. Kimrey, H.D., White, T.L., Bigelow, T.S. and Becher, P.F., Initial results of a high power microwave sintering experiment at ORNL. J Microwave Power, 22 81-82, (1986).

76. Kimrey, H.D. and Janney, M.A., Design principles for high-frequency microwave cavities. Mat Res Soc Symp, 124 367-372, (1988).

77. Meek, T.T., Blake, R.D. and Petrovic, J.J., Microwave sintering of Al_2O_3 and Al_2O_3-SiC whisker composites. Ceram Eng and Sci Proc, 8 [7-8] 861-871, (1987).

78. Borom, M.P. and Lee, M., Effect of heating rate on densification of alumina-titanium carbide composites. Ad Ceram Mat, 1 [4] 335-340, (1986).

79. Meek, T.T. and Blake, R.D., Ceramic-ceramic seals by microwave heating. J Mat Sci Lett, 5 270-274, (1986).

80. Palaith, D., Silberglitt, R. and Libelo,E.L., Microwave joining of ceramic materials. Presented at the 2nd International Conference on Ceramic Materials and Components for Engines, Lubeck-Travemunde, W. Germany. April 14-17, (1986).

81. Palaith, D., Silberglitt, R., Wu, C.C.M., Kleiner, R. and Libelo, E.L., Microwave joining of ceramics. Mat Res Soc Symp, 124 255-266, (1988).

82. Fukushima, H., Yamanaka, T. and Matsui, M., Microwave heating of ceramics and its application to joining. Mat Res Soc Symp, 124 267-272, (1988).

83. Scott, C. and Tran, V.B., Diffusion bonding of ceramics. Am Ceram Soc Bull, 64 1129-1131, (1985).

84. Gyarmati, E., Naoumidis, A. and Nickel, H., Presented at 2nd International Conference on Ceramic Materials and Components for Engines, Lubeck-Travemunde, W. Germany. April 14-17, (1986).

85. Bennett, C.E.G., McKinnon, N.A., and Williams, L.S., Sintering in gas discharges. Nature, 217 1287-88, (1968).

86. Kemer, E. and Johnson D.L., Microwave plasma sintering of alumina. Am Ceram Soc Bull, 64 [8] 1132-1136, (1985).

87. Ji, T-R, and Gerling, J.E., A versatile microwave plasma applicator. Mat Res Soc Symp, 124 353-366, (1988).

88. Jolly, J.A., Financial techniques for comparing the monetary gain of new manufacturing processes such as microwave heating. J Microwave Power, 7 [1] 5-16, (1972).

89. Das, S. and Curlee, T.R., Microwave sintering of ceramics; can we save energy?. Am Ceram Soc Bull, 66 [7] 1093-1094, (1987).

90. Sanio, M.R. and Schmidt, P.S., Economic assessment of microwave and radio frequency materials processing. Mat Res Soc Symp, 124 337-346, (1988).

9

Thin Film Deposition Processes for Electronic and Structural Ceramics

R. C. Budhani [†] and R. F. Bunshah [*]

[*] Department of Materials Science and Engineering, University of California, Los Angeles, Los Angeles, California 90024-1595, USA.

[†] Brookhaven National Laboratory, Upton, New York 11973, USA.

1. INTRODUCTION

Ceramic materials, depending on their electrical conductivity, can be broadly classified as metallic and non-metallic. The metallic ceramics primarily comprise the compounds of transition metals with the elements of group III, IV and V of the periodic table, e.g. boron, carbon, silicon and nitrogen, thus constituting the borides, carbides, silicides and nitrides. The metallic character in these systems arises because of the intermixing between s-p electrons of the non-metals and d electrons of the transition metal. Apart from their high electrical conductivity, this class of ceramics is also characterized by high melting point, extreme hardness and relative immunity to chemical corrosion and oxidation. Furthermore, their relatively close packed structure makes them less susceptible to diffusion of foreign atoms through them. A combination of these properties have made this class of materials very useful in a variety of structural and electronic applications.

The other class of ceramics consists of the oxides of various elements and the compounds involving two elements whose electronic character is decided by s-p electrons only. Excluding some transition metal oxides, most of these compounds are electrically insulating. The predominantly insulating

369

character arises from the strong chemical bonding between the constituents and the mixed ionic-covalent character of the bonds. The wide band gap resulting from such bonding provides these materials their characteristic transparency in the visible and near UV regions of the electromagnetic spectrum. The wide band gap can be modulated via a controlled introduction of non-stoichiometry or dopants. This permits the creation of a variety of ceramic systems with tailored electrical, magnetic and optical responses. Another cluster of properties resulting from the unique bonding character in these materials consists of a very high values for strength, stiffness, melting point and corrosion resistance.

Considering the immense technological potential emanating from the above mentioned properties of ceramics, the past decade has witnessed a considerable scientific and industrial interest in the area of developing thin film/coating deposition techniques for these materials. The large number of coating techniques which exist today and the newer ones in the making, have resulted essentially because of the capability of deposition processes to modulate/tailor a specific property of the material and secondly due to the diversity of applications requiring film thickness ranging from a few nanometres or less to several millimeters. For instance, the deposition process (e.g. plasma spraying) for a thick (300 - 400 μm) thermal barrier coating for a high temperature turbine application, where good adhesion and poor thermal conductivity are the only requirements, can not be used to deposit the gate dielectric of a Metal-Insulator-Semiconductor Field Effect Transistor (MISFET), where extreme dimensionality and a finely tuned electronic defect structure are important.

All coating methods however, consist of three basic steps; synthesis or generation of the coating species or the precursors at the source; transport from the source to the substrate and nucleation and growth of the coating at the substrate. These steps can be completely independent of each other or may be superimposed on each other depending on the coating process. A process in which the steps can be varied independently offers greater flexibility and a large variety of material with tailored properties can be deposited.

Numerous schemes can be devised to classify deposition processes. The scheme used here is based on the dimensions of the depositing species, i.e. atoms and molecules, liquid droplets, atomic clusters or use of surface modification processes. The four major classes of coating processes are

Atomistic deposition	Particulate deposition	Bulk coatings	Surface modification
electrolytic environment	thermal spraying	wetting processes	chemical conversion
electroplating	plasma-spraying	painting	electrolytic
electroless plating	D-gun	dip coating	anodization (oxides)
fused-salt electrolysis	flame-spraying	electrostatic spraying	fused salts
chemical displacement	fusion coatings	printing	chemical-liquid
vacuum environment	thick-film ink	spin coating	chemical-vapour
vacuum evaporation	enameling	cladding	thermal
ion-beam deposition	electrophoretic	explosive	plasma
molecular-beam epitaxy	impact plating	roll bonding	leaching
plasma environment		overlaying	mechanical
sputter deposition		weld-coating	shot peening
activated reactive evaporation		liquid-phase epitaxy	thermal
plasma polymerization			surface enrichment
ion plating			diffusion from bulk
chemical-vapor environment			sputtering
chemical-vapor deposition			ion implantation
reduction			
decomposition			
plasma enhanced			
spray pyrolysis			

Table 1: Coating methods

listed in Table 1.[1]   The ceramic coatings/films used in electronic, optical, photonic and magnetic device applications are primarily deposited by the atomistic processes which have the flexibility to deposit films of nano-dimensions with a precisely controlled electronic and atomic structure. In applications involving wear and corrosion protection, structural applications, thermal barrier coatings, oxidation and corrosion protection coatings, etc. where the coating thickness may vary from a few microns to several thousand microns, the choice of a deposition process is mostly decided by the economics and the degree of preformance anticipated from the coating.   The emphasis of this chapter is on highlighting the important features of the atomistic deposition processes which play a dominant role in deciding the performance of ceramic thin films/coatings.

2. ATOMISTIC DEPOSITION PROCESSES

Figure 1 shows the family tree of the atomistic deposition processes. These are classified into two groups known as Physical Vapor Deposition (PVD) and Chemical Vapor Deposition (CVD) techniques as described in the following sections.

2.1 Physical Vapor Deposition (PVD) Processes

The PVD processes comprise the class of deposition techniques in which one or all constituents of the deposit are generated from a solid source by heat induced vaporization or momentum transfer processes.   These two mechanisms thus lead to "Evaporation" and "Sputtering" being the two fundamental PVD processes.

2.1.1 Evaporation Processes

The evaporation of a material in vacuum requires a vapor source to support the evaporant and to supply the heat of vaporization while maintaining the material at a temperature sufficiently high to produce the desired vapor pressure[2].   The temperature can be obtained by resistive heating of the vapor source, a crucible or a hearth of a suitably chosen material, or by directly heating the evaporant using a laser or electron beam.   For the deposition of multicomponent systems such as alloys,

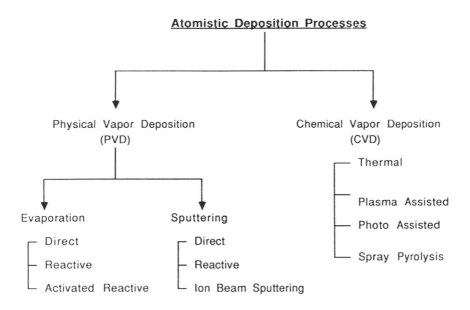

Figure 1: Family tree of atomistic-deposition processes.

intermetallic compounds and ceramics, a single source direct evaporation or
the use of multiple sources for each of the constituents may be made. The
direct evaporation process is extensively used for metals and elemental
semiconductors. In the case of ceramic materials however, the evaporation
occurs with the dissociation of the compound into fragments. Very few
systems such as SiO, MgF_2, B_2O_3, CaF_2 and other group IV divalent oxides
evaporate without any dissociation. The stoichiometry of the deposit in such
a situation depends on the ratio of various molecular fragments striking the
substrate, the sticking coefficient of the fragments (which may be a strong
function of the substrate temperature) and the reaction rate of the fragments
on the substrate to reconstitute the compound. In the case of ceramic
systems, where one of the constituents is a gas in its elemental form, the
films are generally deficient in the gaseous constituent. For example,
direct evaporation of Al_2O_3 results in a deposit deficient in oxygen. The
imbalance of stoichiometry occurs to a lesser degree in ceramic systems where
all the constituents are solids with nearly the same vapor pressure in their
elemental forms. The direct evaporation process, using electron beam or

thermal evaporation techniques, has been successfully used in the case of transition metal carbides, silicides and borides[2-4]. The slight variation in the stoichiometry of the coatings of such systems deposited by a single source evaporation can be avoided if a multi-source evaporation technique is used. A schematic diagram of a two source evaporation process is shown in Figure 2. This technique is used quite extensively for the deposition of transition metal silicides[5] and sulphides and selenides of group II elements (Zn and Cd)[6]. The molecular beam epitaxy (MBE) technique[7], which is used extensively for depositing epitaxial silicides, III-V and II-IV-VI compound semiconductors is also a multi-source thermal evaporation process.

2.1.1.1 Reactive Evaporation: The reactive evaporation process is used to compensate for the loss of the gaseous constituents of a ceramic during its direct evaporation. Thus, for example, stoichiometric Al_2O_3 films can be deposited by direct evaporation of Al_2O_3 in an atmosphere of oxygen. However, due to the high melting point and the extreme reactivity of ceramic melts it is generally difficult to evaporate ceramic materials. To avoid

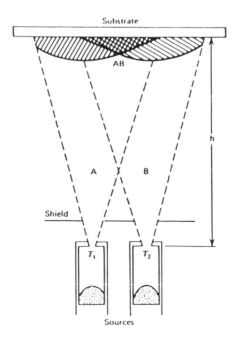

Figure 2: Two source evaporation arrangement yielding variable film composition (72).

these problems, the reactive evaporation process is commonly used in a mode where the metallic constituent of the ceramic is evaporated in a partial pressure of the reactive gas to form a compound in the gas phase or on the substrate as a result of a reaction between the metal vapor and the gas atoms, e.g.:

$$2Al + 3/2O_2 \rightarrow Al_2O_3$$

The reactive evaporation method has been used to synthesize films of SnO_2,[8] $SnInO_3$,[9] In_2O_3,[10] Al_2O_3,[11] SiO_2,[12] $Cu_xMo_6O_8$,[13] Y_2O_3,[14] TiO_2[15] and more recently of $YBa_2Cu_3O_{7-x}$ type perovskite superconductors[16,17].

Since the reactive gas in the 'RE' process is in a molecular state and thus less likely to react with the vapor species, the formation of a well crystallized stoichiometric film requires high thermal activation at the substrate. This problem becomes acute in situations where the reactive gas consists of more than one element. The typical examples are SiH_4 and CH_4 C_2H_2 for the formation of silicides and carbides respectively. The concept of electron impact ionization and excitation of the reactive gas in the substrate-source space, as introduced by Bunshah and Raghuram[18] solved this problem. The ionization and excitation of the gas activates the compound forming reactions and the compound films can be synthesized at a much lower substrate temperature. This process is known as Activated Reactive Evaporation (ARE).

2.1.1.2 Activated Reactive Evaporation: In this process the metal is heated and melted by a high acceleration voltage electron beam. The continuous electron bombardment of the molten material leads to ionizing collisions in the vicinity of the pool, and thus creation of a thin plasma of the metal vapors. The low energy secondary electrons forming the plasma sheath are pulled upwards into the reaction zone by an interspace electrode placed above the pool and biased to a low positive d.c. or a.c. potential (20 to 100V). The low energy electrons have a high ionization cross section, thus ionizing or activating the metal and gas atoms.

In addition, collisions between ions and neutral atoms results in charge exchange processes yielding energetic neutrals and positive ions. It is believed that these energetic neutrals condensing on the substrate along with the ions and other neutral atoms "activate" the reaction, thus increasing the reaction probability between the reactants. A schematic of the Activated

Reactive Evaporation process is shown in Figure 3.

The synthesis of TiC by reaction of Ti metal vapor and C_2H_2 gas atoms with a carbon/metal ratio approaching unity was achieved with this process[18]. Moreover, by varying the partial pressure of either reactant, the carbon/metal ratio of carbides could be varied at will. The ARE process has also been recently applied to the synthesis of all the five different Ti-O oxides[19]. The authors noted that in the ARE process (i.e. with a plasma) as compared to the RE process (i.e. without a plasma) a higher oxide formed for the same partial pressure of O_2 thus demonstrating a better utilization of the gas in the presence of a plasma. The same observation was noticed by Granier and Besson for the deposition of nitrides[20].

A variation of the ARE process using a resistance heated source instead of the electron beam heated source has been developed by Nath and Bunshah[21] and is particularly useful for evaporation of low melting metals such as indium and tin where electron beam heating can cause splattering of the molten pool. The plasma is generated by low energy electrons from a

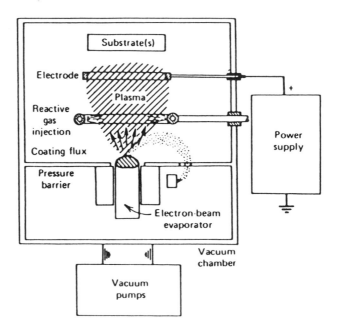

Figure 3: Schematic of the Activated Reactive Evaporation Process (72).

thermionically heated filament and pulled into the reaction zone by an electrical field perpendicular to the evaporation axis. The ionization probability is further enhanced by a superimposed magnetic field which causes the electrons to go into a spiral path. This process has been used to deposit indium oxide and indium tin oxide transparent conducting films. The ARE process has several other variations, as shown in Figure 4.

1. If the substrate is biased in the ARE process, it is called the 'biased activated reactive evaporation process' (BARE). The bias is usually negative to attract the positive ions in the plasma. The BARE process has been reinvented and called Reactive Ion Plating by Kobayasi and Doi[22].

2. 'Enhanced ARE' is the conventional ARE process using electron beam heating but with the addition of a thermionic electron emitter (e.g. a tungsten filament) for the deposition of refractory compounds at lower deposition rates as compared to the basic ARE process. The low energy electrons from the filament sustain the discharge.

3. Using electron beam evaporation sources, the electric field may be generated by biasing the substrate positively instead of using a positively biased interspace electrode. In this case, the technique is called 'low pressure plasma deposition' (LPPD)[23]. However, this version has a disadvantage over the basic ARE process in that there is no freedom of choice concerning grounding the substrate, letting it float or biasing it negatively as in the BARE process.

4. A plasma electron beam gun, instead of the thermionic electron beam gun, can be used to carry out the ARE process. The hot hollow cathode gun has been used by Komiya et al.[24] to deposit TiC films whereas Zega et al.[25] used a cold cathode discharge electron beam gun to deposit titanium nitride films. A plasma assisted deposition process designated 'RF reactive Ion Plating' was developed and used by Murayama[26] to deposit thin films of In_2O_3, TiN and TaN. A resistance or electron beam heated evaporation source is used and the plasma is generated by inserting an RF coil electrode of aluminum wire in the region between the evaporation source and substrate.

The ARE process has also been used in a dissociative mode where instead of using an elemental evaporation source one uses a low melting compound of

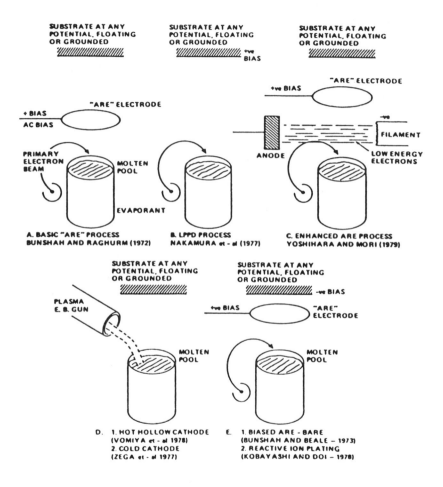

Figure 4: Basic ARE process and later variations (72).

the elements.   For example cubic boron nitride films can be deposited by
evaporating boric acid in an ammonia plasma[27].

The reader is referred to a recent review of the ARE process by R.F.
Bunshah and C.V. Deshpandey.   (Physics of Thin Films Vol. 13, 1987, p 59 -
Academic Press) as well as for a list of ceramic compounds deposited by this
method and their applications.

2.1.1.3 Ion Plating Processes: The beneficial aspects of ion bombardment on
the growth and adhesion of thin films are also realized in the Ion Plating
Process introduced by Mattox[28].   This process differs from ARE in the sense
that the plasma is created by applying a heavy negative bias on the substrate
instead of using a positively biased probe adjacent to the evaporation
sources as in ARE.   In a simple diode ion plating process, the plasma cannot
be supported at pressures lower than about 1.3 Pa.   The higher partial
pressures required to sustain a plasma, on the other hand, may lead to porous
deposits.   Therefore, an auxillary electrode adjusted to a positive low
voltage, as originally conceived in ARE process, is used to initiate and
sustain the plasma (Figure 5).

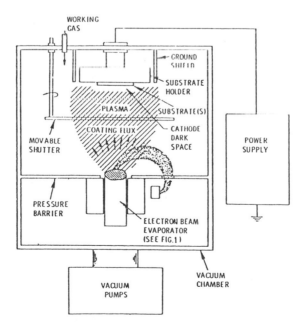

Figure 5: Ion Plating Process (72).

In order to understand the effects of ion bombardment on the microstructure and growth morphology of the films, it is necessary to discuss the kinetics of condensation in some detail. The process of condensation of vapor species on the substrate is a strong function of their energy and the temperature of the substrate. If the energy of the condensate or the substrate temperature, or both, are high, the condensing atoms (adatoms) take time to thermalize on the substrate. The higher thermalization time allows growth of a crystallographically ordered structure. The critical time and energy required for the formation of a crystalline deposit is a strong function of the chemical nature of the adatoms and the crystallographic symmetry of the structure to be deposited. Metals with high symmetry structures (like fcc) always condense in a crystalline phase even at temperatures as low as 4.2K. Covalently bonded, low symmetry structures, on the other hand, condense in amorphous forms. This problem is prevalent in ceramics because of the geometrical constraints of their bonding and their low symmetry structures. The mobility of the adatoms can be increased by subjecting the growing film to energetic ion bombardment. The biased ARE is a typical case where the film properties are improved by ion bombardment.

The capability of thermal evaporation processes to allow high deposition rate and the beneficial effects of ion bombardment have also been merged together in the Ionized Cluster Beam Deposition (ICBM) process[29]. The principle of ICBD is shown in Figure 6. Clusters of atoms are created by condensation of supersaturated vapor atoms produced by adiabatic expansion thorough a small nozzle into a high vacuum region. The clusters are then ionized with the help of electron bombardment and accelerated by a voltage applied between the regions of ionization and the substrate. Ceramic coatings can be made by introducing reactive gases in the chamber.

2.1.2 Sputter Deposition Processes

The basic sputter deposition process involves removal of atoms from the surface of a solid or liquid by energetic ion bombardment and collection of the sputtered species on a solid surface. The basic principle of the sputtering process is shown in Figure 7. A target, consisting of the material to be deposited, is held at a negative potential ranging from a few hundred volts to a few kilovolts. For a critical value of the chamber presure (1×10^{-3} Torr to -1 Torr), the application of the voltage strikes a discharge in the vicinity of the target. The target, because of its negative potential, is bombarded by the ions present in the plasma. The discharge is sustained by the stochastic ionization of the gas

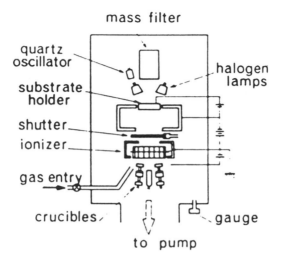

Figure 6: Schematic diagram of the ionized cluster beam deposition equipment (139).

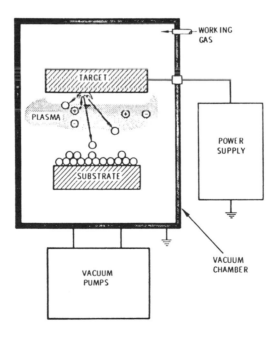

Figure 7: Schematic representations of sputter deposition process (30).

atoms/molecules by the secondary electrons emanating from the target.   In situations when the gas ions are sufficiently heavy, the bombardment of the target leads to sputtering of the target surface by a momentum transfer process.   Argon, because of its higher atomic number and non-reactivity, is the commonly used sputtering gas.   Since the sputtering rate is directly dependent on the number of ions striking the target, its magnitude can be increased by increasing the ion density in the vicinity of the target.   The higher ion densities are realized by application of magnetic fields.   This modification of the sputtering process is known as magnetron sputtering.   If the target material is electrically conducting, a d.c. voltage can be used for sputtering. In the case of insulating targets an RF potential must be applied.

2.1.2.1 Reactive Sputtering Process: The d.c. magnetron sputtering technique has been successfully used to deposit films of transition metal silicides and borides by using a composite ceramic target[30,31].   As in the case of evaporation, however, sputtering also leads to the dissociation of the target material into atoms and molecular fragments and results in the deficiency of the gaseous constituents of the material.   The problem of off-stoichiometry can be eliminated if one adds the constituent gas in the plasma along with the argon.   This modification of the process is called reactive sputtering.

Although a single ceramic target RF or d.c. reactive magnetron sputtering has been used to deposit films of a large class of ceramic materials (Table 2), it essentially suffers from the following disadvantages:

1.   The poor thermal conductivity of ceramics does not allow an effective cooling of the target during sputtering, which results in local hot spots and consequent spitting of the material and the formation of particulates in the films.   The ceramic targets also develop massive cracks after prolonged usage.

2.   In the case of multicomponent targets, such as those used for deposition of high-T_c oxide superconductors and "tungsten bronzes", the preferential sputtering of one of the components leads to off-stoichiometry in the films.   The variation in the film composition can be seen in the case of $La_xSr_{1-x}CuO_4$ superconducting system (Figure 8)[32].

The problems associated with ceramic targets can be eliminated by

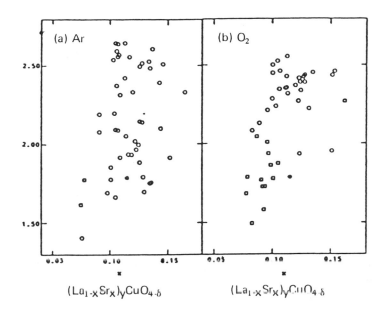

Figure 8: Metallic Composition of the films prepared by the sputtering of $(La_{0.985}Sr_{0.115})_{1.78}CuO_{4-\delta}$ target in (a) Ar and (b) Ar-O$_2$ mixtures, with (○) and without (○) substrate heating. cf: •Target Composition (32).

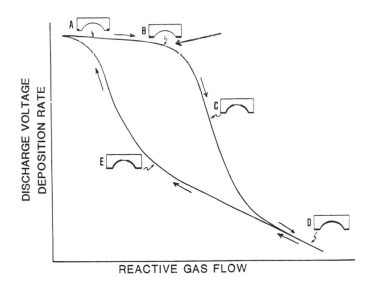

Figure 9: Dependence of discharge voltage and deposition rate on the reactive gas flow (34).

Ceramic Material	Sputtering System Type	Target	Gases	Representative References
AlN	RF Planar	AlN	Ar + N_2	84
BN	" " "	BN	Ar + N_2	85
Si_3N_4	" " "	Si_3N_4	Ar + N_2	86
SiC and $a\text{-}Si_xC_{1-x}$:H	" " "	SiC	Ar + N_2	87,88
In_2O_3	" " "	In_2O_3	Ar/O_2	89,90
SnO_2	" " "	SnO_2	" " "	91–93
ZnO	" " "	ZnO	" " "	94–96
Y_2O_3	" " "	Y_2O_3 ceramic	" " "	97
TiO_2	" " "	TiO_2 ceramic	" " "	98
Ta_2O_5	" " "	Ta_2O_5	" " "	99
Al_2O_3	" " "	Sapphire	" " "	100–102
SiO_2	" " "	Quartz	" " "	103
$Pb(ZrTi)O_3$	" " "	$Pb(ZrTi)O_3$ ceramic	100% O_2	104
$BaTiO_3$	" " "	$BaTiO_3$ ceramic	Ar/O_2 (80/20)	105
$YBa_2Cu_3O_7$	DC Magnetron	$YBa_2Cu_3O_7$ ceramic	Ar	106,107
$La_xSrO_xCuO_4$	60 Hz AC	$La_xSrO_xCuO_4$	Ar	32

Table 2: Ceramic films prepared by reactive sputtering of a compound target.

Ceramic material	Sputtering System Type	Target	Gases	Representative References
AlN	RF Planar	Aluminium	Ar/N_2	108,109
GaN	" " "	Gallium	" " "	110
Si_3N_4	DC Planar	Silicon	$Ar/N_2/H_2$	111,112
SiC	RF Planar	Silicon	$Ar + CH_4$	113
Si_xC_{1-x}	" " "	Silicon + Graphite	$H_2 + Ar$	114
In_2O_3	DC Planar	Indium	Ar/O_2	115
SnO_2	" " "	Tin	" " "	116,117
ZnO	" " "	Zinc	" " "	118
TiO_2, Ta_2O_3	" " "	Titanium	" " "	40
Refractory Silicides	" " "	Refractory Metals	Ar/SiH_4	45
Al_2O_3	RF Planar	Aluminium	Ar/O_2	119
Refractory Nitrides	DC Planar	Refractory Metals	Ar/C_2H_2	120
Refractory Carbides	" " "	" " "	" " "	120
$YBa_2Cu_3O_7$	RF Planar	Y, Ba and Cu Metals	Ar/O_2	121,122

Table 3: Ceramic films prepared by reactive sputtering of elemental targets.

sputtering the constituent metal or metals in a reactive sputtering mode. Table 3 summarizes important ceramic compounds deposited by reactive sputtering. The reactive sputtering of metals, particularly in oxygen environments, suffers from a major problem which is commonly known as target poisoning as discussed below.

When a reactive gas is introduced during the sputtering of metals it is observed that as the flow rate is increased the deposition rate decreases by a factor of three or more, at a particular flow rate termed the critical flow rate f_c. The critical flow rate is a function of power to the target, pressure and target size. The relationship between the film properties and gas injection rate is generally non-linear and it is believed to be due to a complex dependence of the stricking coefficient on the growth rate, composition, structure and temperature of the growing film. A typical dependence of discharge voltage and deposition rate on gas flow rate is shown in Figure 9. The hysteresis effect for the discharge voltage is attributed to the process of formation and removal of a compound layer on the target surface. A similar hysteresis behavior is also observed for the deposition rate, system pressure and RF negative bias (in case of RF sputtering).

It can be seen from the above discussion that target-gas plasma interaction during reactive sputtering leads to the formation of a compound layer on the target which is termed as target poisoning. Numerous papers have been published illustrating target poisoning in reactive sputtering[33-39].

In recent years, many different approaches have been proposed to get around the problem of target poisoning (see Figure 10). Schiller et al.[40] have proposed a technique which initially involved maintaining a gradient in the composition of the reactive gas by injecting argon near the target and oxygen near the substrate as well as by arranging additional gettering surfaces in the chamber to maintain a gradient in composition of the reactive gas. They have reported deposition rates as high as 1.05 μm min^{-1} for Ta_2O_5 and about 0.68 μm min^{-1} for TiO_2 using this technique at a source-to-substrate distance of 5 cm. Westwood et al.[41] have proposed the use of geometric baffles to maintain a pressure differential between the substrate and target to avoid target poisoning. Scherer and Wirz[42] have used similar techniques to produce metal oxide films at high deposition rates. They also introduce a positively biased anode near the substrate to intensify the plasma adjacent to the substrate and increase the reactivity at the

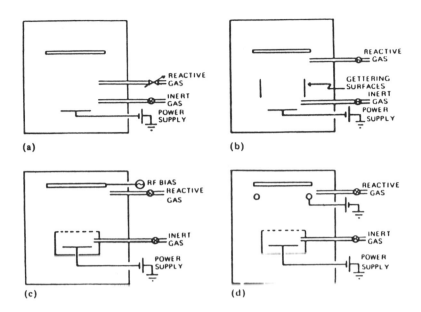

Figure 10: Some techniques used to avoid target poisoning (34).

substrate.   Sproul[43] provided an alternative approach which initially involved pulsing a gas in a controlled manner into the deposition system during magnetron sputtering of metals thus giving a chance for the target to be "depoisoned" during the off-period of the reactive gas injection cycle. In a later development, deposition of compound film was achieved by Sproul[44] without pulsing the gas using on-line mass spectroscopic analysis of the gas phase with feed-back control.   The operating point for this process is just below point B in Figure 9.   Sproul has reported sputter deposition rates equal to those of the corresponding metal by totally avoiding target poisoning.

A major modification of the reactive sputtering process has been introduced by Budhani et al.[45] for deposition of transition metal silicide films.   The technique involves d.c. magnetron sputtering of the metal of interest in a silane diluted argon plasma.   Unlike the sputtering from a composite target, the process allows deposition of the silicides over a wide range of composition.   This method differs from the conventional reactive sputtering in the sense that one of the products of dissociation of the gas is a solid.

In situations where the interaction of the plasma with the growing film may have adverse effects on film properties and where in situ analysis of the growth are required, the direct sputtering methods have limitations. These plasma related problems can be avoided by using ion beam sputtering techniques as described in the following section.

2.1.2.2 Ion Beam Sputtering: In the basic ion beam sputtering process, an energetically well characterized beam of inert gas ions, generated from an ion source, bombards the target to be sputtered. In the case of electrically insulating targets, neutral beams are used. The sputtered species are deposited on a substrate which are situated in a relatively high vacuum. If a metal target is used for the growth of a ceramic film, the reactive gas can be introduced directly in the discharge chamber of the ion gun or a separate gun may be used[46] as shown in Figure 11. Table 4 lists the ceramic systems that have successfully been deposited using ion beam and reactive ion beam deposition techniques.

2.1.2.3   Laser Assisted Vapor Deposition Processes: A variant of the evaporation process which has recently been introduced for deposition of superconducting[139] and diamond or diamond like carbon films[140] is the use

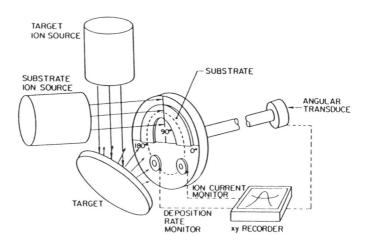

Figure 11: Dual ion beam system for thin film deposition under controlled ion bombardment (75).

Ceramic Compound	Beam Energy (KeV)	Target Material	Gases	Representative References
AlN	Dual Beam Ar^+(1.5), N_2^+(0,8)	Aluminium	Ar and N_2 (separately)	46
	Single Beam (0.8~2.0)	Aluminium	$N_2 + H_2$	123,124
BN	Ion Source of $B_3N_3H_6$	No Target	$B_3N_3H_6$	125
Si_3N_4	$Ar^+ + N_2^+$ (0.5)	Silicon	$Ar + N_2$	126
$a-Si_xN_{1-x}$:H	$Ar^+ + N_2^+$ (1.0~0.5)	Silicon	$Ar + N_2 + H_2$	127,128
AlO_xN_y	Ar and N_2 Neutral Beam	Aluminium	$Ar + N_2 + O_2$ leak	129
SiO_xN_y	Ar and O_2 Neutral Beam	Silicon	$Ar + O_2$	130
$In_2(Sn)O_3$	Ar^+ (0.5)	In/Sn alloy	$Ar + O_2$	131,132
SnO_2	Ar^+ (~)	Tin	$Ar + O_2$	133
Al_2O_3 Refractory Oxides	Ar^+ (1.0)	Aluminium	$Ar + O_2$	134
Y_2O_3 TiO_2 Ta_2O_5	Ar^+ (~)	Refractory Metal	$Ar + O_2$	135
Refractory Nitrides (NbN)	$X_e^+ + N_2^+$ (9.5)	Nb	$X_e + N_2$	136
Lead zirconium titanate (PZT)	Ar^+ (2.0)	Pb, Zr, Ti Oxide Ceramic	$Ar + O_2$	137
$YBa_2Cu_3O_7$	Ar^+ (1.8)	$YBa_2Cu_3O_7$ Oxide Ceramic	$Ar + O_2$	138

Table 4: Ceramic films prepared by reactive ion beam sputtering.

of a laser as a heat source for evaporation.   The evaporation target is either a ceramic compound or metal when the process is used in the reactive mode.   The process also employs as a variant a positively-based ARE type electrode above the target to generate a plasma for reactive deposition. Photonic excitation of a plasma is also believed to be an asset for reactive deposition.   The small area of the deposited film is one of the current limitations of the process.

2.2 Chemical Vapor Deposition Processes

The constituents of the precursors for most of the ceramics exist in the form of heavy molecular gases or volatile liquids such as halides, hydrides, organo-metallics, hydrocarbons and ammonia complexes, which can be dissociated into highly reactive fragments by providing photon, electron or thermal excitation.   These fragments can easily react with the second component of the desired ceramic and form a solid product.   Most of the chemical vapor deposition processes work on this basic principle.   They are known as Thermally Assisted CVD, Plasma Assisted CVD and Photo CVD for thermal, electron and photon excitation respectively.

2.2.1 Thermally Assisted Chemical Vapor Deposition Processes

The principle of the thermal dissociation process or conventional CVD, as shown in Figure 12, requires substrate temperatures in excess of 600°C in most cases for any significant deposition to occur[47].   The high substrate temperatures have both advantages and adverse effects on the properties of the films.   The higher substrate temperature promotes the growth of a dense and well crystallized structure with the minimum of impurities trapped in. Also, since the dissociation reactions occur on the surface of the substrate itself, the conventional CVD process does not have the line of sight limitation of the PVD processes.   With CVD, all parts of an irregularly shaped substrate are coated uniformily provided the temperature and gas flow conditions are the same everywhere.   A wide class of oxides, borides, silicides, nitrides and carbides have been deposited by the conventional thermal CVD process. Some important CVD reactions for the deposition of ceramics are listed below:

$$Si(CH_3)Cl_3(g) \rightarrow SiC(s) + 3HCl(g)$$

$$SiCl_4(g) + CH_4 \rightarrow SiC(s) + 4HCl(g)$$

REACTOR

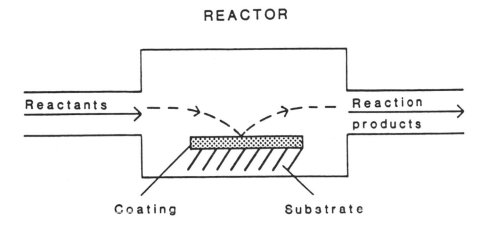

Figure 12: The principle of chemical vapour deposition.

$$3SiH_4(g) + 4NH_3(g) \rightarrow Si_3N_4(s) + 12H_2$$

$$SiH_4(g) + 2H_2O(g) \rightarrow SiO_2(s) + 4H_2$$

$$2AlCl_3(g) + 3CO_2(g) + 3H_2(g) \rightarrow Al_2O_3(s) + 3CO(g) + 6HCl(g)$$

$$2Al(CH_3)_3(g) + 9O_2(g) \rightarrow Al_2O_3(s) + 6CO(g) + 9H_2O(g)$$

$$Al(CH_3)_3(g) + NH_3(g) \rightarrow AlN(s) + 3CH_4(g)$$

$$BCl_3(g) + NH_3(g) \rightarrow BN(s) + 3HCl(g)$$

$$SnCl_4(g) + O_2(g) \rightarrow SnO_2(s) + 2Cl_2$$

The higher deposition temperatures required in a CVD process have adverse effects in situations where the substrate is susceptible to temperature induced irreversible structural/electrical changes. This type of problem is encountered most frequently while depositing dielectric coatings on previously processed electronic devices/substrates. However, low temperature chemical vapor deposition of ceramic coatings can be realized by plasma or photon assisted chemical vapor deposition processes.

2.2.2 Plasma Assisted Chemical Vapor Deposition

The basic design of a plasma CVD reactor is shown in Figure 13. It essentially consists of two parallel plates one of which is connected to a d.c. or RF power supply.   RF is preferred in situations when the film is insulating.   If the gas pressure in the chamber is properly adjusted, a glow discharge can be created between the two plates. At low frequencies, (low KHz range), some degree of secondary electron emission is necessary from the electrodes in order to maintain the discharge. At higher frequencies, enough electrons gain energy to ionize the molecules in the gas to maintain the discharge without secondary emission[48].   Electron energy is lost through momentum transfer, vibrational excitations, dissociation and ionization processes in decreasing order[49].

One of the most common examples of the PACVD processes is the growth of silicon nitride by using a mixture of SiH_4 and NH_3 gases.   For this particular case, the electron impact dissociation of silane and ammonia leads

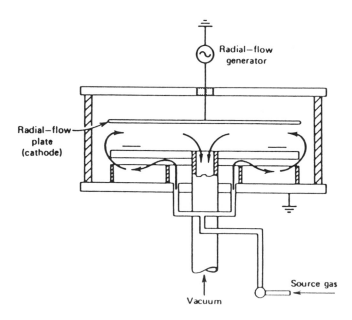

Figure 13: Schematic representation of a radial flow parallel plate plasma CVD reactor (48).

to the following fragments:

$$SiH_4 + e^- \rightarrow SiH, SiH_2, SiH_3, Si, H, + e^-$$

$$NH_3 + e^- \rightarrow NH_2, HN, N, H, + e^-$$

The reaction between the nitrogen and hydrogen containing species in the plasma results in a solid deposit which is commonly written as a - Si_xN_{1-x}:H. The physical properties (electrical conductivity, optical absorption, refractive index), Si/N ratio and the stress state of the plasma CVD silicon nitride are highly sensitive to the amount of bonded hydrogen in the material and the degree of ion bombardment during the film growth. These effects have been studied by Budhani et al.[50-52] and several other workers[53-55]. Apart from silicon nitride, plasma assisted CVD technique has been used to deposit a wide range of other ceramic materials.

In order to avoid the ion bombardment of the growing film in a parallel plate PACVD process, Lucovsky et al.[55] have introduced a modification in which the substrates are placed away from the plasma. The technique is known as Remote PACVD.

2.2.3 Photo CVD

A relatively new entry in the field of CVD processes is Photo CVD. In this case the CVD precursors are dissociated by a high energy photon flux incident from an ultraviolet lamp or a high frequency laser[56,57]. The Photo CVD processes are particularly important from the standpoint of avoiding any ion bombardment of the growing film which invariably occurs in the case of Plasma CVD processes. The Laser CVD also allows the deposition of films with fine line resolution. The main draw back of this process is the poor dissociation efficiency of the precursors and hence the lower deposition rates.

2.2.4 Spray Pyrolysis

Spray pyrolysis involves the spraying of a solution, usually aqueous, containing soluble salts of the constituent atoms of the desired compounds, onto heated substrates. A typical example of the spray pyrolysis process is deposition of tin oxide films by thermal dissociation of an alcoholic solution of $SnCl_4$[58]. Whether or not the process can be classified as CVD depends on whether the liquid droplets vaporize before reaching the substrate or react on it after splashing. Several workers have used preheating (temperature about 200-500°C) of the sprayed droplets to ensure vaporization of the reactants before they undergo a heterogeneous reaction at the substrate[59-63]. The technique is very simple and is adaptable for mass production of large-area coatings for industrial applications. Various geometries of the spray set-ups are employed, including an inverted arrangement in which larger droplets and gas phase precipitates are discouraged from reaching the substrate, resulting in films of better quality. This technique has been used extensively to deposit films of $ZnO^{[64,65]}$, $In_2O_3^{[66,67]}$, $SnO_2^{[68,69]}$, $CdS^{[70]}$, $Al_2O_3^{[71]}$ and other ceramic systems.

3. MICROSTRUCTURE AND CRYSTALLINITY OF THE DEPOSITS

The microstructure and the degree of crystallinity in the films deposited by physical and chemical vapor deposition processes and their plasma and photonic modifications depend on several factors. The important ones are the nature of the chemical bonding in the material to be deposited, the substrate temperature, the crystallographic nature of the substrate with respect to the structure of the material to be deposited, vacuum conditions, impurities present in the deposition environment, deposition rate and the extent of electron and ion bombardment of the growing film.

As stated in previous sections, due to the geometrical constraints imposed by bonding and complex crystal structures, most of the ceramic systems condense into an amorphous phase when deposited on unheated substrates. This tendency is promoted further by the presence of gaseous impurities in the deposition environment. It is generally observed that deposition of crystalline films of a particular ceramic by CVD requires higher substrate temperatures as compared to PVD processes. The reason being the presence of gaseous by-products of dissociation in the deposition environment.

Requirements on the extent of crystallinity in a film/coating is decided by the application involved. In uses where the broad structural inhomogeneties such as grain boundaries, dislocations, stacking faults and other high energy sites in the material may have deleterious effects, amorphous coatings are preferred. Some typical examples where amorphous coatings/films perform better are corrosion resistance applications, implantation and passivation applications in microelectronics, decorative applications and optical coatings. Materials such as silicon nitride, silicon dioxide, boron nitride, aluminum nitride, magnesium fluoride and tin oxide are most frequently used in these applications.

In situations where the performance is dependent on the uniqueness of the crystal structure of the material, such as the metal-insulator transition in vanadium oxide, acousto-optic and acoustic-electric responses of AlN, ZnO, PZT, $BaTiO_3$ and $SrTiO_3$, superconducting properties of copper oxide based perovskites, A-15 silicides, and NbN, wide band gap and dopability of SiC, transistor action in $Si/NiSi_2(CoSi_2)/Si$ epitaxial layers etc., it is important to optimize the deposition conditions for the growth of a well crystallized film.

Although the use of ceramic coatings in metallurgical applications (wear and tribology, heat resistant coatings, aerospace structural applications) does not put a stringent demand on the degree of crystallinity in the material, a well crystallized and structurally dense coating always gives better performance[1,72]. The evolution of microstructural features in metallurgical coatings can be well described in the framework of a structural model proposed by Movchan and Demchishin[73].

The microstructure and morphology of thick single phase films have been extensively studied for a wide variety of metals, alloys and refractory compounds. The structural model was first proposed by Movchan and Demchishin[73]. Figure 14 was subsequently modified by Thornton[74] (Figure 15). Movchan and Demchishin's diagram was arrived at from their studies on deposits of pure metals and did not include the transition zone of Thornton's model, Zone T. This is not prominent in pure metals or single phase alloy deposits but becomes quite pronounced in deposits of refractory compounds or complex alloys produced by evaporation and in all types of deposits produced in the presence of partial pressure of inert or reactive gas, as in sputtering or ARE type processes.

The evolution of the structural morphology is as follows: at low temperatures, the surface mobility of the adatoms is reduced and the structure grows as tapered crystallites from a limited number of nuclei. It is not a fully dense structure but contains longitudinal porosity of the order of a few tens of nanometres width between the tapered crystallites.

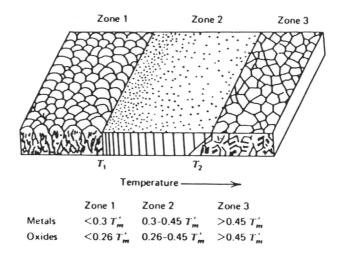

	Zone 1	Zone 2	Zone 3
Metals	$<0.3\ T_m'$	$0.3\text{-}0.45\ T_m'$	$>0.45\ T_m'$
Oxides	$<0.26\ T_m'$	$0.26\text{-}0.45\ T_m'$	$>0.45\ T_m'$

Figure 14: Structual zones in condensates as proposed by Movchan and Demchishan (73).

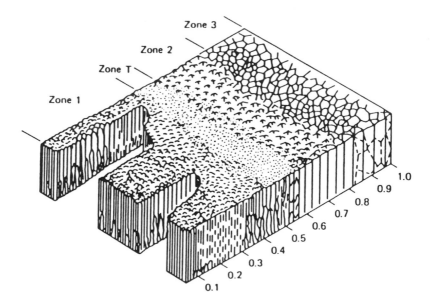

Figure 15: Modified model of Thornton (74).

It also contains a high dislocation density and has a high level of residual stress. Such a structure has also been called "Boitryoidal" and corresponds to Zone 1 in Figures 14 and 15.

As the substrate temperature increases, the surface mobility increases and the structural morphology first transforms to that of Zone T, i.e. tightly packed fibrous grains with weak grain boundaries and then to a full density columnar morphology corresponding to Zone 2 (Figure 14).

The size of the columnar grains increases as the condensation temperature increases. Finally, at still higher temperatures, the structure shows an equiaxed grain morphology, Zone 3. For pure metals and single phase alloys, T_1 is the transition temperature between Zone 1 and Zone 2 and T_2 is the transition temperature between Zone 2 and Zone 3. According to Movchan and Domohishin's original model, T_1 is 0.3 T_m for metals, and 0.22-0.26 T_m for oxides, whereas T_2 is 0.45-0.5 T_m. (T_m is the melting point in K).

Thornton's modification shows that the transition temperatures may vary significantly from those stated above and in general shift to higher temperatures as the gas pressure in the synthesis process increases.

The morphological results reported by Movchan and Demshishin for nickel titanium, tungsten, Al_2O_3 and ZrO_2 have been confirmed for several metals and compounds.

An intense ion bombardment of the substrate during deposition can suppress the development of an open zone 1 structure at low T/T_m. This has been demonstrated for both conducting and non-conducting deposits. Coatings deposited under these conditions have a microstructure similar to the zone T type. This effect appears to be due to the ion induced creation of new nucleation sites and sputtering effects. At this juncture, it should be stated that applicability of the above structural model is only strictly valid in situations where the flux of arriving species is sufficiently low. Or, in other words, the time scale for thermalization of adatoms is sufficiently small as compared to the rate of incoming atoms.

4.   STRESS AND GASEOUS CONCENTRATION

Thin films and coatings, irrespective of the deposition techniques used, always have internal stresses. The stresses are caused either by the thermal expansion mismatch between the coating and substrate or due to the non-equilibrium nature of the process which puts atoms out of position with respect to the minimums in the interatomic force fields. These two contributions are thermal and intrinsic stresses respectively. Small variations in the bond lengths and bond angles in covalently bonded ceramics lead to large internal stresses which have broad, structural and electronic consequences. Structurally, the stresses may lead to failure of adhesion or development of microcracks in the film. Since the intrinsic stresses increase with the coating thickness, failure of adhesion is observed frequently in the case of thick films. Electronically, the stresses affect the pizoelectric properties, optical absorption and optoelectronic properties and electronic dopability of the film or the substrate.

The bombardment of the growing film by energetic ions has a pronounced effect on intrinsic stresses. In the case of evaporated films, a transition from highly tensile to compressive behavior has been observed with increasing ion bombardment[72]. The substrate bias in evaporation and diode sputtering processes is an important tool for tuning[30] the level of stress in the films. In ion beam sputtering processes, bombardment of the growing film with a secondary ion gun can be used for stress tuning. In plasma CVD processes, the RF frequency used for excitation of the plasma is an important parameter in deciding the stress state of the films. At lower RF frequencies, the film is exposed to higher ion bombardment and therefore is generally found in a state of compressive stress[53]. A compressive to tensile change may occur at the higher RF frequencies. In the case of photo and thermal CVD processes, the degree of stress in the films is primarily decided by the gas composition and the substrate temperature.

In situations where a particular application requires exposure of the coating to higher temperatures, it becomes important to study the temperature dependence of stress in the films. Very few studies have so far been reported on the temperature dependence of stress in technologically important ceramic coatings[51,53].

Apart from intrinsic stress, another important factor that plays a critical role in deciding the electrical, optical and structural properties, is the magnitude of gaseous impurities incorporated in the films. These gases are either the impurity gases present in the deposition chamber or are deliberately introduced ones, such as argon in sputtering or the gaseous by-products of the dissociation reactions. In the case of the planar d.c. sputtering process, the amount of sputtering gas trapped in the film increases with the RF bias on the substrates[75,76]. The ion bombardment of the growing film realized by RF or -d.c. bias on the substrate, also tends to decrease the concentration of low stricking coefficient gaseous impurities, such as nitrogen, in the films[77]. In a reactive sputtering process, if the reactive gas consists of two or more elements, even the undesired element is incorporated in the films. A typical example is the incorporation of hydrogen in transition metal silicide films deposited by using SiH_4 as the reactive gas(78).

A profound effect of gaseous impurities is observed in the physical proportios of films deposited by photo and plasma CVD process. For example, silicon nitride films deposited by glow discharge dissociation of SiH_4 and NH_3 or SiH_4 and N_2 gases, contain 30 - 40 atomic percent hydrogen chemically bonded with Si and N atoms in the matrix[51,54]. The presence of bonded hydrogen in the films can be easily seen by infrared transmission measurements (Figure 16). Similarly, films of silicon oxynitride, boron nitride, aluminum oxide, aluminum nitride, titanium oxide, made by photo or plasma CVD contain a large number of chemically bonded H and OH species. A typical example of the presence of OH radicals in CVD deposited aluminum nitride films, as observed by IR spectroscopy, is shown in Figure 16. The concentration of loosely bonded OH radicals decreases on increasing the substrate temperature.

The chemically bonded hydrogen in dielectric films have both beneficial and deleterious effects on the structural and electrical responses of the material. In the case of silicon nitride films deposited by CVD, at temperatures > 700°C, the hydrogen concentration is less than about 5 atomic percent. The hydrogen compensates dangling bonds in the covalently bonded network and thus removes the defect density of states from the band gap and band edges[79,80]. These films are extensively used in Metal-Nitride-Oxide-Semiconductors (MNOS) memory devices[81]. The much higher hydrogen concentration in low temperature (about 30°C) deposited plasma CVD films destroys these properties. The PACVD films, however, act as efficient diffusion barriers against water vapor and alkali ions, which make them

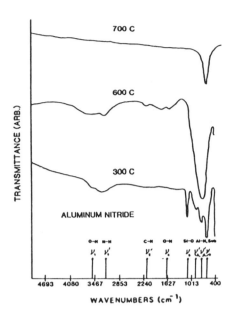

Figure 16: Infrared transmission spectra for aluminium nitride films deposited by chemical vapour deposition process (40).

useful for encapsulation of electronic devices[82,83]. In situations where the dielectric films are subjected to higher temperatures, such as implant activation caps on III-V semiconductors and aerospace applications, the use of heavily hydrogenated films/coatings is not recommended. For a typical case of PACVD silicon nitride, the films start loosing hydrogen above 600°C. The loss of hydrogen is reflected by the decreases in the intensity of infrared absorption bands of N-H and Si-H bonds (Figure 17)[51]. The hydrogen outdiffusion results in a severe microfragmentation of the films.

5. RECENT DEVELOPMENTS

The ultimate ceramic film - Diamond Film - is an excellent example of the influence of the process on the structure and properties of the film. The CVD and PACVD processes using CH_4 and H_2 as reactants require a high deposition temperature (600-1000°C) and the resulting films are very coarse grained, highly faceted and essentially opaque due to internal reflection of

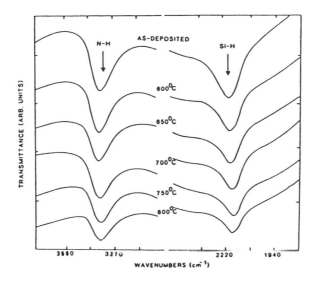

Figure 17: Infrared transmission spectra of as-deposited and annealed Si_3N_4 films deposited by PACVD process.   N-H and Si-H absorbtion bands have been marked in the figure (51).

light from the faceted interfaces in the microstructure[141].   On the other hand, the ARE process evaporates graphite with an electron beam in a hydrogen plasma.   The diamond films are deposited on an unheated substrate with the maximum substrate temperature measured at 350°C from the thermal radiation from the hot graphite target.   The films are very fine grained (grain size less than 100 Å), transparent and with a very smooth surface topography[142,143].   Thus, for the various applications of diamond films, the process has to be matched to the application.

6. CONCLUDING REMARKS

In brief, we have presented an introductory review of currently available atomistic processes for deposition of ceramic thin films and coatings.   It has been shown that the conventional evaporation or sputtering of a bulk ceramic leads to non-stoichiometric deposits which are not suitable for most of the microelectronic and photonic applications. The reactive variations of evaporation and sputtering processes facilitate control over the stoichiometry.   The presence of a gaseous plasma in both of these processes has the additional advantage of activating the compound forming reaction.   The electrons and the reactive and neutral gas ions present in the plasma zone can also be used to modify the microstructure and stress state of the films.   The flux and the energetics of the ion bombardment can be controlled precisely in a dual ion beam reactive deposition processes.   In PACVD, the extent of ion/electron bombardment can be varied with the substrate bias and the plasma excitation frequency. Laser CVD processes are developing rapidly to meet the demands for low temperature deposition of ceramics with restricted dimensionality.   Applications for the various nitride, oxide, oxynitride, carbide, silicide and boride ceramics are wide-ranging. A summary of their most important uses is provided in Table 5.

Ceramic Systems	Deposition Techniques Used	Applications	
		Magnetic,Electronic, Optoelectronic	Structural
AlN	RS, RIBS, ARE, CVD, Photo CVD, Plasma CVD	Implantation mask, heat sink, surface acoustic wave devices	High strength and corrosion resistant coatings
AlN_xO_{1-x}	RIBS, CVD, Photo CVD	Incapsulation layers gate dielectric	
BN	RS, ARE, CVD, Plama CVD	Implantation mask, X-ray lithography mask	Corrosion resistant applications, wear resistant coating, high temperature applications
Si_3N_4	RS, ARE, RIBS, CVD, Photo CVD	Incapsulation, diffusion barrier, gate dielectric applications and solar cells	Aerospace structural material
Transition metal nitrides	RS, ARE, RIBS, CVD	Diffusion barriers, superconducting S-I-S junctions	Decorative coatings hard coatings
Transition metal silicides	Co-sputtering, Evaporation CVD Plasma CVD, RS	Integrated circuit metallization	Heating elements
Transition metal carbides	RS, ARE, CVD		Hard coatings, corrosion resistant coatings

(continued)

Ceramic Systems	Deposition Techniques Used	Applications	
		Magnetic, Electronic, Optoelectronic	Structural
SiC and $a\text{-}Si_xC_{1-x}$:H	RS, ARE, CVD, Plasma CVD, RIBS	High temperature semiconductor devices, solar cells	High temperature structural applications, cutting tool hard coatings
Transition metal borides and rare-earth borides	Sputtering, CVD	Diffusion barriers and metallization in microelectronics	High temperature structural applications
Alkali and alkaline fluorides	Evaporation	Optical filters and allied optics	
In_2O_3, SnO_2, ZnO and Cd_2SnO_4 as transparent conductors	RE, ARE, RS, RIBS, CVD spray pyrolysis	Solar cells, heat mirrors, display devices, gas sensors	Wear resistant coatings
$YBa_2Cu_3O_{7-x}$ and $La_xSr_{1-x}CuO_4$ type perovskites	RE, RS, screen printing	High-T_c superconductors, electromagnetic and magneto-mechanical devices	
$SrTiO_3$, $BaTiO_3$, $Pb(Zr,Ti)O_3$	RS, RIBS	Piezoelectric and ferroelectric devices	
ZnO	RS, RIBS	Surface acoustic wave devices, optical waveguides	
Al_2O_3	ARE, RS, RIBS, CVD, PACVD, Photo CVD	Optical waveguides and optics	Important structural ceramic
SiO_2	RS, CVD, PACVD, Photo CVD	Active and passive applications in microelectronics, optical waveguides	
Diamond and diamond-like carbon	CVD, PACVD, ARE	Protective coatings for IR and UV optics AR coatings, high thermal conductivity substrates, microelectronic devices	Fine particle superabrasives, cutting tool for aluminium alloys, wear coatings for computer memory discs

Table 5: Applications of ceramic coatings.

REFERENCES

1. Bunshah, R.F., Encyclopedia of Chemical Technology, 20, 38 (1982)

2. Maissel, L.I. and Glang R., Handbook of Thin Film Technology (McGraw-Hill, New York, 1970).

3. Budhani, R.C., Memerian, M., Doerr, H.J., Deshpandey, C.V. and Bunshah, R.F, Thin Solid Films 118, 293 (1984).

4. Yokotsuka, T., Narusawa, T., Uchida, Y. and Nakashima, H., Appl. Phys. Lett. 50, 591 (1987).

5. Hunter, W.R., Ephrath, L., Godoman, W.D., Osburn, C.M., Crowder, B.L., Cramer, A. and Luhn, H.E., IEEE Transactions on Electron Devices, ED 26, 353 (1978).

6. Kazmerski, L.L, Ayyagari, M.S. and Samborn, G.A., J. Appl. Phys., 46, 11 (1975).

7. Cho, A.Y. and Arthur, J.R., in Progress in Solid State Chemistry, Somorhai, G. and McCaldin, S. (Eds) Pergamon, New York (1975).

8. Spence, W., J. Appl. Phys. 38, 3767 (1967).

9. Habermeir, H.V., Thin Solid Films, 80, 157 (1981).

10. Laser, D., Thin Solid Films, 90, 317 (1982).

11. Budhani, R.C, Prakash, S. and Bunshah, R.F. (unpublished work).

12. Ritter, E., Opt. Acta 9, 197 (1962).

13. Chi, K.C., Dillon, R.O. and Bunshah, R.F., Thin Solid Films 54, 259 (1978).

14. Colen, M. and Bunshah, R.F., J. Vac. Sci. Technol. 13, 536 (1975).

15. Grossklaus, W. and Bunshah, R.F., J. Vac. Sci. Technol. 12, 593 (1975).

16. Chaudhari, P. et al., Phys. Rev. Lett. 58, 2684 (1987).

17. Laibowitz, R.B., Koch, R.H., Chaudhari, P. and Gambino, R.J., Phys. Rev. B35, 8821 (1987).

18. Bunshah, R.F. and Raghuram, A.C., J. Vac. Sci. Technology 9,1385 (1972).

19. Grossklaus, W., M.S. Thesis UCLA, 1974.

20. Granier, J. and Besson, J., Proc. 9th Plansee Seminar, Reutte, Austria, May 23-26, 1977.

21. Nath, P. and Bunshah, R.F., Thin solid Films 69, 63 (1980).

22. Kobayasi, M. and Doi, Y., Thin Solid Films 540, 57 (1978).

23. Nakamura, K., Inagawa, K., Tsuruoka, K. and Komiya, S., Thin Solid Films 40, 155 (1977).

24. Komiya, S., Umeza, N. and Narusawa, T., Thin Solid Films, _54_, 51 (1978)

25. Zega, B., Kornmann, M. and Amiguel, L., Thin Solid Films _45_, 577 (1977).

26. Murayama, Y., J. Vac. Sci. Tech. _12_ 818 (1975).

27. Chopra, K.L., Argarwal, V., Vankar, V.D., Deshpandey, C.V. and Bunshah, R.F., Thin Solid Films, _126_, 307 (1985).

28. Mattox, D.M., J. Appl. Phys. _34_, 2493 (1963).

29. Takagi, T., Kunori, I.Y.M. and Kobiyama, S., Proc. Int. Conf. Ion Sources 2nd Vienna, 1972, 790 (1972).

30. Thornton, J.A., in Deposition Technologies for Films and Coatings, Bunshah, R.F. (Ed.) Noyes, New Jersey (1982).

31. Mohammadi, F., Saraswat, K.C. and Meindl, J.D., Appl. Phys. Lett. _35_, 529 (1979).

32. Kawasaki, M., Funabashi, M., Nagata, S., Fueki, and Koinuma, H., Jpn. J. Appl. Phys. _26_, L388 (1987).

33. Westwood, W., Prog. Surf. Science _7_, 71 (1976).

34. Bunshah, R.F. and Deshpandey, C.V., _27_, 1 (1986).

35. Geraghly, K.G. and Donaghey, L.F., J. Electrochem. Soc., _123_, 1201(1978).

36. Maniv, S. and Westwood, W.D., J. Appl. Phys. _57_, 718 (1980).

37. Abe, T. and Yamashina, T., Thin Solid Films _30_, 19 (1974).

38. Shinoki, F. and Itoh, A., J. Appl. Phys. _46_, 381 (1975).

39. Natarajam, B.R., Etoukly, A.H., Green, J.E. and Barr, I.L., Thin Solid Films _69_, 201 (1980).

40. Schiller, S., Heisig, V., Stinfelder, K. and Strumpfer, J., Thin Solid Films _64_, 455 (1979).

41. Maniv, S., Miller, C. and Westwood, W.D., J. Vac. Sci. Technol. _A1_, 1370 (1983).

42. Scherer, M. and Wirz, P., Thin Solid Films _119_, 203 (1984).

43. Sproul, W.D., U. S. Patent 442811, 1984.

44. Sproul, W.D., Thin Solid Films, _118_, 279 (1984).

45. Budhani, R.C., O'Brien, B.P., Doerr, H.J., Deshpandey, C.V. and Bunshah, R.F., J. Appl. Phys. _57_, 5477 (1984).

46. Harper, J.M.E., Cuomo, J.J. and Hentzell, H.T.G., Appl. Phys. Lett., _43_ 547 (1983).

47. Carlsson, J.A., J. Vac. Sci. Technol. (1982).

48. Bonifield, T.D., in Deposition Technologies for Films and Coatings, Bunshah, R.F. (Ed.) p 365, Noyes, New Jersey (1982).

49. Garscadden, A., Duke, G.L. and Bailey, W.F., Appl. Phys. Lett. 43, 1012 (1983).

50. Budhani, R.C., Prakash, S., Doerr, H.J. and Bunshah, R.F., J. Vac. Sci. Technol. A5, 1644 (1987).

51. Budhani, R.C., Bunshah, R.F. and Flinn, P.A., Appl. Phys. Lett. 52, 284 (1988).

52. Budhani, R.C., Prakash, S. and Bunshah, R.F., Materials Research Society Spring Meeting, Anaheim, CA 1987.

53. Hess, D.W., J. Vac. Sci. Technol. A2, 244 (1984).

54. Chow, R., Landford, W.A., Wang Ke-Ming and Rosler, R.S., J. Appl. Phys. 53, 5630 (1982).

55. Tsu, D.V., Lucovsky, G. and Mantini, M.J., Phys. Rev. B33, 7069 (1986).

56. Allen, S.D., J. Appl. Phys. 52, 6501 (1981).

57. Mayo, M.J., Solid State Technology, April 1986, p. 144.

58. Chopra, K.L., Major, S. and Pandya, D.K., Thin Solid Films, 102, 1 (1983).

59. Manifacier, J.C., de Murcia, M., Fillard, J.P. and Vicario, E., Thin Solid Films, 41, 127 (1977).

60. Manifacier, J.C., de Murcia, M., Fillard, J.P. and Vicario, E., Mater. Res. Bull, 10, 1215 (1978).

61. Maudes, J.S. and Rodriguez, T., Thin Solid Films 69, 183 (1980).

62. Ishigiro, K., Sasaki, T., Arai, T. and Imai, I., J. Phys. Soc. Jpn., 13, 296 (1958).

63. Grosse, P., Schmitte, J.E., Frank, G. and Kostlin, H., Thin Solid Films 90, 309 (1982).

64. Nobbs, J. Mck, and Gillespie, F.C., J. Phys. Chim. Solids. 31, 2353 (1970)

65. Aranovich, J., Oritz, A. and Babe, R.M., J. Vac. Sci. Technol., 16, 994 (1979).

66. Raza, A., Agrihotri, O.P. and Gupta, B.K., J. Phys. D10, 1871 (1977).

67. Groth, R., Phys. States Solidi, 14, 69 (1966).

68. Shanthi, E., Dutta, V., Banerjee, A. and Chopra, K.L., J. Appl. Phys., 51, 6243 (1980).

69. Shanthi, E., Banerjee, A., Dutta, V. and Chopra, K.L., J. Appl. Phys. 51, 1615 (1982).

70. Chopra, K.L. and Das, S.R., "Thin Film Solar Cells", Academic Press, New York (1983).

71. Chopra, K.L., Kainthala, R.C., Pandya, D.K. and Thakoor, A.P., Phys. Thin Films 12, 167 (1982).

72. Bunshah, R.F., in Deposition Technologies for Films and Coatings, Bunshah, R.F., (Ed.) Noyes, New Jersey, (1982).

73. Movchan, B.A. and Demchishin, A.V., Fizika Metall. 28, 653 (1969).

74. Thornton, J.A., J. Vac. Sci. Technol. 12, 830 (1975).

75. Harper, J.M.E., Cuomo, J.J., Gambino, R.J. and Kaufman, H.R., in Ion Beam Modification of Surfaces: Fundamentals and Applications, Auciello, O. and Kelly, R., p127, Elsevier, Amsterdam (1984).

76. Cuomo, J.J. and Gambino, R.J., J. Vac. Sci. Technol 14, 152 (1977).

77. Winters, H.F. and Kay, E., J. Appl. Phys. 43, 794 (1972).

78. Bardin, T.T., Pronko, J.G., Budhani, R.C. and Bunshah, R.F., J. Vac. Sci. Technol. A4, 3121 (1986).

79. Robertson, J., Phil. Mag. B44, 215 (1981).

80. Lucovsky, G. and Lin, S.Y., J. Vac. Sci. Technol. B3, 1122 (1985).

81. Hezel, R., Blumenstock, K. and Schorner, R., J. Electrochem. Soc. 131, 1679 (1984).

82. Sinha, A.K., Levinstein, H.J., Smith, T.E., Quintano, G. and Haszko, S.E., J. Electrochem. Soc., 125, 601 (1978).

83. Harris, J.S., Eisen, F.H., Welch, B.M., Pashley, R.D., Sigund, D. and Mayer, J.W., Appl. Phys. Lett. 21, 601 (1972).

84. Shuskus, A.J., Reeder, T.M. and Paradis, E.L., Appl. Phys. Lett. 24, 151 (1974).

85. Seidel, K.H., Reichelt, K., Schaal, W. and Dimigen, H., Thin Solid Films, 151, 243 (1987).

86. Xin, S.H., Schaft, W.J., Wood, C.E.C. and Eastman, L.F., Appl. Phys. Lett. 41, 743 (1982).

87. Wasa, K., Nagai, T. and Hayakawa, S., Thin Solid Films 31, 235 (1976).

88. Henmesch, H.K. and Roy, R. Eds., Mater. Res. Bull. 4 (1969).

89. Vassen, J.L., RCA Rev. 32, 289 (1971).

90. Growth, R., Phys. States Solidi, 14, 69 (1966).

91. Peaker, A.R. and Horsley, B., Rev. Sci. Instrum. 42, 1825 (1971).

92. Takao, T., Wasa, K. and Hayakawa, S., J. Electrochem Soc. 123, 1719 (1976).

93. Goodchild, R.G., Webb, J.B. and Williams, D.F., J. Appl. Phys. 57, 2308 (1985).

94. Paradis, E.L. and Shuskus, A.J., Thin Solid Films 38, 131 (1976).

95. Webb, J.B., Williams, D.F. and Buchanan, M., Appl. Phys. Lett., 39, 640 (1980).

96. Barnes, J.D., Leary, D.J. and Jordan, A.G., J. Electrochem. Soc., 7, 1636 (9180).

97. Goldstein, R.M. and Wigginton, S.C., Thin Solid Films 3, R41 (1969).

98. Wasa K. and Hayakawa, S., Microelectron. and Reliab., G, 216 (1967).

99. Young, P.L., Fehler, F.P. and Whitman, A.J., J. Vac. Sci. Technol. 14, 176 (1977).

100. Pratt, I.H., Solid State Technol., 12, 49 (1969).

101. Nowicki, R.S., J. Vac. Sci. Technol., 14, 127 (1977).

102. Salma, C.A.J., J. Electrochem. Soc., 117, 913 (1970).

103. Morrison, D.T. and Robertson, T., Thin Solid Films 27, 19 (1975).

104. Krupandhi, S.B., Maffei, N., Sayer, M. and El-Assad, K., J. Appl. Phys. 54, 6601 (1983).

105. Nagatomo, T., Kosaka, T., Omori, S. and Omoto, O., Ferroelectrics 37, 681 (1981).

106. Lee, S.J., Rippert, E.D., Lin, B.Y., Song, S.N., Hwu, S.J., Poeppelmeier, K. and Ketterson, J.B., Appl. Phys. Lett. 51, 1194 (1987).

107. Shah, S.I. and Carcia, P.F., Appl. Phys. Lett., 51, 2146 (1987).

108. Pearce, L.G., Gunshor, R.L. and Pierret, R.F., Appl. Phys. Lett., 39, 878 (1981).

109. Shiosaki, T., Yamamoto, T., Oda, T., Harada, K. and Kawabata, A., Ultrason. Symp. 451 (1980).

110. Matsusita, K., Matsuno, Y., Hanu, T. and Shibata, Y., Thin Solid Films 80, 243 (1981).

111. Meaudre, K. and Tardy, T., Solid State Commun., 48, 117 (1983).

112. Smith, G.J. and Milne, W.I., Phil. Mag. 47, 419 (1983).

113. Matsumoto, S., Suzuki, H. and Ueda, R., Jpn. J. Appl. Phys. 11, 607 (1972)

114. Shimada, T., Katayama, Y. and Komatsubara, K.F., J. Appl. Phys. 50, 5530 (1977).

115. Liberman, M.L. and Medrud, R.C., J. Electrochem. Society, 116, 242 (1969).

116. Hecg, M. and Portier, E., Thin solid Films 9, 341 (1972).

117. Karim, A., M. Tech. Thesis, UCLA (1988), unpublished.

118. Hata, T., Noda, E., Morimoto, O. and Hada, T., Appl. Phys. Lett. 37, 633 (1980).

119. Deshpandey, C. and Holland, L., Thin Solid Films 96, 265 (1982).

120. Sproul, W.D., Thin Solid Films 33, 133 (1987).

121. Char, K., Kent, A.D., Kapitulnik, A., Beasley, M.R. and Geballe, T.H., Appl. Phys. Lett., 51, 1370 (1987).

122. Chi, C.C., Tsui, C.C., Yee, D.S., Cuomo, J.J., Laibowitz, R.B., Koch, R.H., Braren, B., Srinivasan, R. and Plechaty, M.M., Appl. Phys. Lett., 51, 1951 (1987).

123. Erler, H.J., Reisse, G. and Weissmantel, C., Thin solid Films 102, 345 (1983).

124. Bhat, S., Ashok, S., Fonash, S.J. and Tongson, L., J. Electronic Mat., 14, 405 (1985).

125. Shanfield, S. and Wolfson, R., J. Vac. Sci. Technol. A1, 323 (1983).

126. Bradley, L.E. and Sites, J.R., J. Vac. Sci. Technol. 16, 6 (1979).

127. Singh, Jagriti and Budhani, R.C., Appl. Phys. Lett. 51, 978 (1987).

128. Singh, Jagriti and Budhani, R.C., Solid State Commun. 64, 349 (1987).

129. Sites, J.R., Thin Solid Films 45, 47 (1977).

130. Burdoritsin, V.A., Thin Solid Films, 105, 197 (1983).

131. Dubow, J.B., Burk, D.E. and Sites, J.R., Appl. Phys. Lett. 29, 495 (1976).

132. Cheek, G., Genis, A., Dubow, J.B. and Pai Verneker, V.A., Appl. Phys. Lett. 33, 643 (1978).

133. Giani, E. and Kelley, R., J. Electrochem. Soc. 12, 394 (1974).

134. Hebard, A.F., Blonder, G.E. and Suh, S.Y., Appl. Phys. Lett. 44, 11 (1984).

135. Cole, B.E., Moravec, J., Ahonen, R.G. and Ehlert, L.B., J. Vac. Sci. Technol A2, 372 (1984).

136. Takei, K. and Nagai, K., Jpn. J. Appl. Phys. 20, 5 (1981).

137. Castellano, R.N. and Feinskin, L.G., J. Appl. Phys. 50, 6 (1979).

138. Madakson, B., Cuomo, J.J., Yee, D.S., Roy, R.A. and Scilla, G., J. Vac. Sci. Technol. (in Press).

139. Dijkkamp, D. et al., Applied Phys Lett. 51 [8] 619 (1987).

140. Krishnaswamy, J., Rengan, A. and Narayan, J., Proc 1st Int. Symp. on Diamond and Diamond like Carbon Films, Electrochemical Society, Los Angeles, May 7-12, 1989.

141. Spear, K.E., J. Am. Ceram. Soc. 72 [2] 171 (1989).

142. Deshpandey, C., Doerr, H.J. and Bunshah, R.F., US Patent 4,816,291, March 1989.

143. Deshpandey, C., Doerr, H.J., Bunshah, R.F. and Radhakrishnan, M.C., - to be published.

Index

411